THE CRYOSPHERE AND GLOBAL ENVIRONMENTAL CHANGE

Environmental Systems and Global Change Series

Series Editor: Professor Anthony Orme

The development of this new series of advanced undergraduate and graduate textbooks has been stimulated by three widely recognized trends in the teaching of earth and environmental sciences at university level.

First, the systems' approach is now well established in university physical geography and earth/environmental science curricula around the world, both at undergraduate and graduate levels.

Second, concerns about the pace and extent of global change have increasingly informed – and given an urgent social relevance to – a wide range of course offerings in these subjects.

Last, implicit in the environmental systems approach is the importance of integrating findings and methodologies from a wide range of disciplines, including ecosystems science, geomorphology, hydrology, geophysics, oceanography, climatology, archaeology, and environmental planning.

The ESGC series is explicitly designed to reflect these educational trends. It is an ambitious new venture resulting from the merging of two existing publishing initiatives – Blackwell's Environmental Systems and Pearson's Understanding Global Environmental Change Series – and its objectives may be simply stated:

- to create an awareness and understanding of the way key environmental systems operate and interact.
- to explore the pace and extent of global (and regional) environmental change and to show how environmental systems respond to change over a variety of scales in time and space.
- to attract students from a range of disciplines and to encourage students to think in new ways that transcend traditional discipline boundaries.
- to underline the relevance of these studies to social/environmental problems, and to encourage students to bring a scientific approach to solving such problems.

Books in the series are aimed at advanced undergraduates and graduates taking degree courses in physical geography, earth science, environmental science, ecology, and archaeology. Titles in the series will have an international relevance, with examples and case studies taken from varied environments around the world.

1 *The Cryosphere and Global Environmental Change*, Olav Slaymaker and Richard E.J. Kelly

Forthcoming
2 *The Pace of Environmental Change*, Anthony Orme
3 *Oceans and Global Environmental Change*, Tom Spencer
4 *Water in a Changing World*, John Pitlick

THE CRYOSPHERE AND GLOBAL ENVIRONMENTAL CHANGE

by

Olav Slaymaker and Richard E.J. Kelly

University of British Columbia, Vancouver, Canada
and
University of Waterloo, Ontario, Canada

Blackwell Publishing

BLACKWELL PUBLISHING
350 Main Street, Malden, MA 02148–5020, USA
9600 Garsington Road, Oxford OX4 2DQ, UK
550 Swanston Street, Carlton, Victoria 3053, Australia

First published 2007 by Blackwell Publishing Ltd

1 2007

Library of Congress Cataloging-in-Publication Data

Slaymaker, Olav, 1939-
  The cryosphere and global environmental change / by Olav Slaymaker and Richard E. J. Kelly.
     p. cm. — (Environmental systems and global change series)
  Includes bibliographical references and index.
  ISBN-13: 978-1-4051-2976-3 (pbk. : alk. paper)
  ISBN-10: 1-4051-2976-X (pbk. : alk. paper)
  1. Cryosphere—Textbooks. 2. Global environmental change—Textbooks. 3. Frozen ground—Textbooks. 4. Climatic changes—Textbooks. 5. Global warming—Textbooks. I. Kelly, Richard E. J. II. Title.

QC880.4.C79S57 2007
551.31—dc22
                                                                          2006032075
A catalogue record for this title is available from the British Library.

Set in Meridien Roman 10.5/12.5
by NewGen Imaging Systems Pvt Ltd., Chennai, India
Printed and bound in Singapore
by Markono Print Media Pte Ltd

For further information on
Blackwell Publishing, visit our website:
www.blackwellpublishing.com

CONTENTS

Color Plates appears between pages 142 and 143

LIST OF PERMISSIONS

FIG. 1.1 Reprinted with permission from (Opportunities in the Hydrologic Sciences) © (1991) by the National Academy of Sciences, courtesy of the National Academies Press, Washington, DC

FIG. 1.6 Redrawn with permission from (Glaciers and Climate Change) © 2001 by Taylor and Francis Group, London

FIG. 1.8 Reprinted with permission from (Climate Change 2001: Impacts, Adaptation and Vulnerability) © 2001 by Cambridge University Press, Cambridge

FIG. 1.9 Redrawn with permission from (Impacts of a Warming Arctic) © 2004 by Cambridge University Press, Cambridge

FIG. 1.10 Redrawn with permission from the Annals of Glaciology © 1995 by the International Glaciological Society

FIG. 1.11 Redrawn with permission from (Impacts of a Warming Arctic) © 2004 by Cambridge University Press, Cambridge

FIG. 1.12 Redrawn with permission from Nature © 1997

FIG. 1.14 Reprinted with permission from (Climate Change 1995: the Science of Climate Change) © 1996 by Cambridge University Press, Cambridge

FIG. 1.15 Reprinted with permission from (Climate Change 2001: Impacts, Adaptation and Vulnerability) © 2001 by Cambridge University Press, Cambridge

FIG. 1.16 Redrawn with permission from (Sea Ice Climatic Atlas: 1971–2000) © 2002 by Canadian Ice Service, Environment Canada, Ottawa

FIG. 1.17 Reprinted with permission from (Impacts of a Warming Arctic) © 2004 by Cambridge University Press, Cambridge

FIG. 1.18 Redrawn with permission from (Les Milieux Polaires) © 1999 by Armand Colin Press, Paris

FIG. 1.19 Reprinted with permission from Burneing Images Custom Photographics, Squamish, B.C

FIG. 1.20 Redrawn with permission from (Arctic Human Development Report) © 2004 by Stefansson Arctic Institute, Akureyri

FIG. 1.21 Redrawn with permission from (Les Milieux Polaires) © 1999 by Armand Colin Press, Paris

FIG. 3.1 Redrawn with permission from (Snow Ecology) © 2001 by Cambridge University Press, Cambridge

FIG. 3.2 Reprinted with permission from (Boundary Layer Climates) © 1987 by Taylor and Francis Group, London

FIG. **3.3** Reprinted with permission from the Journal of Glaciology © 1989 by the International Glaciological Society

FIG. **3.4** Redrawn with permission from (Snow Ecology) © 2001 by Cambridge University Press, Cambridge

FIG. **3.5** Redrawn with permission from Journal of Geophysical Research © 2001 by the American Geophysical Union

FIG. **3.6** Redrawn with permission from Journal of Geophysical Research © 1984 by the merican Geophysical Union

FIG. **3.7** Redrawn with permission from the Annals of Glaciology © 1990 by the International Glaciological Society

FIG. **3.11** Redrawn with permission from Bulletin of the Hydrological Sciences © 1978 by Blackwell Scientific Publications, Oxford

FIG. **3.12** Reprinted with permission from (Mass Balance of the Cryosphere) © 2004 by Cambridge University Press, Cambridge

FIG. **3.13** Redrawn with permission from the Journal of Glaciology © 1991 by the International Glaciological Society

FIG. **3.14** Redrawn with permission from (Snow Ecology) © 2001 by Cambridge University Press, Cambridge

FIG. **3.17** Reprinted with permission from (Snow Ecology) © 2001 by Cambridge University Press, Cambridge

FIG. **3.19** Reprinted with permission from (The Avalanche Handbook) © 1993 by Mountaineers Books, Seattle

FIG. **3.20** Reprinted with permission from (The Avalanche Handbook) © 1993 by Mountaineers Books, Seattle

FIG. **3.23** Reprinted with permission from (Snow Ecology) © 2001 by Cambridge University Press, Cambridge

FIG. **3.24** Reprinted with permission from (Snow Ecology) © 2001 by Cambridge University Press, Cambridge

FIG. **3.25** Reprinted with permission from the Journal of Glaciology © 1962 by the International Glaciological Society

FIG. **3.26** Reprinted with permission from EOS (May 15, 1984) by the American Geophysical Union

FIG. **3.27** Reprinted with permission from EOS (July 17, 1984) by the American Geophysical Union

FIG. **3.28** Redrawn with permission from (River-Ice Ecology) © 2000 by the National Water Research Institute, Environment Canada, Saskatoon. (no written response to request)

FIG. **3.29** Redrawn with permission from (River-Ice Ecology) © 2000 by the National Water Research Institute, Environment Canada, Saskatoon. (no written response to request)

FIG. **5.2** Redrawn with permission from Nature© 1990

FIG. **5.4** Redrawn with permission from Nature © 1988

FIG. **5.5** Redrawn with permission from (Ice Age Earth) © 1992 by Routledge, Taylor and Francis

FIG. **5.6** Redrawn with permission from (Ice Age Earth) © 1992 by Routledge, Taylor and Francis

FIG. 5.7 Reprinted with permission from Canadian Journal of Earth Sciences © 1996 by CISTI

FIG. 5.8 Redrawn with permission from Royal Geographical Society © 1970; (Earth Rheology, Isostasy and Eustasy) © 1980 by Wiley, Chichester

FIG. 5.9 Redrawn with permission from Journal of the Geological Society of London © 1995

FIG. 5.10 Redrawn with permission from Paleooceanography © 1992 by the American Geophysical Union

FIG. 5.11 Redrawn with permission from Earth and Planetary Science Letters © 2000 by Elsevier

FIG. 5.12 Redrawn with permission from (Ice Sheets and Late Quaternary Environmental Change) © 2001 by Wiley

FIG. 5.13 Redrawn with permission from Annals of Glaciology © 1990 by the International Glaciological Society

FIG. 5.14 Redrawn with permission from Global and Planetary Change © 1991 by Elsevier

FIG. 5.15 Reprinted with permission from Journal of Quaternary Science © 2004 by Wiley

FIG. 5.16 Reprinted with permission from Geografiska Annaler © 2004 by Blackwell

FIG. 5.17 Redrawn with permission from Quaternary Research © 1999 by Elsevier

FIG. 5.18 Redrawn with permission from (The Last Great Ice Sheets) © 1981 by Wiley, New York; Quaternary Science

Reviews © 1990 by Elsevier and Quaternary Research © 1994 by Elsevier

FIG. 5.19 Redrawn with permission from (Late Quaternary Environments of the Soviet Union) © 1984 by Longman

FIG. 5.20 Redrawn with permission from (Geocryology) © 1979 by Edward Arnold and from (Late Quaternary Environments of the United States) © 1983 by Longman

FIG. 5.22 Redrawn with permission from (Glacial and Pleistocene Geology) © 1957 by Wiley

FIG. 5.23 Reprinted with permission from Geografiska Annaler © 1966 by Blackwell and from Boreas © 2005 by Blackwell

FIG. 5.24 Redrawn with permission from Boreas © 1996 by Edward Arnold

FIG. 5.25 Reprinted with permission from Boreas © 2005 by Blackwell

FIG. 5.26 Redrawn with permission from Geological Survey of Canada © 1972.

FIG. 5.27 Reprinted from Bulletin of the Geological Society of America © 1972 and from Nature © 1989

FIG. 6.1 Redrawn with permission from Glacial Geology © 1983 by Elsevier

FIG. 6.2 Redrawn with permission from Quaternary International © 1994 by Elsevier

FIG. 6.3 Redrawn with permission from Geografiska Annaler © 1977 by Blackwell

FIG. 6.4 Redrawn with permission from Geomorphology ©1994 by Elsevier

FIG. 6.5 Redrawn with permission from Revue de Geologie Dynamique et de

Geographie Physique © 1986 by Societe Geologique de Belgique (no written response to request)

FIG. 6.6 Redrawn with permission from Erdkunde © 1948

FIG. 6.7 Reprinted with permission of the Province of British Columbia

FIG. 6.8 Redrawn with permission from Quaternary Science Reviews © 2002 by Elsevier

FIG. 6.9 Reprinted with permission from Geological Survey of Canada © 1972

FIG. 6.10 Redrawn with permission from Geological Society of America © 1987

FIG. 6.11 Reprinted with permission from (Mountain Geomorphology) © 2004 by Edward Arnold

FIG. 6.12 Redrawn with permission from Bulletin of the Geological Society of America © 1993 and from Glaciers and Landscape © 1976 by Edward Arnold

FIG. 6.13 Reprinted with permission from National Air Photo Library, Ottawa and from 8th edition (Landforms and Surface Materials of Canada) © 1996 by J.D.Mollard and Associates Limited, Regina

FIG. 6.14 Reprinted with permission from National Air Photo Library, Ottawa and from 8th edition (Landforms and Surface Materials of Canada) © 1996 by J.D.Mollard and Associates Limited, Regina

FIG. 6.15 Reprinted with permission from National Air Photo Library, Ottawa and from 8th edition (Landforms and Surface Materials of Canada) © 1996 by J.D.Mollard and Associates Limited, Regina

FIG. 6.16 Reprinted with permission from National Air Photo Library, Ottawa and from 8th edition (Landforms and Surface Materials of Canada) © 1996 by J.D.Mollard and Associates Limited, Regina

FIG. 6.17b Redrawn with permission from (Advances in Hillslope Processes) © 1996 by Wiley

FIG. 6.18 Redrawn with permission from (Facies Models: Response to Sea Level Change) © 1992 by Geological Association of Canada

FIG. 6.19 Redrawn with permission from Sedimentary Geology © 1993 by Elsevier

FIG. 6.20 Reprinted with permission from National Air Photo Library, Ottawa and from 8th edition (Landforms and Surface Materials of Canada) © 1996 by J.D.Mollard and Associates Limited, Regina

FIG. 6.21 Reprinted with permission from Geographie Physique et Quaternaire, © 1991 by Les Presses de líUniversite de Montreal

FIG. 6.22 Reprinted with permission from Canadian Journal of Earth Sciences © 1994 by CISTI

FIG. 6.23 Reprinted with permission from National Air Photo Library, Ottawa and from 8th edition (Landforms and Surface Materials of Canada) © 1996 by J.D.Mollard and Associates Limited, Regina

FIG. 6.24 Redrawn with permission from Geografiska Annaler © 1989 by Blackwell

FIG. 6.25 Redrawn with permission from (Glaciers and Glaciation). © 1998 by Edward Arnold

FIG. 6.26 Reprinted with permission from Zeitschrift fur Geomorphologie © 1990

FIG. 6.27 Reprinted with permission from (Mountain Geomorphology) © 2004 by Edward Arnold

FIG. 7.1 Reprinted with permission from (Panarchy) © 2002 by Island Press

FIG. 7.2 Redrawn with permission from (Collapse) © 2005 by Penguin Books

FIG. 7.4 Reprinted with permission from (The Blue Planet) © by Wiley

FIG. 7.5 Redrawn with permission from Physical Geography © by V.H.Winston and Son and from (The Periglacial Environment) © by Addison, Wesley Longman

FIG. 7.6 Redrawn with permission from Geophysical Research Letters © by American Geophysical Union

FIG. 7.7 Redrawn with permission from Arctic Pollution 1998 © 1998 by Arctic Monitoring Assessment Programme (no written response to request)

FIG. 7.8 Redrawn with permission from AMAP Assessment 2002 © 2004 by Arctic Monitoring Assessment Programme (no written response to request)

FIG. 7.9 Reprinted with permission from (Panarchy) © 2002 by Island Press

FIG. 7.10 Redrawn with permission from AMAP Assessment 2002 © 2004 by Arctic Monitoring Assessment Programme (no written response to request)

FIG. 7.11 Redrawn with permission from AMAP Assessment 2002 © 2004 by Arctic

Monitoring Assessment Programme (no written response to request)

FIG. 7.12 Redrawn with permission from Arctic Pollution 2002 © 2003 by Arctic Monitoring Assessment Programme (no written response to request)

FIG. 7.13 Redrawn from Arctic Human Development Report © 2004 by Stefansson Arctic Institute, Akureyri

FIG. 7.14 Redrawn with permission from Arctic Marine Transport Workshop © 2004 by US Arctic Research Commission

FIG. 7.15 Redrawn with permission from Arctic Marine Transport Workshop © 2004 by US Arctic Research Commission

FIG. 7.17 Reprinted with permission from (Canadaís Cold Environments) © 1993 by McGill.Queens University Press

PLATE 1.3 Reprinted with permission from (Remote Sensing in Northern Hydrology) © 2005 by American Geophysical Union

PLATE 1.4 Reprinted with permission from (Mass Balance of the Cryosphere) © 2004 by Cambridge University Press and from Science © 2002 by AAAS

PLATE 1.5 Reprinted with permission from Geology © 1994 by GSA

PLATE 3.1 Reprinted with permission from (Mass Balance of the Cryosphere) © 2004 by Cambridge University Press

PLATE 3.3 Reprinted with permission from Bulletin of the Geological Society of America © 1999 by GSA

PLATE 5.1 Reprinted with permission from American Scientist © 1999 by AAAS

PLATE 5.2 Reprinted with permission
of the Minister of Public Works
and Government Services Canada
© 2006 and courtesy of Natural Resources
Canada, Geological Survey of
Canada

PLATE 5.3 Redrawn with permission from
Quaternary Research © 1999 by Elsevier

PLATE 7.1 Reprinted with permission from
Contes Rendus de líAcademie des Sciences
© 1999

PREFACE

This book attempts to deal integratively with all elements of the cryosphere in the context of a changing global environment. Not only is that global environment changing with respect to climate, but the accelerating pressures on the environment from anthropogenic activity are complicating our understanding of cryospheric change. The cryosphere is an essential member of the globe's environmental systems. Relatively few recognize the sheer extent of snow, ice, and permafrost environments, much less the critical regulatory function of the cryosphere subsystem within the global environmental system. We therefore offer this book as a contribution to the Environmental Systems and Global Change Series and aim at meeting the objectives that are common to that series, as follows:

1   to create an awareness and understanding of the way in which the cryosphere is responding, has responded, and continues to interact with the changing global environment;

2   to explore the pace and extent of global environmental change and to show how the cryosphere responds to change over a variety of scales of time and space;

3   to attract upper level undergraduate students from a range of disciplines and to encourage them to think in new ways that transcend traditional disciplinary boundaries;

4   to underline the relevance of these changes to social and environmental problems and to encourage students to

bring a scientific approach to solving such problems.

The topic of snow, ice, and permafrost response to global environmental change and the implications for landscapes and human livelihoods has become a central concern over the past decade. Both within the deliberations of the Intergovernmental Panel on Climate Change (IPCC) and as the primary focus of reports such as the Arctic Climate Impact Assessment (ACIA 2004), the accelerating rate of change of the cryosphere has been emphasized. There are many specialized treatises dealing with individual components of the cryosphere, such as Siegert (2001) on ice sheets or even two of the components, such as Bamber and Payne (2004) on ice sheets and sea ice. None, as far as we are aware, deal with all six components of ice sheets, sea ice, snow, river and lake ice, glaciers, and permafrost. The reason has become evident in the course of preparation of this book. There has had to be sacrifice of considerable depth of discussion of each component. However, we are convinced that the task of synthesis, the comprehensive completion of which still lies ahead, is just as urgently needed as the in-depth analysis of each component of the cryosphere.

In order to avoid repetition within the text, we have decided to list the major findings of the IPCC and the ACIA, which are specifically relevant to our theme. Changes identified by the IPCC with intermediate to high confidence levels and which specifically affect or are affected by the cryosphere (McCarthy et al. 2001) are the following:

A Changes documented from available data:

    a. In the Arctic, increase in air temperature over land of as much as 5°C has occurred during the 20th century but only a slight warming over sea ice between 1961 and 1990 has been documented.

    b. Arctic sea ice extent has decreased by 2.9% per decade over the 1978–96 period.

    c. Regions underlain by permafrost have been reduced in area and a general warming of ground temperatures has occurred.

    d. A decrease in spring snow extent over Eurasia has occurred since 1915.

    e. A warming trend in the Antarctic Peninsula is evident.

    f. Between the mid-1950s and the early 1970s, Antarctic sea ice retreated south by 2.8° of latitude, but no significant change occurred between 1973 and 1996.

    g. Surface waters of the Southern Ocean have warmed and become less saline.

B Major changes predicted by climate models under a doubling of CO_2 scenario:

    a. Increased melting of arctic glaciers and the Greenland Ice Sheet will occur. Most of the Antarctic Ice Sheet is likely to thicken as a result of increased precipitation.

    b. Warmer water in the Southern Ocean will intensify biological activity and growth rates of fish.

    c. Summer ice in the Arctic Ocean could shrink by 60% and Antarctic sea ice volume could decrease by 25%.

    d. Thickening of the active layer will lead to widespread thermokarst and damage to infrastructure.

    e. There will be a weakening of the global thermohaline circulation as a result of increased flux of freshwater from the Arctic Ocean and consequent risk of rapid climate change, especially in Europe and the North Atlantic region.

    f. Changes in sea ice will alter the seasonal distribution, geographic ranges, patterns of migration, nutritional status, reproductive success, and abundance and balance of species.

    g. Polar regions are major sources and sinks for carbon dioxide and methane; projected climate change will increase contributions to greenhouse gases (GHGs).

    h. The Arctic is extremely vulnerable and major ecological, sociological, and economic impacts are predicted.

    i. Habitat loss for certain species of seal, walrus, and polar bear is anticipated.

    j. Loss of sea ice in the Arctic will provide increased opportunities for new sea routes, fishing and new settlements but also for wider dispersal of pollutants.

    k. Traditional life styles will be seriously affected. Maintenance of self-esteem, social cohesion, and cultural identity of communities will be the greatest threat.

Major recommendations of the Arctic Climate Impact Assessment Report (2004; ACIA), whose findings are generally consistent with those of the IPCC above, include ways of improving future assessments:

A improvement of subregional scale impacts, perhaps at the local level, where an assessment of impacts has the greatest relevance for residents;

B socioeconomic impacts in oil and gas production, mining, transportation, fisheries, forestry, and tourism need to be quantitatively determined;

C vulnerability or the degree to which a system is susceptible to adverse effects of multiple interacting stresses requires better understanding of the capacity of the system to adapt. In ecological context, vulnerability is commonly expressed as a function of sensitivity (response time following disturbance) and resilience (ability to absorb the effects of a disturbance).

But in order to achieve priorities A, B, and C, improvements in long-term monitoring, process studies, modeling, and analyses of impacts on society will be needed.

Much of this book is concerned with the identification of the state of our understanding of past and present interactions of cryosphere and changing environments. We have deliberately downplayed the prediction of future changes on the pragmatic and philosophical grounds that the future is, in principle, unknowable. But we do consider the very real probability of future collapse of both cryosphere and related socioeconomic systems.

In Chapter 1, we identify some of the most important systemic and cumulative changes in snow, ice, and permafrost that are likely to occur at global and circumpolar scales under the influence of climate and humankind. In Chapter 2, we address the measurement and monitoring problem; paying particular attention to ways of integrating observations over larger areas. In Chapter 3, we enquire into the role of the cryosphere as energy regulator and water store at local and micro-scales and examine the spatial variability of the cryosphere at global and circumpolar scales. In Chapter 4, we illustrate the dramatic ways in which remote sensing and satellite imagery have revealed spatial patterns of the cryosphere. In Chapter 5, we review our understanding of temporal variability of the cryosphere, with particular emphasis on the critical transition from glacial to postglacial conditions. In Chapter 6, we show the spatially variable imprint of the cryosphere on landscape. We view the landscape as a palimpsest,

made up of successive layers of erosional and depositional evidence. In the final chapter, we discuss possible future transitions of the society–environment relation, the sustainability of the cryosphere, and the potential for cryospheric and societal collapse. The key themes that unify the book are:

A the unique sensitivity of the cryosphere as an indicator of change at all spatial and temporal scales;

B the transient nature of environmental responses of the cryosphere to disturbance by climate and/or human activities;

C transient facies, landform, and landscape responses to changing cryospheric conditions and the nature of the resistances to change, whether inherent to the cryosphere or ecologically, socioeconomically, politically, or culturally induced.

The senior author spent the early part of his career investigating snow quantity and quality variations in relation to runoff and sediment sources in the Coast Mountains of British Columbia. More recently, he has become impressed with the importance of the meta-problem of global environmental change in both its natural science and social science formulations. It therefore seemed appropriate to connect these two sets of interests at widely differing spatial scales. It became apparent that the area of remote sensing and geomatics would be essential and it was most fortunate that the junior author was willing to lend his expertise to this project. Without his coauthorship, this book would have been less scientifically credible.

ACKNOWLEDGMENTS

The senior author acknowledges the beneficial influence of a year as Distinguished Scholar in Residence at the Peter Wall Institute for Advanced Studies during calendar year 2005. The inspiration provided by the 2005 cohort of distinguished scholars

(Drs. Dom Lopes, Ken Carty and Lawrence Ward), by the Peter Wall Distinguished Professor, Dr. Brett Finlay, and by the Director of the Institute, Dr. Dianne Newell, is deeply appreciated.

The senior author acknowledges Margaret, whose love and patience over 40 years has been unfailing. The junior author would like to express his grateful thanks to his colleagues at NASA Goddard Space Flight Center in Maryland, USA, particularly Drs. Dorothy Hall and James Foster, for providing an incredible breadth of opportunities to learn about the cryosphere and the science of measurement. He would also like to acknowledge the senior author, who has demonstrated a clear and inspirational pathway to such an integrative project. Finally, the junior author would like to thank Sasha, Hannah, and Oscar for their deep and sustaining support; without them, none of his work could be possible.

The coauthors acknowledge their special indebtedness to Eric Leinberger and Dr Dori Kovanen. Eric is a superb cartographer who has redrawn and improved each of the maps and line drawings; Dori understands the science and has corrected many points of detail. Without Eric and Dori, this book would not have been so attractively illustrated and accessible.

We also offer our appreciation to the staff at Blackwell Publishing: Ian Francis, Delia Sandford, and Rosie Hayden. Their collective confidence in the original project proposal has been sustained even as their patience has been strained.

Finally, to two anonymous reviewers we are deeply grateful. One North American and one European expert took the time to save us the embarrassment of omitting some critical references and one or two errors of judgment. Nevertheless, remaining errors are the sole responsibility of the authors.

We would also like to acknowledge the following people for providing figures: Donald Cavalieri (NASA/GSFC), Christopher Shuman (NASA/GSFC), H. Jay Zwally (NASA/GSFC), Claude Duguay (University of Alaska, Fairbanks), John Kimball (University of Montana), Frank Weber (BC Hydro, Canada), and Mike Church (University of British Columbia), J. Ross Mackay (University of British Columbia).

Richard E.J. Kelly
University of Waterloo, Kitchener and
Olav Slaymaker
University of British Columbia
Vancouver.

THE EVIDENCE FOR CRYOSPHERIC CHANGE

1.1 INTRODUCTION

The cryosphere (the subsystem of the Earth characterized by the presence of snow, ice, and permafrost) plays a vital and central role in changes occurring in the Earth's environment. Several General Circulation Models have predicted that global warming will be first obvious and most extreme in polar regions. This is because snow cover, glaciers, and sea ice will all tend to diminish, and this will produce further warming because of the decrease in albedo associated with the greater extent and duration of dark surfaces (i.e. the Earth's surface absorbs more solar radiation). At the same time, the thawing of permafrost will release methane and greenhouse gases (GHGs) which will further warm the globe.

These feedback effects are fundamentally caused by some unique properties of water at the surface of the Earth that make it particularly susceptible to change. It is particularly important to the dynamics and energetics of the Earth's system that all three phases of water (solid, liquid, and vapor) coexist over the range of the Earth's temperatures and pressures. The sensitivity of water to the normal range of the Earth's temperature and pressure conditions is such that change of phase from liquid water to solid ice and vice versa occurs frequently. On no other planet within the solar system

is this the case (Fig. 1.1). This particular change of phase leads to a dramatic change in surface cover characteristics which can be observed, measured, and sensed remotely. Further, this change of phase introduces a different set of energy and mass exchanges at the surface that may in turn produce a positive feedback that reinforces environmental changes induced by the phase change (see Chapter 3).

However, a single focus on climate is likely to be counterproductive in the interpretation of environmental change (Slaymaker 2000, 2001; Dowlatabadi 2002). Land use change is the other major driver of global environmental change

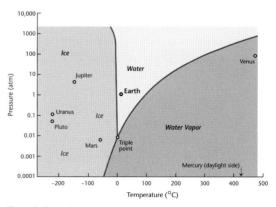

FIG. 1.1 Planetary positions on the phase diagram of water (from National Academy of Sciences 1991).

(Adger & Brown 1994). Such human activity also leads to dramatic changes in surface cover characteristics, thereby causing new energy and mass balances leading to land degradation. In fact, any surface cover change, whether natural or anthropogenic, causes changes in energy and mass balance that can have dramatic effects on the land-scape. In permafrost environments, such effects are magnified throughout the active layer, which is that depth of soil, etc. which overlies the permafrost and melts on an annual basis. Road and runway construction on massive ice-rich permafrost leads to warming, land subsidence, and destruction of the transport corridor.

The impacts of changes are more difficult to understand and predict when several processes of change happen simultaneously. They are still more challenging when the systems are near a threshold (Slaymaker 1990). In this sense, we need to consider the fact that different populations of physical systems, ecosystems, and human systems are at varying distances from thresholds. The most vulnerable subpopulations in polar regions, whether permafrost, plants, penguins, polar bears, or people, are teetering on the edge of viability. Within the cryosphere, the critical threshold is the freezing point of water; in the ecosystem and human system, the presence or absence of snow and ice may be only one of several thresholds that needs to be considered.

We normally think of the following compartments of the cryosphere: seasonal snow, mountain glaciers and ice caps, ice sheets and ice shelves, permafrost, seasonal frozen ground, river and lake ice, and sea ice. Seasonal frozen ground has the largest area of any component of the cryosphere: at its maximum in the northern hemisphere winter it covers over 5.2 (10^7) km^2 (or about 50% of the northern hemisphere land mass; National Snow and Ice Data Centre (NSIDC) new data set; Zhang et al. 2005); ice sheets contain the largest proportion of the volume of the cryosphere at about 3.3 (10^7) km^3 (Fig. 1.2). These areal and volumetric data are important, but because of our interest in changes, it is the sensitivity of the various compartments to environmental change that is most important (Table 1.1). From this perspective, snow, lake and river ice, sea ice, ice caps and glaciers, seasonal frozen ground, and permafrost are more important than the 98.5% of the ice that is stored in the polar ice sheets and ice shelves. Snow is sensitive to individual weather events as well as seasonal fluctuations and climatic trends over decades and longer; river, lake, and sea ice are sensitive to seasonal fluctuations and longer trends; permafrost and glaciers are sensitive to decadal and longer climatic trends; and ice sheets are sensitive to millennial and geological timescale events. The precise sensitivities are not known in general and the question of threshold exceedances is a lively research topic.

Global environmental change is defined as environmental change that consists of two components, namely systemic and cumulative change (Turner et al. 1990). Systemic change refers to global scale, physically interconnected phenomena whereas cumulative change refers to unconnected local to intermediate scale actions which have a significant net effect on the global system. The coupled ocean–atmosphere system is an example of the former; land cover and land use changes produce cumulative change. Systemic changes are conceptually and intuitively obvious, but difficult to measure. Cumulative change is relatively easy to measure at a specific site or within a region, but the global impacts are often not simply additive. Feedback effects, both positive and negative, and variations in the levels of threshold exceedance complicate the calculation.

Our interest then goes beyond the topic of climate change but is more circumscribed

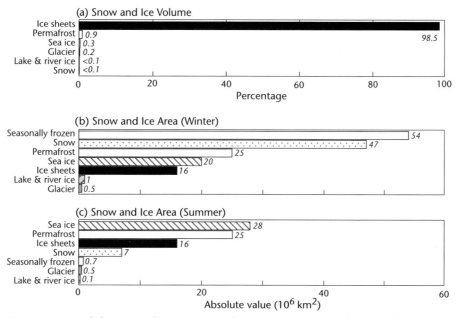

Fig. 1.2 Components of the cryosphere expressed as (a) percentage of total and (b) and (c) absolute values (Fitzharris 1996).

than the generic term "global change." The study of global change includes the full range of globalization processes, economic, social, cultural, and political, whereas global environmental change includes only the biogeochemical processes and the extent to which they are being modified by human activity (Slaymaker & Spencer 1998).

Table 1.1 Sensitivities of cryosphere compartments (after McGuffie & Henderson-Sellers 1997).

I Component	Response time
Sea ice	Days to centuries
Snow and surface ice	Hours
Lakes and rivers	Days
Soil and vegetation	Days to centuries
Glaciers	Decades to centuries
Ice sheets	Millennia
Mantle's isostatic response	Millennia

Global environmental change then, for the purpose of this book, occupies an intermediate slot between the almost limitless topic of global change and the more narrowly constrained field of global climate change.

There are several reasons for a new emphasis on global environmental change and these reasons emerged during the last four decades of the 20th century. In the 1970s, we first saw the Earth from space. We perceived an island Earth surrounded by the atmosphere and oceans and an Earth overwhelmingly dominated by water (the Blue Planet). This awareness forced an adjustment of our scientific scale of enquiry from site, plot, and watershed scale toward a global system scale. In the 1980s, we rediscovered the 19th century understanding that all environmental and human systems are interlinked at different temporal and spatial scales (von Humboldt 1849). Ecologists like H.T. Odum (1983) and Holling (1986) provided much of the leadership and renewal of these ideas. In the 1990s, for the first time, we came to the realization that anthropogenic (cumulative) processes are quantitatively as pervasive as biogeochemical (systemic) processes in modifying our global environment (Turner et al. 1990; Vitousek et al. 1997). And finally in the 21st century, the rapid deployment of remote sensing systems and the computational capacity to store, manipulate, and present spatially referenced data have made the visual representation of complexly interacting human and biogeochemical systems more realistic and compelling.

The three major cryosphere regions of the world are Antarctica, the Arctic Ocean, and the extra-polar snow and mountain environments. Antarctica is a continent surrounded by ocean (Fig. 1.3). East Antarctica contains 77% of the Earth's ice by volume and West Antarctica contains 10%. A further

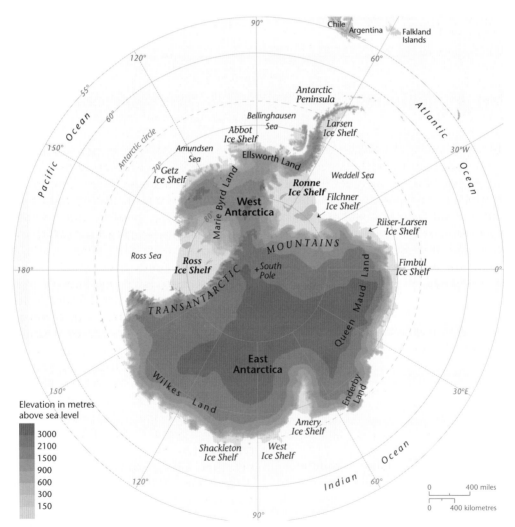

FIG. 1.3 The Antarctic cryosphere.

2.5% is found in the Antarctic ice shelves. The ice cover of Antarctica evolved in response to the opening of the Southern Ocean, first when it separated from Australia about 50 Ma but more dramatically when it finally separated from South America about 20–22 Ma. It is thought that East Antarctica has remained ice-covered throughout this period, including warmer episodes, such as the Pliocene (2–3 Ma). However, the West Antarctic Ice Sheet has probably decayed and regrown several times. In Chapter 5, we will return to this topic in more detail.

The Arctic (Fig. 1.4) is an ocean surrounded by land. Much of the surface consists of sea ice or pack ice, which circulates slowly in a clockwise direction, in response to underlying ocean currents. Most of the ice in the Arctic is contained in the Greenland Ice Sheet (9% of Earth's ice by volume) and in permafrost (0.9%). Less than 10% of Canada, Siberia, and northern Alaska is covered by glaciers and ice caps. Much of the ice-free terrain is underlain by permafrost.

The extra-polar regions (Fig. 1.5) contain only 0.5% of the Earth's ice and snow by volume. Yet, these small ice caps (Plate 1.1), mountain glaciers, snow, and alpine permafrost are extremely important in that their interaction with people is more intense than that in the circumpolar regions. Indeed, the fact that $c.5.2 \ (10^7) \ km^2$ of the extra-polar region is affected by seasonal frozen ground from time to time is a better index of the importance of the cryosphere in this region than the volumetric data. The case will be advanced in this book that, in the short- and medium term, changes of the cryosphere in the extra-polar regions are the most important ones to monitor, for at least two reasons: in these regions the cryosphere is closest to its threshold value and the anthropogenic impact is most pronounced.

We will first discuss the extra-polar cryosphere and potential effects on ocean circulation in the Arctic; then we will look at meltwater sources in the polar regions from the major ice sheets and the cumulative effect of these processes on global sea level changes and sea ice conditions; and finally we will look at major ecological impacts and some of the more urgent socioeconomic implications of cryosphere change.

1.2   THE GEOMORPHIC AND HYDROLOGIC EFFECTS OF CRYOSPHERIC CHANGE

The case for the importance of the climatic roles of snow and ice has been well and comprehensively documented. The case is made in two broad ways: first, in relation to the fundamental physical properties of snow and ice that modulate energy exchanges between the Earth's surface and the atmosphere (albedo, thermal diffusivity, and latent heat as well as surface roughness, emissivity, and dielectric characteristics) and second, in relation to the concept of residence time (flux/storage) of water within the cryosphere. Water with short residence times participates in the fast response regime of the climate system, whereas long residence time components act to modulate and introduce delays into the transient responses. Nevertheless the threat of abrupt changes in the slow response components must also be taken seriously.

The case for the importance of the geomorphic and hydrologic roles of snow and ice has also been comprehensively documented, but rarely with the same global sense as that of the climatic effects. The fact is that similar rationales can be provided for climatic, hydrologic, and geomorphic implications of cryospheric change. With respect to the fundamental physical properties of

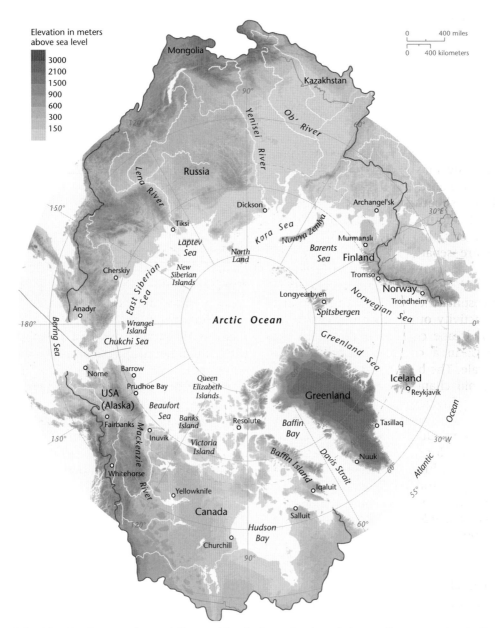

Fig. 1.4 The Arctic cryosphere, delineated by drainage basins of rivers tributary to Arctic Ocean.

snow and ice which affect hydrologic systems, it is for example, the strength, the crystal structure, the density, and the liquid water content of snow and ice that modulate both the mass and energy exchanges. In geomorphic terms, the mass exchanges of sediment are modulated by the ratio of the mean magnitude of barriers to change (lithology and available sediments) to the mean magnitudes of the disturbing forces (e.g. snowmelt runoff, jokulhlaups, or glacial erosional power) (Brunsden 1993).

Under the topic of residence times, the issue of flux to storage ratios has become a central topic, most recently sensitively discussed by Church (2002) in the context of the paraglacial concept. The distinctive characteristic of geomorphic systems is the way in which the storage term tends to dominate the mass balance. But by direct analogy with abrupt changes in the behavior of ice sheets, the history of glacial/ nonglacial alternating morphogenetic systems suggests that short periods of rapid change are separated by long periods of relative quiescence.

The practical importance of this point is that cryospheric change induced either by human activity or by climate can generate rapid landscape changes both at the local scale and, in a cumulative sense, at the global scale. The best documented example is the rapid change in discontinuous permafrost landscapes under the influence of infrastructure construction and/or progressive warming. But more dramatic are the effects of sandur landscape transformation under the influences of jokulhlaups in Iceland (Björnsson 2004) and elsewhere.

In a brilliant and compelling introduction to his book *Landscapes of Transition* (Hewitt et al. 2002), Hewitt has introduced the notion of "transition" to address the sense in which landscape development is not merely chronological and linear, or simply a "lagged" response to climatic and tectonic changes:

There are diachronous episodes of (incomplete) readjustment to the cessation of past conditions, and towards later conditions, of which those at present are only one set. There are distinctive spatial and temporal patterns of adjustment, including self adjustment specific to the earth surface processes at work. The paraglacial is a classic example. It is suggested that such temporal and spatial responses in earth surface processes apply much more generally as part of landscape transformation in the Quaternary. (Hewitt 2002, p. 2)

We have found this seminal idea helpful in interpreting cryospheric change. There are parts of the cryosphere that are in incomplete transition from past conditions, notably the Antarctic and Greenland ice sheets and Siberian, Alaskan, and Canadian permafrost. At depth, these ice sheets and this permafrost are still responding to impetus from past extremes of cold, whereas at the surface they are driven by contemporary mass and energy budgets.

Few aspects of the landscape respond instantaneously to mass and energy inputs. Some aspects of the cryosphere, such as "wet snow," do, but the majority of the components of the cryosphere resists change until subjected to stresses sufficient to deform or remove them. In Chapter 6, we shall explore the implications of complex transitions of the cryosphere on the landscape.

1.3 SUBARCTIC AND ALPINE HYDROLOGY

The major Arctic rivers originate in more temperate latitudes, where human population densities are markedly greater than those within the Arctic region proper. The melting of glaciers in temperate alpine source areas, the changing regional hydrology of the subarctic, and the degradation of permafrost, especially in the discontinuous permafrost zone, will dominate the discharge regime of the major rivers. In addition, the land use impacts associated with accelerating resource development in the subarctic affect discharge and sediment transport regimes in medium-size basins.

Most developmental and industrial activities have a higher potential for producing significant pollution in the Arctic than in more temperate regions. Environmental sensitivity associated with freeze–thaw processes; the temperature-dependent rates of chemical and biological processes that degrade pollutants; and the fragility and low

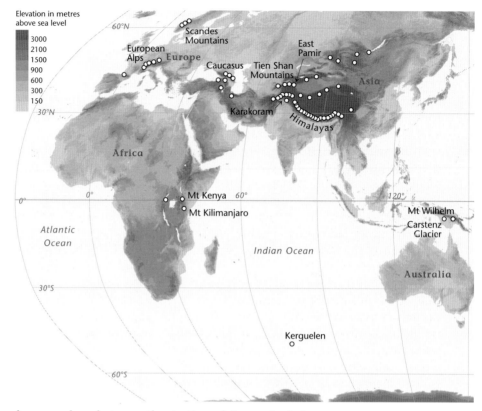

Fig. 1.5 The extra-polar cryosphere between the Arctic and Antarctic circles.

assimilative capacity of northern ecosystems may lead to cumulative and negative synergistic effects. Much pollution in the North originates from southerly latitudes and may be carried north not only by large river systems, but also by ocean currents and atmospheric processes (see Chapter 7).

The average rise in winter temperatures north of 60°N is expected to be 2–2.4 times the global average, resulting in temperature increases of 6–7°C at about 70°N (Roots 1990). Most warming is likely to occur in winter in continental locations, but only a small summer temperature increase of perhaps 1°C is anticipated. The zones of highest temperature increase should occur at the snow, ice/snow, and permafrost boundaries where the albedo changes dramatically.

Although evaporation rates may not change significantly, precipitation is expected to increase along the Arctic mainland coast and in the Arctic archipelagos up to five-fold (Roots 1990). The warmer conditions that are predicted for much of the circumpolar region will reduce the duration of winter. Seasonal snow accumulation could increase in higher-elevation zones and increased summer storminess may reduce melt at intermediate elevations as a result of increased cloudiness and summer snowfall (Woo 1996). At lower elevations, however, rainfall and rain-on-snow melt events will probably increase. There will be a shift from nival toward more pluvial runoff regimes.

Overall, present runoff to the Arctic Ocean is approximately twice that produced by precipitation minus evaporation over the Arctic Ocean. Of the total runoff to the Arctic Ocean, 70% is provided by the Ob, Yenisei, Lena, and Mackenzie rivers.

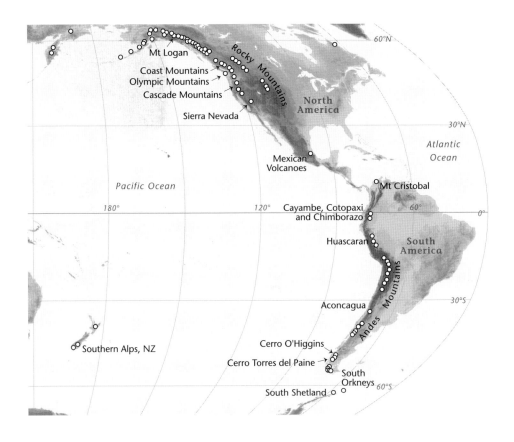

Models are predicting a total annual inflow increase to the Arctic Ocean of 10–20% with an atmospheric CO_2 doubling (Plate 1.2). Recent work on Siberian rivers emphasizes the importance of major dams, topography, and permafrost conditions in controlling discharge regimes. On the Ob and Yenisei rivers, winter snow accumulation influences summer and fall discharges whereas on the Lena River, which is further east, the winter and spring discharges are most affected (Ye et al. 2003).

The single most important point to note is that export of freshwater from the Arctic Ocean into the North Atlantic is the main coupling mechanism that links northern latitudes to the global thermohaline circulation (see Section 1.5). This export of freshwater seems very likely to increase under all scenarios modeled under a doubling of CO_2 assumption; this, as will be discussed later, will tend to reduce the intensity of that circulation, with potentially abrupt influence on the climate of the North Atlantic region.

1.4 GLACIER LOSS AND MOUNTAIN PERMAFROST

The estimated relative contribution to sea level rise that has been made by glaciers and small ice caps in the extra-polar regions over the past century is disproportionately large (Table 1.2). This is largely because the measurements of mass balances outside the polar regions are unambiguously negative, by contrast with the ambiguities of the mass balances of the major ice sheets. During the 20th century, there has been obvious thinning, mass loss, and retreat of mountain

TABLE 1.2 Estimated contributions to sea level rise over the last century in cm (from Warrick et al. 1996).

Component contributions	Low	Middle	High
Thermal expansion	2	4	7
Glaciers/small ice caps	2	3.5	5
Greenland Ice Sheet	−4	0	4
Antarctic Ice Sheet	−14	0	14
Surface water and ground water storage	−5	0.5	7
Total	−19	8	37
Observed	10	18	25

glaciers (Fig. 1.6). Nine regions (Fig. 1.6a) and nine individual glaciers with long historical records (Fig. 1.6b) demonstrate this trend and also show that such a trend will continue even if the present climatic regime were to continue unchanged. Doubling of CO_2 concentrations in the atmosphere is likely to lead to pronounced reductions in seasonal snow, permafrost, glacier, and periglacial belts of the world and a corresponding shift in landscape forming processes. Disappearance of up to 25% of presently existing mountain glacier mass is anticipated. Tropical alpine glaciers, especially those on Mount Kilimanjaro, have attracted serious concern. Increases in the thickness of the active layer of permafrost and the disappearance of extensive areas of discontinuous permafrost in mountain areas are also predicted. Borehole measurements in the European Alps (Haeberli et al. 2000) show that permafrost is warming in some areas but not everywhere.

More water will be released from regions with extensive glaciers. In temperate mountain regions, reduced duration of snow cover will cause moderation of the seasonal flow regime of rivers so that winter runoff increases and spring runoff decreases. Widespread loss of permafrost over mountain areas is expected to trigger accelerated mass movement, erosion, and sedimentation.

1.5 PERMAFROST

There is a long history of research experimentation on periglacial processes in the Arctic (Fig. 1.7).

Permafrost records temperature changes via the gradient of its subsurface temperature change with depth; it transmits these changes to other components of the environment and it facilitates further climate change by releasing trace gases. The physical expression of warming of the permafrost is the presence of thermokarst, a landscape of subsidence and highly unstable, spatially variable active layer. In unglaciated parts of Siberia, coalescing thaw depressions (known as alases), formed during Holocene warm intervals, occupy areas as large as 25 km². Frozen-ground activity has been designated as a "geoindicator" for monitoring and assessing environmental change (Berger & Iams 1996). It is critical to understand that ground ice content is highly spatially variable and that thaw settlement hazard is controlled by a combination of depth of thaw and ground ice content, not by the presence or absence of permafrost.

Permafrost regions represent 25% of the exposed land area of the northern hemisphere (Fig. 1.8). But because much of those regions is discontinuous permafrost, permafrost actually underlies 13–18% (Zhang et al. 2000). The interpretation of

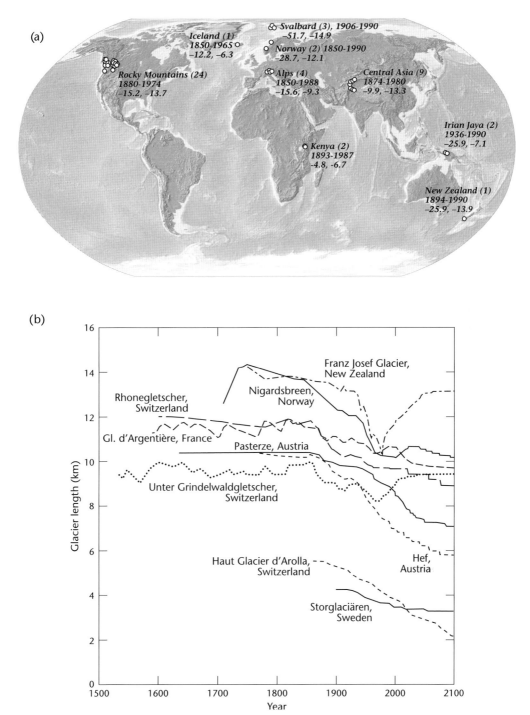

FIG. 1.6 (a) Results of an analysis of glacier length changes in km, summarized for nine regions. For each region are indicated the number of glaciers, the period of observations, the mean rate of retreat, and the mean scaled rate of retreat. The scaling effects maritime and low gradient glaciers differentially because of their greater sensitivity (after Oerlemans 2001). (b) Records of nine historic glacier lengths and projections into the future: Franz Josef Glacier (New Zealand); Glacier d'Argentière (France); Haut Glacier d'Arolla (Switzerland); Hintereisferner (Austria); Nigardsbreen (Norway); Pasterze (Austria); Rhonegletscher (Switzerland); Storglaciaren (Sweden); Unter Grindelwaldgletscher (Switzerland) (after Oerlemans 2001).

FIG. 1.7 The camp of the Polish research expedition in Arfersiorflk Fjord, West Greenland, just at the edge of ice sheet. July 1937 (photograph by the late Alfred Jahn).

these permafrost maps is a guide to those areas that are most susceptible to change as a result of predicted warming. Both thawing index and depth of permafrost are used to predict the level of hazard. It is anticipated that the area of continental permafrost may be reduced by 12–22% of its current extent under CO_2 doubling scenarios, and Smith and Burgess (1999) have predicted eventual disappearance of half of Canada's permafrost under such a scenario. The time lag involved in the thawing of thick permafrost may allow some relict permafrost to persist for millennia.

1.6   THE CARBON BALANCE OF THE CRYOSPHERE

A series of questions concerning the changing carbon balance resulting from melting permafrost and wetland hydrology under warming temperatures has arisen. In the Southern Ocean, it is projected that there will be a reduced uptake of GHGs. A coupled climate model estimate (Matear & Hirst 1999) suggests a reduced cumulative uptake by 2100 of 56 Gt. This is equivalent to a 4% a^{-1} increase in CO_2 emissions over the next century (Fig. 1.9). Whether the

Arctic will be a net source or sink of CO_2 depends largely on the direction of hydrological change and the rate of decomposition of exposed peat in response to temperature change (McKane et al. 1997a, b). Tundra ecosystems have large stores of nutrients and carbon bound in permafrost, soil, and microbial biomass and have low rates of CO_2 uptake because of low net primary production (Callaghan & Jonasson 1995). The net effect of complex processes is the likelihood that GHG emissions will increase with warming temperatures.

It is also possible that natural gas will be released to the atmosphere as a result of destabilization of gas hydrates. In the northern seas, gas hydrates are sometimes deposited in the near-bottom zone and their decomposition is likely with a comparatively small increase in temperature. Methane hydrate destabilization has been documented during Quaternary interstadials (Kennett et al. 2000).

Predicting how northern peat carbon stocks may respond to a warming Arctic climate is a complex problem that remains intractable to date. One scenario is that warming will fuel an appreciable new CO_2 source, should currently frozen or waterlogged peats experience warmer temperatures, permafrost degradation, decreased water table elevation, and enhanced aerobic decomposition. Such fluxes are potentially large: Assuming no enhanced carbon uptake by the biosphere or ocean, complete oxidation of west Siberian peatlands over the next 500 years would release sufficient carbon to the atmosphere, so as to boost the present-day rate of atmospheric CO_2 increase by 4%. In terms of net greenhouse forcing, the warming effect of such a release would likely be offset by reduced methane emission but, even in the unlikely event of a total shutdown of methane emission, would still attain at least approximately 80% of its greenhouse warming potential. Evidence for a recent slowdown or stoppage in west

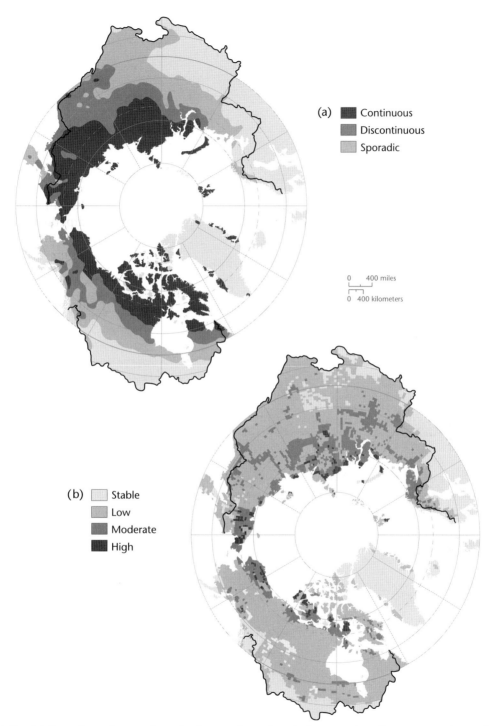

Fig. 1.8 (a) Zonation of permafrost in the northern hemisphere under climate scenario predicted by ECHAM1-A GCM (Cubasch et al. 1992) and (b) Hazard potential associated with degradation of permafrost under ECHAM1 – a climate change scenario. Classification is based on a thaw settlement index calculated as the product of existing ground ice content (Brown et al. 1997) and predicted increases in the depth of thaw (Anisimov et al. 1997; Anisimov & Fitzharris 2001).

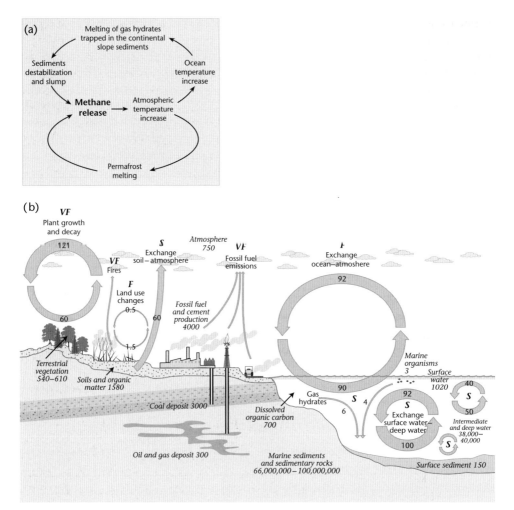

FIG. 1.9 (a) Positive feedback loops in carbon cycling resulting from methane release and melting of frozen gas hydrate. (b) Global carbon cycling, slow (s), fast (f), and very fast (vf) cycling is discriminated. Storage in billions of tonnes and flux of carbon in billions of tonnes per annum (from Arctic Climate Impact Assessment 2004).

Siberian peat accumulation does exist, and detailed studies of contemporary peat accumulation rates are needed to confirm or disprove this possibility (Smith et al. 2004).

1.7   RIVER AND LAKE ICE BREAK-UP AND FREEZE-UP

Chronologies of river and lake ice formation and disappearance provide broad indicators of climate change over extensive lowland areas. Broad scale patterns of freeze-up are available for Russia from 1893 to 1985 (Soldatova 1992). In general, freeze-up in western Russia is 2–3 weeks later now than at the turn of the century, whereas further east there is a slight trend toward earlier freeze-up (Ginzburg et al. 1992). Similar patterns are available for ice break-up dates, with western Russian rivers breaking up 7–10 days earlier now than in the

19th century (Soldatova 1993) but in central and eastern Siberia, break-up is later, giving an overall expansion of the ice season. In North America, records from 1823 to 1994 at six sites on the Great Lakes show that freeze-up came later and break-up was earlier until the 1890s but they have remained constant during the 20th century (Assel et al. 1995; Fig. 1.10). According to Magnusson et al. (2000), freeze-up and break-up dates of ice on lakes and rivers provide consistent evidence of later freeze-up and earlier break-up around the northern hemisphere from 1846 to 1995. They examined 39 time series from Russia, Finland, Japan, Canada, USA, and Switzerland.

Under conditions of overall annual warming, the duration of river ice cover can be expected to be reduced (see Chapter 3). Many rivers within temperate regions

would tend to become ice-free, whereas in colder regions the present ice season could be shortened by up to one month by 2050. Warmer winters would cause more mid-winter break-ups as rapid snowmelt becomes more common. Warmer spring temperatures could affect the severity of the break-up, but the effect is the result of a complex balancing between downstream resistance (ice strength and thickness) and upstream forces (flood wave). Although thinner ice produced by a warmer winter would tend to promote a thermal break-up, this might be counteracted by the earlier timing of the event, reducing break-up severity (Prowse et al. 1990; Fig. 1.11).

The onset of thawing has been imaged over Siberia (Plate 1.3) from 1992 to 1999 and demonstrates the large temporal and spatial variability of this part of the cryosphere.

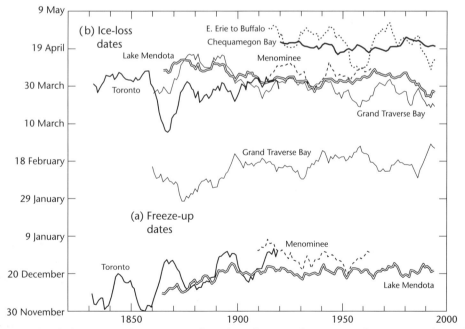

FIG. 1.10 Documented changes in freeze-up and ice loss dates in the Great Lakes region of North America. (a) Chequamegon Bay (1911–94); (b) Grand Traverse Bay (1851–1994); (c) Menominee (1900–63); (d) Lake Mendota (1856–1994); (e) Eastern Lake Erie (1905–94); and (f) Toronto Harbour (1823–1919) (from Assel et al. 1995).

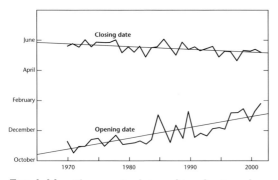

FIG. 1.11 Changes in dates of tundra travel on Alaska's North Slope (from Arctic Climate Impact Assessment 2004).

1.8   OCEAN CIRCULATION

The global thermohaline circulation ("conveyor belt") is driven primarily by the formation and sinking of deep water in the Norwegian Sea (Fig. 1.12). It is controlled by two counteracting forces operating in the North Atlantic: a thermal forcing, from low-latitude heating to high-latitude cooling, and a haline forcing, from high-latitude freshwater gain to low-latitude evaporation (Rahnstorf 1997). In today's North Atlantic, the thermal forcing dominates. When the strength of the haline forcing increases due to excess runoff and/or ice melt, the conveyor belt will weaken or may even shut down. It is predicted from numerical simulations that small changes in the external forcing at high latitudes can switch off the present conveyor belt and force the ocean to approach another equilibrium state in which the Antarctic Circumpolar Front is the only deepwater source. Periodic movement of excessive sea ice from the Arctic Ocean into the Greenland Sea is one mechanism that appears to be responsible for the interdecadal variability of the conveyor belt (Mysak et al. 1990). Modeling of ocean circulation suggests increasing stability of the surface mixed layer, reduction in salt flux, less ocean convection, and less deep water formation. This could lead to abrupt climate change and substantial impacts.

Paleorecords show that there have been millennial oscillations of several degrees of the sea surface temperature in the North Atlantic during the interval 10–80 ka BP. These millennial cycles, called Dansgaard–Oeschger cycles, contribute cumulatively to long-term cooling cycles which culminate in an enormous discharge of icebergs into the North Atlantic (Heinrich events) and followed by an abrupt shift to a warmer climate. The interesting question, which is difficult to resolve, is exactly how abrupt these climate shifts can be, but a decadal timescale is probable (Dansgaard et al. 1993). The regional climate of Europe and the North Atlantic would be directly impacted.

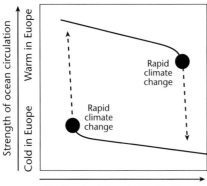

FIG. 1.12 Rapid climate change in Europe as a function of changing freshwater inflow to the North Atlantic Ocean and changing strength of ocean circulation (from Rahnstorf 1997).

1.9   THE MASS BALANCE OF THE POLAR ICE SHEETS

The West Antarctic Ice Sheet is grounded below sea level in many places. It is possible that the ice sheet is buttressed by the surrounding ice shelves and that a small rise in sea level or an increase in the melting rates beneath the ice shelves could trigger ice calving and the collapse of the ice sheet. It has long been argued (Weertman 1974) that this ice sheet may be inherently unstable because the interior grounded ice cannot respond fast enough to changes in thickness of the floating portions at the grounding line. It is now known that the ice sheet is dominated by fast flowing ice streams (Fig. 1.13; Plates 1.4 & 1,5) whose response times to changes in the grounding line appear to be very rapid (Alley & Whillans 1991). There is at present no agreement on whether these dynamic ice streams increase or decrease the stability of the ice sheet. This ice sheet, by contrast with the East Antarctic Ice Sheet, does have a dynamic history but it is thought that if a collapse were to occur, it would be due more to changes in climate of the last 10 ka rather than to contemporary climate change. The probability of the West Antarctic Ice Sheet surging within the next century "may be relatively remote, but not zero."

The response of Greenland to warming is likely to be different. Both the melt rates at the margins and the accumulation rates in the interior should increase. Because the melt rates are likely to be greater than the accumulation rates, the mass balance will become negative. In southwest Greenland there has been a general retreat of outlet glaciers since the end of the 19th century and general trends have been extrapolated together with the net effect of predicted warming over the next two centuries. But the documented retreat has slowed down and a significant number of glaciers is now advancing.

Recent break-ups of the Larsen and Wordie ice shelves in the Antarctic Peninsula, discharges of huge icebergs from the Filchner and Ross ice shelves, and the discovery of major recent changes in some Antarctic ice streams have focused attention on the possibility of collapse of Antarctic ice shelves

within the next century (Cook et al. 2005). But in Antarctica as a whole, temperatures are so low that comparatively little surface melting occurs and the ice loss is mainly by iceberg calving, the rates of which are determined by dynamic processes with long response times.

Mass balance studies have large error estimates because of the sheer vastness of the ice sheet. In principle, summing the balance observations at the surface of the ice and comparing with loss of ice by calving rates should allow a net balance to be calculated. If Antarctica were to warm in the near future, its mass balance would be positive. The rise in temperature would be insufficient to initiate melt but would increase snowfall.

At all events, no one is seriously predicting a collapse of the entire Antarctic ice sheet within the next century. Conventional wisdom has stated that the ice sheet has been stable since 20–22 Ma (Kennett 1978). However, diatom evidence has raised the possibility of unstable behavior in the relatively recent past (2.5–3 Ma) (Barrett 1991; Webb & Harwood 1991) and on that basis the probability of early collapse is very nearly but not quite zero.

1.10 Sea level

There is much debate over the relation between climate change and sea level. Sea level changes at any location are affected by a combination of local, regional, and global factors, but the general trend from six coastal stations with long records is consistent (Fig. 1.14). Global mean sea level has risen 10–25 cm over the last 100 years (Houghton et al. 1996); there has been no detectable acceleration of sea level rise during this period, though the average rise during the last 100 years is significantly higher than the rate averaged over the last several thousand years. Possible climate-related factors

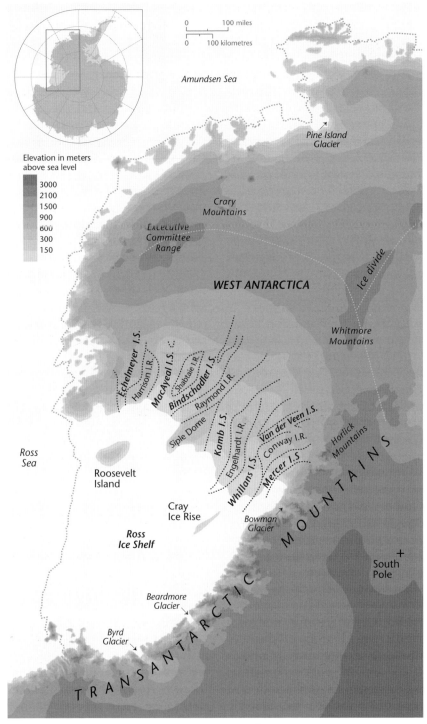

Fig. 1.13 West Antarctic Ice Sheet, Ice Streams (I.S.) and Ice Rises (I.R.).

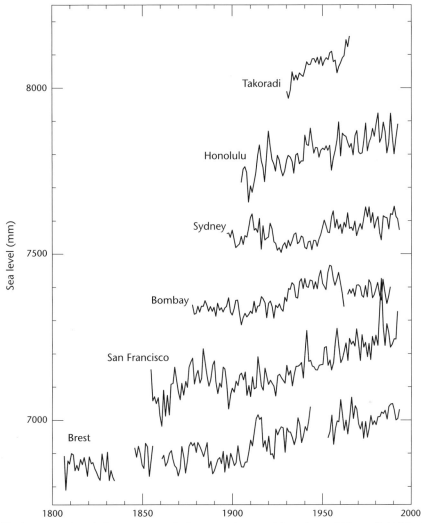

Fig. 1.14 Six long sea level change records from major world regions: Takoradi (Africa), Honolulu (Pacific), Sydney (Australia), Bombay (Asia), San Francisco (North America), and Brest (Europe). The observed trends (in mm yr^{-1}) for each record over the 20th century are, respectively, 3.1, 1.5, 0.8, 0.9, 2.0, and 1.3. The effect of postglacial rebound as simulated by the Peltier ICE-3G model is less than, or of the order of, 0.5 mm yr^{-1} at each site (from Houghton et al. 1996).

contributing to this rise include thermal expansion of the ocean and melting of glaciers, ice caps, and ice sheets. Thermal expansion over the last 100 years may have produced 2–7 cm rise (Table 1.2); general retreat of glaciers can account for 2–5 cm; contributions from the Greenland and Antarctic ice sheets are ambivalent.

Projections of future changes until 2100 are 9–88 cm, with most of this sea level rise due to thermal expansion and melting of glaciers and ice caps (Table 1.3). The contribution of the major ice sheets is assumed to be minor, but they are a major source of uncertainty. Not only is there a lack of understanding of the current mass balance,

TABLE I.3 Estimates of future global sea level rise in centimeters (from Warrick et al. 1996).

Source (emission scenario)	Sea level rise component (cm)				Total rise (cm)*		
	Thermal expansion	Glaciers and icecaps	Greenland Ice Sheet	Antarctic Ice Sheet	Best estimate	Range†	To (year)
IPCC90-A (Warrick & Oerlemans 1990)	43	18	10	−5	66	31–110	2100
Church et al. (1991)‡	25	[. . .]	10 (all ice)	. . .]	35	15–70	2050
Wigley and Raper (1992)§	[. . .]	. . .]	Not specified	. . .]	48	15–90	2100
Wigley and Raper (1993)	25	[. . .]	21 (all ice)	. . .]	46‖	3–124¶	2100
Titus and Narrayanan (1995)**	21	9	5	−1	34	5–77††	2100
IPCC projections, this report‡‡	28	16	6	−1	49	20–86	2100
This report (Section 7.5.3.2)§§	15	12	7	−7	27		2100

* In most cases from 1990.
† No confidence intervals given unless indicated otherwise.
‡ Assumes rapid warming of 3°C by 2050 for best case.
§ For IPCC emission scenario IS92a.
‖ For IPCC (1990) Scenario B, best estimate model parameters.
¶ For IPCC (1990) Scenarios A and C, with high and low model parameters, respectively.
** Incorporates subjective probability distributions for model parameter values based on expert opinion.
†† Represents 90% confidence interval.
‡‡ For the IPCC IS92a forcing scenario, using a climate sensitivity of 2.5°C for the mid projection and 1.5°C and 4.5°C for the low and high projections, respectively. Also see Raper et al. (1996).
§§ For IPCC IS92a forcing scenario, with a constant 2.2°C climate sensitivity.

but there is also considerable uncertainty over the possible dynamic responses on timescales of centuries. The likelihood of a major sea level rise by 2100 due to collapse of the West Antarctic Ice Sheet is considered low, but not zero.

It is instructive to note that sea level rise in northwest England during the catastrophic melting of the Laurentide Ice Sheet about 7.8 ka BP (^{14}C years) ago ranged from 34 to 44 mm·a^{-1} (Tooley 1978). All this uncertainty points to the need for better monitoring of the major ice sheets and a continuing urgent search for information on thresholds of surging glacier behavior.

1.11 IMPORTANCE OF SEA ICE

Sea ice is a thin layer of ice floating on the sea surface that forms when the temperature of the sea falls below its freezing point. The salinity of sea water depresses the freezing point to about −1.9°C. Standard terms as defined by the World Meteorological Organization (WMO) (1970) are listed in Table 1.4. The more recent Sea Ice Glossary (WMO 2005) should also be consulted. Sea ice is a predominant feature of the polar oceans (Fig. 1.15). It has a dramatic effect on the physical characteristics of the ocean surface. The normal exchange of heat and mass between the atmosphere and ocean is strongly modulated by sea ice, which isolates the sea surface from the usual atmospheric forcing. Sea ice also affects albedo, exchange of heat and moisture with the atmosphere, the thermohaline circulation of the ocean, and the habitats of marine life. The complexity of the process of change from open sea water to sea ice, and the importance of albedo changes, is explained by the sea ice terminology of gray, gray white, and white ice.

Warming is expected to cause a reduction in the area covered by sea ice, which allows increased absorption of solar radiation and a further increase in temperature. This sea

TABLE 1.4 Sea ice terminology (adapted from WMO 1970).

General	Technical term	Description
New ice (<10 cm thick)	Frazil ice	Fine spicules or plates of ice suspended in water. Gives water an oily appearance
	Grease ice	Ice crystals coagulate to form a soupy layer
	Slush	Snow which is saturated. A viscous floating mass after snow fall
	Nilas	A thin elastic crust of ice (interlocking fingers)
	Pancake ice	Circular pieces of ice (30 cm–3 m diameter) with raised rims
Young ice (10–30 cm thick)	Grey ice	10–15 cm thick. Forms rafts under pressure
	Grey white ice	15–30 cm thick. Forms ridges under pressure
First year ice (30 cm–2 m thick)	White ice	30–70 cm thick
	Medium ice	70–120 cm thick
	Thick ice	>120 cm thick
Old ice	Second year ice	Has survived one summer's melt. Thicker and less dense than First year ice. Stands higher in the water
	Multiyear ice	Old ice up to 3 m or more thick. Common in the Arctic but mostly confined to the Weddell Sea in the Antarctic

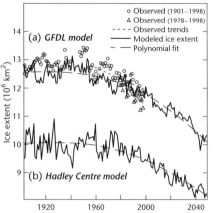

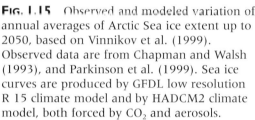

FIG. 1.15 Observed and modeled variation of annual averages of Arctic Sea ice extent up to 2050, based on Vinnikov et al. (1999). Observed data are from Chapman and Walsh (1993), and Parkinson et al. (1999). Sea ice curves are produced by GFDL low resolution R 15 climate model and by HADCM2 climate model, both forced by CO_2 and aerosols.

ice–albedo positive feedback has the potential to produce an ice-free Arctic Ocean that is irreversible.

Arctic sea ice extent decreased by approximately 3% per decade from 1978 to 1996. Summer sea ice extent has shrunk by 20% over the past 30 years in the Atlantic part of the Arctic Ocean but the shrinkage has only been 5% in the Canadian Arctic Sea. But the data are highly variable from year to year and region to region (Fig. 1.16). Overall sea ice extent in April has been reduced in the Nordic seas by 33% over the last 135 years (Vinje 2001). Models predict sea ice extent in the Arctic Ocean to be reduced by 20% by 2050 (Vinnikov et al. 1999). In the Antarctic Ocean, Wu et al. (1999) concluded that sea ice extent has been reduced by 0.4–1.8° of latitude over the 20th century. This generalization hides a high degree of spatial and temporal variability, and no significant trend prevails over the satellite era (1973–96). The sector from 0–40°E shows significant increase

in sea ice cover and the sector from 65–160°W has a significant decrease. Models predict a reduction in sea ice extent of 45% by volume with a doubling of CO_2 (Wu et al. 1999).

1.12 ECOLOGICAL IMPACTS

What will be the response to predicted climate change of the structure of marine communities and the overall productivity of polar oceans? Existing models do not provide quantitative answers. There is some risk of unforeseen collapse of parts of the marine biological system with consequent global effects, particularly on fisheries.

A shift in the global thermohaline circulation would be serious in both the Arctic and Antarctic oceans. Biology through the entire food chain from algae to higher predators will be strongly affected. Warmer water in the Southern Ocean will intensify biological activity and increase growth rates of fish. Changes in sea ice distribution will alter the seasonal distribution, geographic ranges, patterns of migration, nutritional status, reproductive success, and abundance and balance of species. Biological activity, especially in lakes and ponds, will increase. Enhancement of winter flow will mean that streams that currently freeze to their beds will retain a layer of water below the ice and this will be beneficial to invertebrates and fish populations. Less sea ice will reduce ice edges, which are prime habitats for marine organisms. There will be major ecological implications for the Arctic, especially habitat loss for certain species of seal, walrus, and polar bears (Fig. 1.17).

Ice shelves on the Antarctic Peninsula are expected to continue to retreat, expose more bare ground, and cause changes in the ecology, especially with the introduction of exotic plants and animals. Poleward migration of taxa from boreal forest to the Arctic

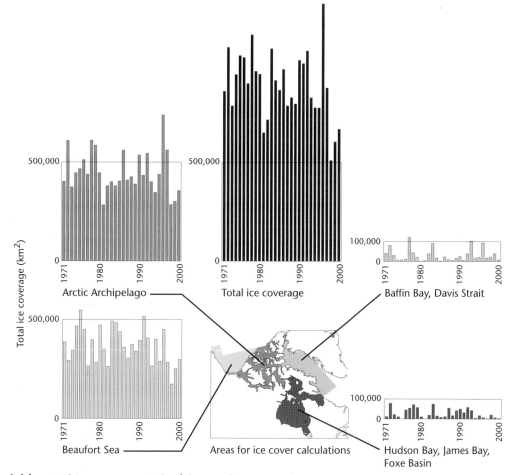

FIG. 1.16   Total ice coverage in km² for northern Canadian waters (September 10) 1971–2000 (from Environment Canada 2002).

tundra (Fig. 1.18) will depend not only on warming climate but also on dispersal rates, colonization rates, and species interactions and, therefore, may exhibit substantial time lags. The Arctic historically has experienced fewer invasions of weeds and other exotic taxa than other regions. Some of the most important changes in diversity in the Arctic may be changes in the abundance of caribou, waterfowl, and other subsistence resources. Reduction in frequency and severity of ice-jam flooding will have a serious effect on northern riparian ecosystems,

particularly the highly productive river deltas, which depend on frequent inundation for their source of rich substrate. Changes in community composition and productivity may be particularly pronounced in the High Arctic, where much of the surface is currently unvegetated and is prone to establishment and expansion of additional vegetation (Wookey et al. 1993; Callaghan & Jonasson 1995). Degradation of permafrost also causes complex response in Arctic and Antarctic biota. Thermokarst in relatively warm, discontinuous permafrost

FIG. 1.17 Polar bears are dependent on sea ice where they hunt seals and use ice corridors to move from one area to another (photograph from Arctic Climate Impact Assessment 2004).

in central Alaska has transformed some upland forests into extensive wetlands (Osterkamp et al. 2000), whereas in the northern foothills of the Brooks Range, plant density may increase and prevent increases in active layer thickness (Walker et al. 1998). Climatic changes observed in the European Alps have been associated with upward movement of some plant taxa of 1–4 m per decade and the loss of some taxa that were formerly restricted to high elevations. In the Coast Mountains of British Columbia (Fig. 1.19), Brink et al. (1967) documented a rising tree line during the first half of the 20th century.

1.13 SOCIOECONOMIC EFFECTS

The expected smoothing of the annual runoff amplitude could be beneficial or adverse, depending on whether the region is relying on energy production during the winter or water supply for irrigation during the summer. The hydroelectric industry may benefit from moderation of the stream flow regime. In some semiarid areas, like Argentina or central Asia, glacial runoff may increase water resources, but in other places, such as interior British Columbia, summer water resources will diminish. Of all the river ice processes, ice jams are the major source of economic damage. Less river ice and a shorter ice season in northward flowing rivers in Canada, Russia, and Siberia should enhance north–south transport. Reduced snow cover and glaciers will detract from the scenic appeal of many alpine landscapes. For temperate mountains, less snow will restrict alpine tourism and limit the ski industry to higher alpine areas than at present. Detrimental socioeconomic impacts on mountain communities that depend on winter tourism will occur. Buildings, other structures, pipelines, and communication links will be threatened by reduced slope stability. Weller and Lange (1999) note that the bearing capacity of permafrost has decreased with warming, resulting in failure of pilings for buildings and pipelines as well as road beds. The extensive development of thermokarst will cause damage to infrastructure. Engineering and agricultural practices will need to adjust to changes in ice, snow, and permafrost distribution. New building codes for roads, railways, and buildings to cope with the effects of permafrost thawing will be needed. It is anticipated that there will be lower operational costs for the oil and gas industry, lower heating costs, and easier access for ship-based tourism. Increased costs will include infrastructure on permafrost and reduced transportation possibilities across frozen ground and water. Thawing of permafrost could lead to disruption of existing petroleum production and distribution systems in the tundra.

Extraction of oil and mineral resources is likely to be the greatest direct human disturbance in the Arctic. Although the spatial

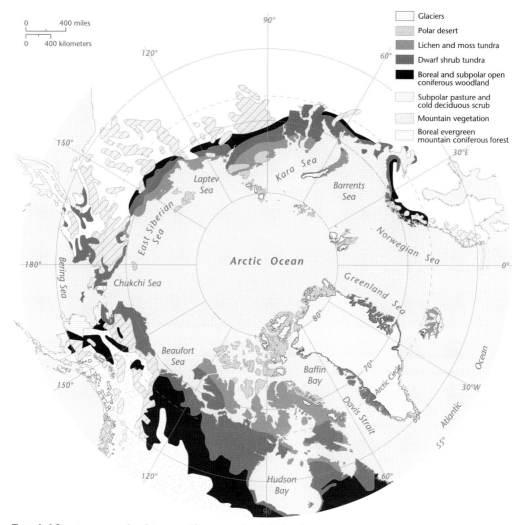

FIG. 1.18 Circumpolar biomes (from Godard & Andre 1999).

extent of these disturbances is small, their impacts can be far reaching because of road and pipeline systems associated with them. Roads, in particular, open previously inaccessible areas to new development either directly related to tourism and hunting or to support facilities for resource extraction.

In alpine ecosystems, grazing, recreation, and other direct impacts will be the most serious. Many of the alpine areas with greatest biodiversity, such as the Caucasus and the Himalayas, are areas where human population pressures may lead to the most pronounced land use impacts in alpine zones. Improved road access to alpine areas often increases human use for recreation, mining, and grazing and increased forestry pressure at lower elevations. Overgrazing and trampling by people and animals may destabilize vegetation, leading to erosion and loss of soils that are the long-term basis of the productive capacity of alpine ecosystems. Alpine areas that are downwind of human population or industrial centers experience

Fig. 1.19 Black Tusk is a volcanic plug located in Garibaldi Provincial Park, British Columbia. At this location there have been scientific investigations of changing tree line (Brink et al. 1967), permafrost conditions, and rates of gelifluction (Mackay & Mathews 1974).

substantial rates of nitrogen deposition and acid rain. Continued nitrogen deposition at high altitudes, along with changes in land use that lead to soil erosion can threaten provision of clean water.

A shift in the global thermohaline circulation would influence the climate of Europe and the North Atlantic with attendant impacts on tourism and local job creation possibilities. Changes in sea ice distribution are more serious for human activity in the Arctic than in the Antarctic. In the Arctic, loss of sea ice will affect indigenous peoples and their traditional life style (Fig. 1.20). There are major uncertainties about the ability of the local communities to adapt. Although most indigenous peoples are highly resilient, the combined impacts of climate change and globalization create new and unexpected challenges. Because their livelihood and economy

increasingly are tied to distant markets, they will be affected not only by climate change in the Arctic but also by other changes elsewhere. Local adjustments in harvest strategies (Fig. 1.21) and in allocation of labor and capital will be necessary. Maintenance of self-esteem, social cohesion, and cultural identity of communities will be the greatest threat. Changes in sea ice, seasonality of snow, and habitat and diversity of food species will affect hunting and gathering practices and threaten long-standing traditions and ways of life. Loss of sea ice in the Arctic will provide increased opportunities not only for fishing and new settlements but also for wider dispersal of pollutants. Collectively, these changes emphasize the need for an adequate infrastructure to be in place before they occur. Disputes over jurisdiction in Arctic waters, sustainable development of fisheries and other marine

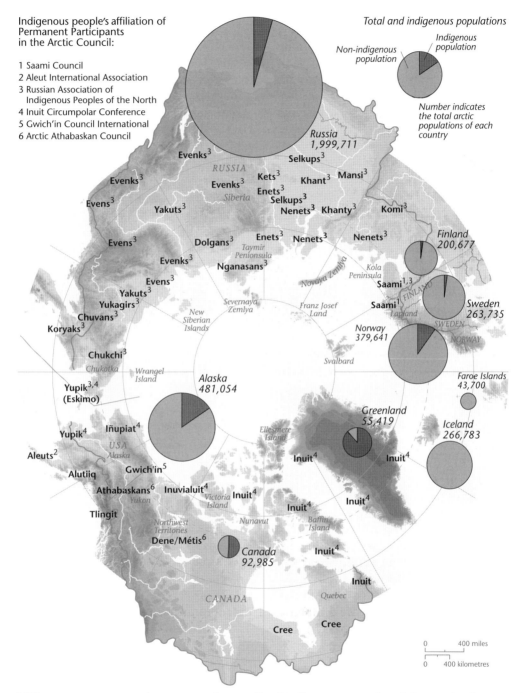

Indigenous people's affiliation of Permanent Participants in the Arctic Council:

1 Saami Council
2 Aleut International Association
3 Russian Association of Indigenous Peoples of the North
4 Inuit Circumpolar Conference
5 Gwich'in Council International
6 Arctic Athabaskan Council

Total and indigenous populations

Non-indigenous population

Indigenous population

Number indicates the total arctic populations of each country

Evenks[3]
Selkups[3]
RUSSIA
Evenks[3]
Evenks[3]
Kets[3] Khant[3] Mansi[3]
Evens[3]
Enets[3]
Selkups[3]
Yakuts[3]
Nenets[3] Khanty[3] Komi[3]
Siberia
Russia 1,999,711
Finland 200,677
Evens[3]
Dolgans[3]
Enets[3] Nenets[3]
Nenets[3]
Taymir Peninsula
Evenks[3]
Nganasans[3]
Novaya Zemlya
Kola Peninsula
Saami[1,3]
Evens[3]
Yakuts[3]
Severnaya Zemlya
Franz Josef Land
Saami[1]
Lapland
Sweden 263,735
Yukagirs[3]
New Siberian Islands
Norway 379,641
Chuvans[3]
Koryaks[3]
Chukchi[3]
Chukotka *Wrangel Island*
Svalbard
Faroe Islands 43,700
Yupik[3,4]
(Eskimo)
Alaska 481,054
Greenland 55,419
Iceland 266,783
Yupik[4] Iñupiat[4]
Ellesmere Island
Aleuts[2]
USA Alaska
Inuit[4] Inuit[4]
Alutiiq
Gwich'in[5]
Athabaskans[6] Inuvialuit[4] Inuit[4]
Victoria Island
Inuit[4] Inuit[4]
Tlingit
Yukon
Northwest Territories
Baffin Island
Nunavut
Dene/Métis[6]
Canada 92,985
Inuit[4]
Inuit
CANADA
Quebec
Cree Cree

0 400 miles
0 400 kilometres

FIG. 1.20 Associations, conferences, and councils of indigenous peoples of the Arctic (from Arctic Climate Impact Assessment 2004).

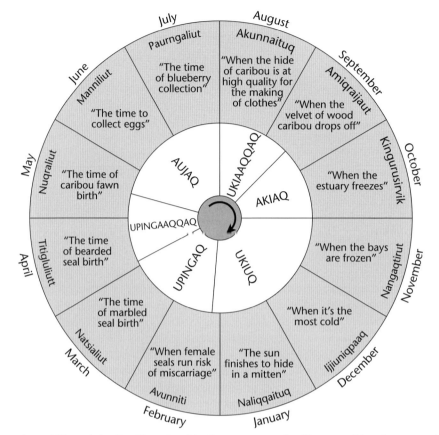

FIG. 1.21 Vulnerability of the livelihoods of indigenous peoples illustrated by the annual calendar of the Inuit at Nunavik in Hudson Bay (from Godard & Andre 1999).

resources, and construction of navigational aids and harbor facilities, as well as problems arising from oil and gas development, including pollution and environmental monitoring will all have to be resolved by polar and associated nations. By contrast, jurisdiction over the Antarctic is likely to be unaffected by environmental change unless some new resources are suddenly more accessible.

1.14 CONCLUSIONS

The cryosphere is one of the most important yet least understood subsystems of the

Earth system. All of the Earth's landscapes that have been influenced by glaciers and ice sheets, permafrost, sea and freshwater ice, and snow cover and snow avalanches have been transformed in both surface and subsurface expression (see Chapter 6). Alternation of glacial and interglacial stages on an approximate 100 ka cycle over the past 2.5 Ma has generated huge fluctuations in the extent of the cryosphere (see Chapter 5). Changes in the cryosphere system at a seasonal and event timescale are also significant (see Chapter 4). In short, the cryosphere is the most sensitive of environmental systems at the Earth's surface. Spatially, the cryosphere is also highly variable in so far as it is controlled by latitude, altitude, continentality, and aspect. Changes in the cryosphere in response to global warming are likely to be observed initially in transition zones: around the edges of glaciers and ice sheets, in the zone of discontinuous permafrost, and in the subalpine–alpine ecotone. Furthermore, midlatitude pockets of the cryosphere are already responding to climate changes with many tropical glaciers facing complete ablation within the next 10–20 years based on current retreat rates.

Many ecological and socioeconomic reasons for focusing on the cryosphere have been identified. Both new opportunities and new challenges will be presented as a result of global environmental change. These issues can be summarized as new opportunities for major resource development companies; new challenges in the development of construction techniques in thawing permafrost and drilling structures in newly ice-free coastal waters; new challenges with respect to environmental security under permafrost degradation and land use changes; new opportunities to learn from environmental and traditional knowledge in informing understandings of the circumpolar world; challenges to, and possible disappearance of, traditional lifestyles; and challenges to sovereignty and nation-state security with competing claims to the Northwest and Northeast Passages. Some of these issues, though by no means all of them, will be pursued in Chapter 7.

It has been claimed that the severe and rapid environmental changes imposed by alternating glacial and interglacial stages were responsible for the present genetic structure of human populations (Hewitt 2000). If we are, in a sense, created by cryospheric change, is it not then one of our highest priorities to preserve and enhance human well-being during a time of rapid cryospheric change?

2

TRADITIONAL *IN SITU* APPROACHES TO THE MONITORING OF CRYOSPHERIC CHANGE

2.1 INTRODUCTION

What sets the cryosphere apart from other hydrological environments is its relative inaccessibility to human investigation simply through its geographical remoteness and extreme thermal regime. As a result, and until relatively recently, fewer studies of the cryosphere have been conducted compared with studies of temperate hydrological processes. Not only are financial costs large for studies of the cryosphere, there are often fewer field researchers available and prepared to subject themselves to subzero temperatures for long periods of time. That being said, and with technological improvements, there have been more measurement studies in the past 30–40 years to help us understand better cryosphere dynamics. With advances in modern remote sensing methods, scientists are now beginning to be able to understand better the nature of key cryospheric parameters. Through careful use of traditional historical ground-based measurement records and sophisticated numerical models, scientists are now beginning to extend the measurement record back through history in order to understand the longer-term variations of important cryosphere variables and how they might affect the Earth system.

In this chapter, we examine how the cryosphere can be "measured." For example, in any given winter season, terrestrial snow cover extent can extend to more than 50% of the northern hemisphere's total land area and contract to less than 1% during summer (Robinson et al. 1993). This rapidly varying cryospheric variable, therefore, requires measurement strategies that account for its dynamic change in extra-polar regions. In other words, measuring strategies must account for variations of key parameters in extra-polar regions that contribute directly to the cryosphere's status. Our approach, therefore, necessarily forces us to consider the cryosphere as an integrated system of the polar and extra-polar regions. If we are to fully understand the nature of the cryosphere, we must be able to observe mass and energy exchanges through time not only in the polar regions but also in the extra-polar regions.

Measuring the cryosphere is not a straightforward undertaking. By the sheer size and remoteness of much of its geographic location, to fully understand its dynamic nature, measurements of key variables have been made using a variety of different traditional approaches. Some methods have been used for a long time while others are undergoing continual refinement. This chapter is separated into sections that describe measurement approaches and data sources that are essential for our understanding of different parts of the cryosphere. Some measurements are considered operational in as much as they are routinely managed through automated networks while most are the result of either concerted field experiments that may be discrete in time, or measurements made by remote sensing instruments many of which were not developed expressly for cryosphere applications. The following sections describe data used for cryospheric studies from *in situ* measurements. In each section, we identify the variable of interest (snow, sea ice, ice sheets, alpine glaciers, permafrost, and river runoff) and the measurement approach taken, including the associated errors and uncertainties of measurement. We also describe efforts to "spatialize" the measurements through regional operational networks of measurements or more locally through significant discrete experiment campaigns and identify how the measurements might be improved either through process improvement or network improvement. Throughout, we outline some of the recent and planned initiatives to expressly enhance our understanding of the cryosphere through internationally coordinated and integrated sustained measurement programs.

It is not our purpose to give an exhaustive account of all measurement methods and subsequent data sources available for cryospheric studies. Several texts exist that describe aspects of measurement approaches to specific components of the cryosphere in detail. For example, in Bamber and Payne (2004), there is a full account of *in situ* measurement, remote sensing, and modeling approaches for ice sheets and sea ice. The chapter by Collins in Goudie (1990) gives an excellent account of alpine glacier hydrology measurement approaches. Gray and Male (1981) is a standard reference text for most snow hydrologists. Our intent, therefore, is to provide an overview of the key data sources that have been compiled for which rigorous analysis of the cryosphere is possible. This chapter, therefore, is an important one for the book as later discussion is based on the data sources described here.

> **A note on data and measurements**: In this chapter we attempt to be consistent about our use of the terms data and measurement. While the difference might seem inconsequential to many, there are significant reasons for this distinction. Hand (2004) suggests a definition of measurement as "quantification: the assignment of numbers to represent the magnitude of attributes of a system we are studying or wish to describe." This definition underlies our discussion here as it implies that measurements are measurement **processes** that are subject to errors introduced by the way the measurement process was designed and executed. The word "represent" is also important because it indicates that recorded numbers from a measurement process are merely representations of the real world state; errors of the measurement processes can bias or distort that representation of the real world. Finally, the term data is also used to indicate assignment of numerical values to the measurement process outcome; data are not synonymous with measurement processes! While to some these definitions might seem obvious, it is important to state them early on because in subsequent sections we discuss cryosphere data sources assigned to different measurement processes that represent cryospheric states. The data sources, therefore, have different heritages and pedigrees (depending on the maturity of the measurement processes) and, therefore, have different biases associated with their representational capabilities.

2.2 IN SITU MEASUREMENTS

The cryosphere can be represented by several different geophysical state variables. Traditionally, in situ measurements have been used to quantify the cryosphere's state both in the northern and southern hemispheres. There are challenges with these measurements primarily because while they may be precise, accurate, and representative of the site-specific location, they may not necessarily be representative of the surrounding region. Great efforts, therefore, are made to understand the representativity of these data and the uncertainty of measurement process. Measurement uncertainties are often the starting point for discussions about the validity of interpretations drawn from the instrument record so it is important to understand whence these uncertainties derive.

The construction of distributed but linked data archives of *in situ* measurements of cryospheric variables has begun to make the analysis of regional to global environmental cryospheric issues more straightforward. For example, through sponsorship by international bodies such as the United Nations, snow accumulation, glacier mass balance, and permafrost depth measurements are all available for *in situ* sites through the United Nation's Global Observing Systems which are comprised of the Global Climate Observing System (GCOS), the Global Terrestrial Observing System (GTOS), and the Global Ocean Observing System (GOOS) (see <http://earthwatch.unep.net/data/g3os.php>). These three observing systems were formulated and are cosponsored by five United Nations organizations: World Meteorological Organization (WMO), the United Nations Environment Programme (UNEP), the International Council for Science (ICSU), Intergovernmental Oceanic Commission (a part of UNESCO), and the Food and Agricultural Organization (FAO). In addition, each of the three observing systems are sponsored by more discipline-specific organizations. For example, GTOS is also sponsored by the United Nations Educational, Scientific and Cultural Organization (UNESCO) and the FAO. Within each program, there are several efforts to organize data to make them available to the scientific community.

2.2.1 LAND SURFACE AIR TEMPERATURE

We begin this section by briefly noting that measurements of land surface temperature are sometimes used to characterize the global cryosphere. The literature on global temperature measurements from in situ sources is large and temperature, specifically air temperature, measured at standard surface meteorological weather stations is one of several important variables used to characterize the cryosphere. Indeed, in situ temperature measurements have been used in the periodic reports of the Intergovernmental Panel on Climate Change (Houghton et al. 2001, see <http://www.ipcc.ch/>). Temperature measurements are the basis of work by several researchers, especially Jones (1994), Hansen and Lebedeff (1988), and Vinnikov et al. (1990). For example, Jones (1994) used 2,961 measurements from weather stations around the world that adhere to World Meteorological Organization (WMO) climate station standards to quantify global temperature trends. A total of 998 measurement stations in this study (at January 1, 2005) are publicly available through GCOS (<http://www.wmo.ch/web/gcos/gcoshome.html>) which acquires data from weather stations around the world (Fig. 2.1). It is interesting to note that in a more recent study by Mann and Jones (2003), "proxy" temperature records derived from the analyses of lake sediments, tree rings, and ice cores agree with the recent instrument record well.

GCOS was established in 1992 to ensure that the observations and information

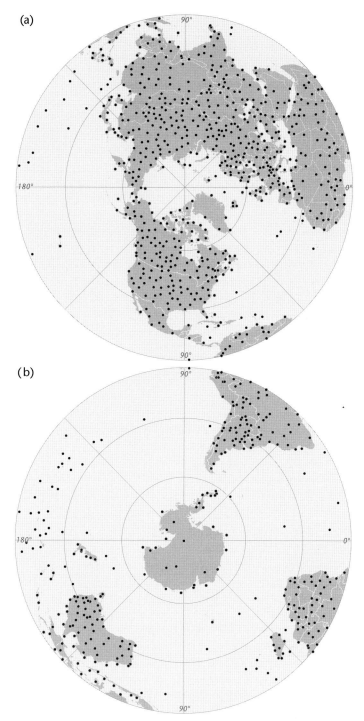

FIG. 2.1 The Global Climate Observing System Terrestrial Network of weather stations used in temperature trend calculations in the Northern Hemisphere (a) and Southern Hemisphere (b).

needed to address climate-related issues are obtained and made available to all potential users. It is cosponsored by the WMO, the Intergovernmental Oceanographic Commission (IOC) of UNESCO, UNEP, and the ICSU. GCOS is intended to be a long-term, user-driven operational system capable of providing the comprehensive observations required for monitoring the climate system, for detecting and attributing climate change, for assessing the impacts of climate variability and change, and for supporting research toward improved understanding, modeling, and prediction of the climate system.

The instrument records from weather stations are often used to provide a reference period against which present-day or pre-20th century measurements are compared. Temperature **anomalies** are variations in the difference between the temperature for a given time increment (e.g. average or minimum daily or monthly) and the average temperature over a reference period of time (e.g. 1951–70 in the case of Jones, 1994). In this way if the anomaly is positive, then temperatures are warmer than the reference average and if negative, then temperatures are cooler than the average. Clearly the reference period is the key to identifying the anomaly. Understanding the uncertainties in the measurement record is also important for these studies and work by Karl et al. (1994) demonstrated that uncertainties in the calculation of historical trends is affected by the actual pattern of temperature change, how nonrandom spatial sampling is dealt with, and the time over which the trend is calculated. There has been much discussion about the use of these data especially when used in conjunction with proxy records and we will return to this topic in Chapter 4.

Discrete limited lifetime programs and experiments, such as NASA's Program for Arctic Regional Climate Assessment (PARCA) and Surface Heat Budget of the Arctic Ocean (SHEBA) have been organized in the recent past to improve our understanding of cryospheric processes. These experiments were conducted in the Arctic regions including Greenland, Antarctica, and the Arctic ocean. While the experiments do not necessarily add to the GCOS time series, they were designed to improve our understanding of regional cryospheric mass and energy dynamics. The experiments generally have relied on, among other things, automatic weather stations (AWS) installed at representative sites throughout well-defined regions. For example, in the PARCA program, Steffen and Box (2001) describe the Greenland Climate Network which contains 18 AWS distributed over the ice sheet. Data are still available for many of the sites through the Cooperative Institute for Research in Environmental Sciences in Colorado. By using these data carefully, regional to local scale variations in the cryosphere's temperature regime can be better understood.

2.2.2 TERRESTRIAL SNOW AND SNOW ON SEA ICE

Snow is an important cryospheric variable and is one of the most dynamically varying components of the global water cycle. Each year, the average maximum terrestrial snow extent in the Northern Hemisphere is approximately 47 million square kilometers, or 50% of the Northern Hemisphere land surface (Robinson et al. 1993). While snow cover extent affects the energy budget of the lower atmosphere, the thickness, mass, and melt characteristics of snow affects the dynamics of terrestrial hydrological systems. Estimation of total snow water equivalent (SWE), therefore, is of great interest not only to climatologists but also to river basin water resource managers who need timely information about snowpack volume and condition (stable or melting)

TRADITIONAL *IN SITU* APPROACHES

for effective runoff prediction. Snow also forms a significant accumulation to Arctic and Antarctic ocean sea ice which on any given day in the year covers about the size of the US. The snow variables of interest are snow cover extent, snow depth, and snow density or SWE which is the columnar height of melted liquid water from a column of snow (e.g. 100 cm snow depth with 0.1 g cm^{-3} density equals 10 cm SWE) and snow wetness or snow liquid water content. Snow stratigraphy is also important for some management applications, especially avalanche risk control but is generally done manually in the field by those with an interest, such as ski patrols.

Gray and Male (1981) provide excellent comprehensive information about snow measurements. Doeksen and Judson (1996) suggest that there are three dominant factors that make the characterization of snow on the ground difficult. First, snow often melts as it lands or as it lies on the ground from warm underlying soil or warm air, wind, or sunshine above, or both, making it difficult to characterize the snow event. Second, snow is easily blown and redistributed and it generally does not land or lie uniformly on the ground. Third, consistent and comparable snow data are only possible if standard procedures are established and followed. Thus, to make accurate and useable measurements of snow, it is important to identify representative locations for measuring snow, and to implement a consistent time interval for repeat measurements.

Snowfall and accumulated snow volume or snow mass are probably the most important snow variables that can be measured *in situ*. Snowfall measurement is not straightforward to measure because the method requires using light attenuation sensors that are calibrated to give less than quantitative snowfall rates. The approach yields categories of snowfall intensity that are of light intensity, reported when the visibility is

greater than or equal to 1.0 km, moderate snow intensity, when the visibility is between 1 and 0.5 km, and heavy snow intensity when the visibility is less than or equal to 0.5 km (National Weather Service 1994). Attempting to quantify these measurements with respect to the liquid water equivalent of snowfall does not yield robust estimates of snow accumulation (Rasmussen et al. 1999) and efforts continue both to improve the estimation of falling snow rate and to automate the measurements for the National Weather Service's Automated Surface Observing System network which services the US aviation requirements. Globally, these measurements are not available.

Direct measurements of snow depth or accumulated SWE on the ground are by far the most commonly used methods of characterizing instantaneous snow mass on the ground. It is worth noting that both approaches, de facto, provide information on the presence or absence of snow for snow extent mapping. Snow depth tends to be measured using a vertically placed ruler and a snow board. The observer measures the accumulated snow by inserting vertically the ruler and reading off the snow depth. To measure fresh snow accumulation, the flat snow board is placed on top of the old snow and new snow accumulations are measured after a set time interval. Historically, snow depth measurements are made manually and are not automated but may well be recorded at meteorological stations and reported with the ensemble of standard AWS data. More recently, sonic snow depth sensors are being used at automatic snow monitoring sites such as those in the United States Department of Agriculture (USDA) and National Resource Conservation Service (NRCS) SNOwpack and TELemetry (SNOTEL) network. These sensors measure snow depth as a function of the time delay for a signal to travel from the sensor, mounted above ground on an AWS, to the snow surface and back again. With respect to direct measurements of snow mass, snow gauges, or AWS rain gauges, can record accumulated snowfall in a heated receptor. This approach requires careful gauge calibration and gauge shielding. Precipitation gauges designed to measure rain often undermeasure snowfall because the gauge acts as an obstacle to the airflow with slower falling snowflakes tending to follow the airflow around the gauge rather than fall into the gauge. Two types of shield are commonly used: the Nipher Shield, a flared metal device that attaches to the precipitation gauge, and the Alter type which consists of 32 free swinging galvanized metal leaves, or baffles, attached to a steel ring 4 ft in diameter and supported by 3 or 4 galvanized pipes. Although the wind shields improve the catch, significant undermeasurement (50% or more) may still result particularly during high wind speed events (Goodison 1978). Figure 2.2a shows an example of a Nipher shielded gauge. To compensate for undermeasure by precipitation gauges, a recent study by Yang et al. (2005) demonstrated a method to correct the biases in a 31 year record of precipitation from more than 4,000 gauges in northern regions. The correction factors applied varied spatially and temporally and the resulting corrected data enhance the observed monthly precipitation trends between 5 and 20%.

Snow pillows, a preferred method of measuring SWE directly, convert the weight of snow overlying a pressure-pillow into a SWE measurement. Figure 2.2b illustrates the principle in general terms. Snow pillows are part of the standard measurement suite in the SNOTEL network. Correct and safe siting of snow pillows is important because they should be located where extraneous objects (such as windblown vegetation) do not accumulate on the pillow thus biasing the measurement.

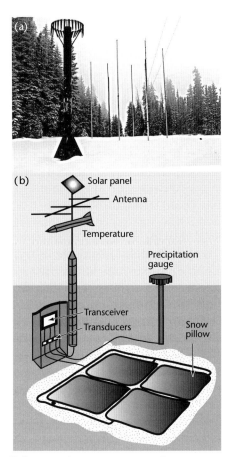

FIG. 2.2 (a) Alter snow gauge for a SNOTEL site in the western USA (<http://www.crh.noaa.gov/git/images/towerssnotel102302b.jpg>). (b) Illustrates the design of a snow pillow system (<http://www.nationalatlas.gov/articles/climate/a_snow.html>).

For all *in situ* methods of snow measurement, representative siting is very important. As synoptic scale snow accumulation patterns change under a changing climate, this representativity is becoming especially critical. Automatic and nonautomatic snow measurement networks have been developed to spatially represent the snow depth or SWE but if these were sited deliberately, which is not always the case, they would have been based on snow climatologies that may now be changing. Table 2.1 gives a

TABLE 2.1 Data sources for operational snow depth (SD) and SWE measurements globally and in Canada and the USA.

Data set name	Global summary of the day (GSOD)*	NWS Cooperative Stations (COOP)†	SNOw TELemetry network (SNOTEL)‡	Meteorological services of Canada§
Source	NOAA NCDC	NOAA NCDC	USDA NRCS	CRYSYS/CCIN
Spatial coverage	Global	USA	Western USA and Alaska	Canada
Temporal cover	1973–present although most complete from 1982	1890–present	1970s–present	1950s–present
Measurements (SD/SWE)	SD	SD	SD and SWE	SD and SWE
Measurement frequency	Daily	Daily	Daily	Weekly, biweekly, and monthly
Total number of stations	8,000 report meteorological variable in total but far fewer reporting SD (<1,500)	11,000 COOP stations in total but around 3,000 report snow at maximum winter snow extent	920 at August 2005	SD: 800 stations per year (max) with 84 stations providing continuous records back to mid-1940s. SWE: 2,128 stations per year (max) with 14 stations providing continuous records back to mid-1940s

*Data are normally available about 4 days after the end of the data week. For some periods, one or more countries' data may not be available due to data restrictions or communications problems.
†The measurements are representative of where people live, work, and play but do not necessarily represent snow conditions in more remote locations.
‡Managed by the United States Department of Agriculture's Natural Resource Conservation Service (NRCS). SNOTEL data constitute the most comprehensive operational snowpack measurements in the USA.
§Compiled by 20 agencies and MSC snow depth observing stations. A pronounced peak in the number of stations is found in the 1976–85 decade.

summary of several different *in situ* operational snow monitoring data sets that are publicly available. NOAA represents the National Oceanic and Atmospheric Administration and NCDC represents the National Climate Data Center.

Data from the sources identified in Table 2.1 is shown in Fig. 2.3. The WMO global sites shown are not necessarily for snow depth but include the entire network of stations. Locations where snow is measured only submit their data when snow is present. For the North American examples, both Canada and the USA have experienced significant funding shortfalls in their observing networks which have caused a general decline in the availability of snow depth and SWE data. Data quality and representativeness is an important issue with many of these operational data products. Serreze et al. (2000) provides an account of how to screen erroneous data from SNOTEL sites and the methodology can be extended to COOP measurements for which observing sites often are located on private property. Moreover, many Canadian snow depth sites are biased toward lower elevations and lower latitudes because of the scarcity of mountain and high-latitude sites (Brown & Braaten 1998).

Examples of *in situ* terrestrial snow depth and SWE measurements made operationally can also be found for Finland, Norway, Sweden, and in other Alpine countries that rely on snowmelt runoff for domestic and industrial water supply. Some of these networks are available to researchers but many are not as they constitute data used by private commercial power generation companies. Other data networks are no longer operational. The Former Soviet Union Hydrological Surveys spanned the period from 1966 to 1996 and covered much of Russia and Siberia (Krenke 1998). Furthermore, it is unlikely that the networks currently available will be improved or

augmented in any time soon. Hydrologists are looking to remote sensing and models to provide improved spatially and temporally comprehensive cover.

In addition to various operational data sets, there have been many finite-duration field experiments to measure snow parameters. The 2002–3 NASA-funded Cold Land Processes eXperiment (CLPX) organized 18,000 snow depth measurements to be made during the experiment period and the excavation of 144 additional snow pits. The BOReal Ecosystem-Atmosphere Study (BOREAS) experiment also conducted significant snow field measurements. Details of many of these experiments can be found at the National Snow and Ice Data Center (NSIDC) in Boulder, Colorado (<http://www.nsidc.org>) and at the Canadian Cryospheric Information Network (CCIN, <http://www.ccin.ca>).

The measurement of snow on sea ice is reliant on ad hoc field experiments. Very few experiments have measured snow in this context and recent work has been done with the express purpose of validating remote sensing observations from satellite instruments (Markus & Cavalieri 1998). Few other experiment measurement campaigns have been conducted.

2.2.3 SEA ICE

Sea ice is an element of the global energy and mass transfer system that is of key importance to the cryosphere. For example, it has a strong insulation capacity that restricts energy, mass, and momentum transfer between the atmosphere and ocean. It also influences the saline content, and, therefore, the ocean density of the upper ocean layers which, in turn, can influence ocean circulation and bottom water formation. Sea ice extent changes seasonally in the polar regions; at its minimum extent in the northern and the southern hemisphere, sea ice covers $6-8 \times 10^6$ km^2 and $2-3 \times 10^6$ km^2 respectively and at its maximum, it covers $15-16 \times 10^6$ km^2 and $18-19 \times 10^6$ km^2 respectively (Cavalieri et al. 1997b). The key variables of interest for sea ice characterization are extent, concentration, ice type, and thickness. In the Arctic, some sea ice persists year after year, termed multi-year ice (MYI) and some as first year ice (FYI) which is "seasonal ice," meaning it melts away and reforms annually. Almost all Southern Ocean or Antarctic sea ice is FYI in character. Characterization of sea ice variations is challenging because not only does the extent of sea ice change seasonally, but also sea ice does not remain in one place; it drifts and moves as the total extent expands and shrinks seasonally.

Attempts at measuring sea ice variables using *in situ* techniques are limited at best because the hostile polar climate tends to constrain human activity. Thus, measurements are usually spatially and temporally restricted to experiment campaigns and in many cases these follow the tracks of ship cruises; no spatially comprehensive or consistent measurement networks are in existence to measure sea ice variables. The closest to an operational monitoring program are the measurements made by the Former Soviet Union over the Arctic Basin which consists of sea ice and snow measurements associated with the Former Soviet Union's historical Sever airborne and North Pole drifting station programs. These surveys took place in 1937, 1941, 1948–52, and 1954–93 (Konstantinov & Grachev 2000). In addition, the Danish Meteorological Institute (DMI) produced monthly sea ice charts of ice distribution in spring and summer on the basis of coastal and commercial ship reports. Vinje (2001) has provided a consistent data set of these charts back to 1864 and these data are available from <http://acsys.npolar.no/ahica/intro.htm>.

Furthermore, Walsh and Chapman (2001) have produced a synthesis of these data plus recorded ship and coastal observation data from the 20th century held at the UK Meteorological Office and the US National Ice Center (www.natice.noaa.gov). The US SCientific ICE EXpeditions (SCICEX) civilian submarine program of 1993–97 made upward looking sonar (ULS) measurements along multiple transects in the Arctic Ocean (Rothrock et al. 1999). ULS measurements gauge the distance from the submarine sensor to the base of the sea ice and since the position of the submarine is known precisely and the ocean surface is known, the ice thickness can be calculated. In an effort to tie the SCICEX data to a longer term record of intermittent summer cruises between 1958 and 1976, Rothrock et al. (2003) produced a data set that has been used to successfully analyze seasonal variability of sea ice thickness.

In Antarctica and the Weddell Sea, a measurement network of 14 moored ULS was deployed by the Alfred Wegener Institute (AWI) to measure water pressure, water temperature, and ice draft (Harms et al. 2001). Figure 2.4 shows the position of the 14 sites that were deployed during the 1992–98 timeframe.

In addition to the AWI network, a one-season experiment was executed in 1992 called Ice Station Weddell. This successful experiment studies ocean and atmospheric dynamics (including sea ice kinematics) in the western Weddell Gyre of Antarctica. *In situ* measurements were made on an ice floe in the southwestern Weddell as the floe drifted to the north. Measurements were taken at the ice camp, on snowmobiles, from helicopters which flew up to 100 miles from the camp, and onboard Russian and US icebreakers (Gordon et al. 1993). Research results from this experiment have been reported more recently by Geiger and Drinkwater (2005). Further *in situ* data

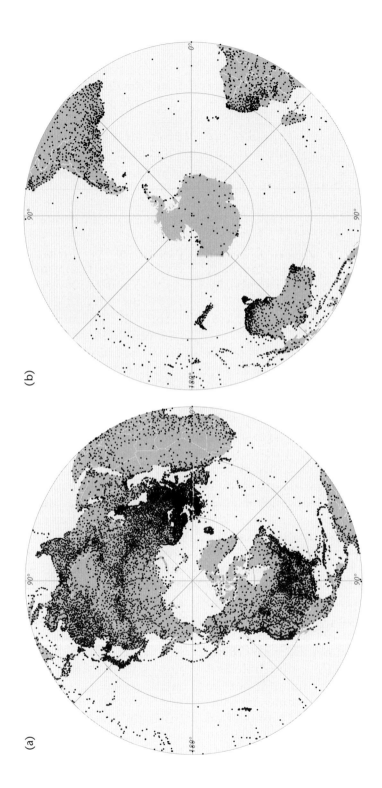

(a)

(b)

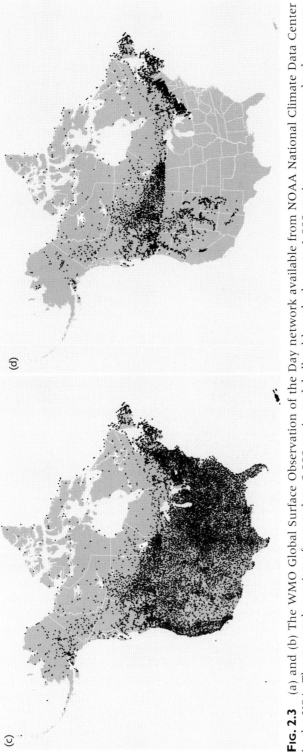

Fig. 2.3 (a) and (b) The WMO Global Surface Observation of the Day network available from NOAA National Climate Data Center in the USA. The network consists of more than 8,000 stations globally although closer to 1,000 stations report snow depth measurements. (c) Snow depth *in situ* measurements sites in North America. The sites in the USA constitute the National Weather Service Cooperative network of observers many of which measure snow depth. In Canada, the locations represent places where snow depth was measured during the long-term data set available through the Canadian Cryospheric Information Network. (d) SWE measurement stations in the USA (SNOTEL) and snow course sites in Canada during the maximum monitoring of 1976–85.

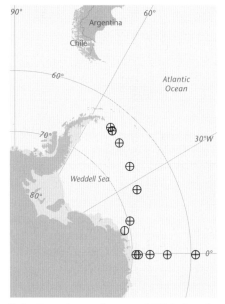

FIG. 2.4 Locations of moored ULS sites from the AWI data set.

sources can be found through the British Antarctic Survey, the US National Ice center, and the Russian Federal Service for Hydrometeorology and Environmental Monitoring (<http://www.mecom.ru>).

Other data sets related to the Arctic can be found at the ARCtic System Science (ARCSS) Data Coordination Center (ADCC) at NSIDC. For example, the Coordinated Eastern ARctic EXperiment (CEAREX) was a multiplatform field program conducted in the Norwegian Seas and Greenland north to Svalbard from September 1988 through May 1989. Canada, Denmark, France, Norway, and the United States participated in the experiment. Sea ice measurements were made during the EUrasian Basin EXperiment (EUBEX) in 1981, the Marginal Ice Zone EXperiment (MIZEX) in 1983, 1984, and 1987, and the Seasonal Ice Zone EXperiment (SIZEX) in 1989. Additionally, the Surface HEat Budget of the Arctic Ocean (SHEBA) experiment between 1995 and 2002 was an important experiment that

has aimed to determine the ocean–ice–atmosphere processes that control surface albedo and cloud-radiation feedback mechanisms over Arctic pack ice. Ultimately, it is hoped that the measurements can be used to improve models of Arctic ocean–atmosphere–ice interaction thereby improving our understanding of the Arctic climate. An intensive measurement approach was taken during the SHEBA experiment including the deployment of a network of automatic buoys for monitoring ice kinematics throughout the Arctic Basin. Details of this experiment can be found at <http://sheba.apl.washington.edu>.

In situ sea ice measurements are important to help understand ocean–atmosphere energy and mass exchanges. In an experimental context, data from these experiments allow us to develop increasingly robust numerical models of sea ice and ocean–atmosphere exchanges. They are also very important for validating remote sensing approaches (see Chapter 4). However, like terrestrial snow cover, undertaking comprehensive studies of sea ice throughout the cryosphere using ground surveys is impossible. Thus, experiments, through necessity, are generally highly focused to address specific science modeling questions or remote sensing needs.

2.2.4 ICE SHEETS AND ALPINE GLACIERS

Of great interest to Earth system scientists is the mass balance of glaciers, ice sheets, and ice caps because these water stores contribute directly to sea level and ocean circulation, and potentially to climate change (Meier 1998). For example, it is estimated that the Greenland and Antarctic ice sheets together contain enough water to raise sea level by almost 70 m (Houghton et al. 2001). Clearly, quantifying ice sheet and glacier mass balance processes is of great interest globally.

According to Paterson (1994), the mass balance at any time is "the change in mass (expressed as the equivalent volume of water) per unit area relative to the previous summer surface." A glacier, ice cap, or ice sheet gains mass by accumulating snow which is transformed into ice by densification processes. It loses mass (ablation) mainly by melting at the surface or base with subsequent runoff or evaporation of the meltwater. Meltwater may refreeze within the snow instead of being lost, and some snow may sublimate from the surface back to the atmosphere. Ice may also be removed by flowing into a floating ice shelf or glacier tongue, from which it is lost by basal melting and calving of icebergs. In general, net mass accumulation at higher altitudes and net ablation at lower altitudes are balanced by the downhill flow of ice under internal deformation (since ice is a plastic material) and by the sliding and bed deformation at the base. This balance is expressed usually as the rate of change of the equivalent volume of liquid water, in cubic meter per year; for a steady state the mass balance is zero. Mass balances are computed for both the whole year and individual seasons (winter and summer) with the specific mass balance being the net summer and winter mass balances averaged over the surface area, in meter per year.

Ice-covered regions are dynamic environments that are characterized by forcing responses at variable timescales which may not always be synchronous with external weather and climate forcing factors. While seasonal accumulation and ablation processes might balance in approximate terms, internal ice dynamical processes occurring at both shorter and longer timescales contribute to balance inequalities. For example, the average annual solid precipitation falling onto Greenland and Antarctica is equivalent to 6.5 mm of sea level, this input being approximately balanced by loss from melting and iceberg calving. However, the balance of these processes is not the same for the two ice sheets, on account of their different climatic regimes. Antarctic temperatures are so low that there is virtually no surface runoff; the ice sheet mainly loses mass by ice discharge into floating ice shelves, which experience melting and freezing at their underside and eventually break up to form icebergs. On the other hand, summer temperatures on the Greenland Ice Sheet are high enough to cause widespread melting, which accounts for about half of the annual ice loss, the remainder being discharged as icebergs or into small ice-shelves. Understanding ice sheet dynamics in the context of global environmental change, therefore, requires us to characterize the present state of the ice sheets and glaciers and to establish the historical and prehistorical trends of ice masses in order that we can understand variations in global and regional mass balance behaviors.

Current state characterizations can be made through specific *in situ* measurements of ice mass balance processes. These measurements can be linked to cryospheric climate forcing variables derived from *in situ* measurements or from coupled atmospheric–ocean global climate models. Braithwaite (2002) provided an excellent historical perspective and summary of glacier mass balance measurements over the last 50 years. In his paper, he charts the progress of balance monitoring from the work of H.W. Ahlman, a Swedish glaciologist who made glacier balance measurements in the 1920s and 1930s (Ahlmann 1948). Hagen and Reeh (2004) recently produced an excellent and comprehensive synthesis of the different measurement approaches of land ice taken by glaciologists and geophysicists. Interestingly, in their paper, they include aerial photography in the discussion of *in situ* techniques noting that this method has been used for 50 years in conjunction with field studies. They make a useful categorization of measurement approaches into two groups: (i) direct estimation of the glacier or ice sheet volume over time from integrated surface topography measurements, assuming knowledge of the subglacial bedrock topography; and (ii) measurements of specific components of mass balance equations, particularly mass accumulation and ablation. We also use this distinction but emphasize Braithwaite's observation that mass balance measurements, especially those that attempt to quantify specific components of the mass balance equations, can be uncertain with large errors. This is because field conditions often add significant difficulty to the measurement process.

For volume estimates, the best way is to calculate the change in volume with time through the construction of digital elevation models (DEMs). Direct measurements of elevation change have a long history of application on alpine glaciers through the use of field survey methods. In mountain terrain, theodolites and/or electronic distance measurers survey marked locations on the ice surface from fixed, stable sites located off the ice on the surrounding valley walls. Precise measurements, to centimeter accuracy, are possible provided that the markers, usually in the form of embedded stakes, remain intact during the survey period. The outcome of this approach is that three-dimensional DEMs can be produced enabling annual changes in glacier volume to be made. For large ice sheets, however, it is generally not possible to survey from peripheral stable sites. Instead, repeat leveling along predetermined transects is performed. This gives the relative height of the level sites and can be calibrated to absolute benchmark heights. Leveling experiments, however, are subject to uncertainties which can prevent the detection of long-term trends. In addition,

while the full three-dimensional survey may be spatially comprehensive for mountain glaciers, this will not apply for ice sheets; the challenge is to derive representative transect profiles that can be repeated from year to year.

Traditional cartographic surveys have also been undertaken to directly estimate mass balance changes from year to year. Elevation contours derived from surveys can be produced to provide an estimate of surface topography. Repeated from year to year, it is possible to compare changing glacier volume as an indicator of mass balance variation. Maps can also be derived from aerial photography, and, using stereo photography methods, DEMs can be constructed directly, assuming stable ground control points can be located from year to year. The accuracy of cartographic methods, however, is 1–2 m on account of both repeated height inaccuracy and horizontal uncertainty of stable targets. This precision might be good enough to estimate general trends but it is inadequate for direct measurement of ablation or accumulation variations on a year to year basis. A more recent approach to mapping is through the use of global positioning system (GPS) surveys. Very accurate measurements of the glacier surface in three dimensions are possible with this method and profiles of GPS receivers mounted on "fixed" stake have been used to characterize the changing mass balance of both small glaciers to large ice caps. This approach has the advantage of being automated and also measuring the ice surface velocity, thereby providing a measurement of the glacier kinematics. Finally, a recent method tested by Hulbe and Whillans (1994) has been proposed that measures the vertical velocity and snow accumulation simultaneously. The submergence rate of a "coffee can" collector is measured and is combined with the accumulation rate to give the rate of thickening.

In Greenland and Antarctica, the uncertainty of this method is estimated to be about 0.02 m yr^{-1}. Again, however, for many volume estimates of glacier mass, comparisons of one year to the next are not straightforward on account of the measurement errors in producing the volume estimates. If taken over several years, this method becomes more reliable in estimating trends in glacier mass balance variations.

The measurement of specific components of mass balance equations, particularly mass accumulation and ablation, is the second classification used by Hagen and Reeh (2004). They contend that it is easier to measure *in situ* volume changes through time and compute the mean annual net balance than obtain spatially comprehensive annual or seasonal data. The approaches available for large glaciers and ice sheets, however, are generally separate than those that have been developed for smaller glaciers.

Measurement of accumulation and ablation rate can be achieved by repeatedly surveying stakes that have been partially drilled into the glacier to determine the accumulation and ablation of snow and ice relative to the stake during the year (Østrem & Brugman 1991). This is repeated for a series of stakes comprising the "stake network." To ensure a comprehensive identification of the accumulation area, "snow probing" can be undertaken. Snow density measurements from snow pits or cores must also be measured and when coupled with the stake measurements, an estimate of net surface accumulation or ablation can be obtained. By undertaking a survey at the end of the winter and summer seasons, the maximum mass accumulation and maximum mass ablation can be determined respectively. If a distinct "summer surface" is present over the glacier, the winter and summer balances can be measured at the end of the spring provided that the previous

year's stakes are intact. Paterson (1994) expresses the net balance of a glacier as:

$$B_n = \int_{S_c} b_n dS + \int_{S_a} b_n dS$$

where b_n is the net balance at measurement points in the accumulation area S_c and the ablation area S_a. Figure 2.5 gives an example of the specific net mass balance for Hardangerjøkulen glacier in southern Norway for 1995–96. In this instance, the annual net balance was negative over all elevations of the glacier.

Contour maps of lines of equal balance can be constructed for the glacier. This method works well for midlatitude mountain glaciers where well-defined accumulation and ablation patterns are observed. For higher-latitude glaciers, however, this approach may be subject to greater errors on account of the presence of superimposed ice layers and all-round snow accumulation.

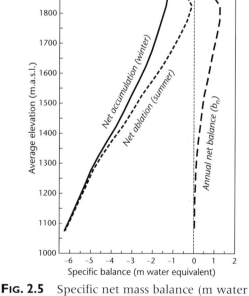

FIG. 2.5 Specific net mass balance (m water equivalent) for Hardangerjøkulen, Norway for 1995–96 balance year (courtesy of the Norwegian Water Resources and Energy administration).

The spatial representativity of these measurements is an important aspect of their use in mass balance studies. Cogley and Adams (1998) showed that for small valley glaciers, a spatial density of 1 stake km^{-2} is required. Over larger glaciers, the spatial dependency of accumulation profile is greater so the sampling spatial density can be relaxed.

A second method for obtaining representations of accumulation in mass balance equations is by undertaking stake measurements over an "index area" that is a highly representative of the glacier as a whole. This is similar to the first method except that it is applicable for much smaller glaciers and does not require the construction of glacier mass balance contours.

For long-term studies (up to 250 k years) and to reconstruct previous accumulation rates, snow pit studies and ice core measurements have been used widely. Snow pits (Fig. 2.6) are used to quantify vertical variations in snow density and to identify accumulation layers in the snow that can be used to calculate densification rates. This process controls the rate at which snow is converted to ice. Ice core studies are used to reconstruct past climate variations. O^{18}/O^{16} ratios from ice cores can be used to identify temperature variations over the past 250 k years. However, analysis of the oxygen isotopic record enables the identification of discrete layers in the profile at depth that can be used to assess the accumulation rate over the glacier or ice cap (Fig. 2.7). It should be remembered, however, that these methods are *in situ* in scope and are spatially limited. To extend the areal coverage of these measurements, accumulation layers can also be obtained from sled-based ground penetrating radar (GPR) measurements. Such instruments resolve accurately the accumulation layers for areas that are accessible from sled-mounted systems (e.g. Spikes et al. 2004). GPR penetration depths may be limited depending on the snow and

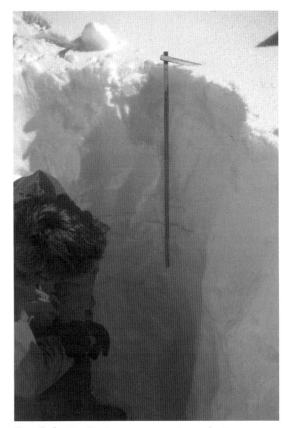

Fig. 2.6 Preliminary snow pit analysis on Hardangerjøkulen, Norway in 1995 (photograph by R.E.J. Kelly).

ice stratigraphy but several surveys have been conducted in this way (e.g. US International Trans-Antarctic Scientific Expedition described by Mayewski & Goodwin 1997). Figure 2.8 shows a GPR survey result from the US International Trans-Antarctic Scientific Expedition (ITASE) in 2000.

Ablation terms are less straightforward to measure directly. Successive ice height measurement against a graduated stake is one approach and requires that the stake remains stable and fixed in the ice. This is the dominant measurement approach taken. For glaciers that discharge ice into the ocean, ice calving is generally defined as

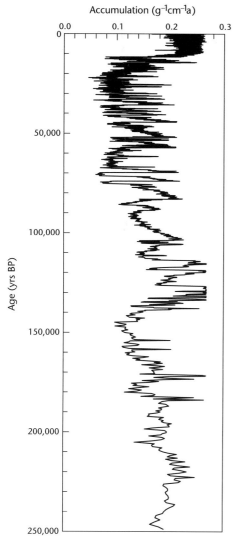

FIG. 2.7 Ice core stratigraphy from a 3-km core, revealing accumulation rate in the vertical column at GRIP (see Dalh-Jensen et al. 1993; Greenland Summit Ice Core CD-ROM 1997). Accumulation is the snow accumulation rate in $g^{-1}cm^{-1}a$.

the mass of ice loss per unit time. The dimensions of the calved ice can be estimated either from the ablation area ice velocity and thickness or by surveying overturned recently calved icebergs.

The World Glacier Monitoring Service (WGMS), a part of the Commission of Cryospheric Sciences of the International Union of Geodesy and Geophysics, makes available information on mass balance measurements for many glaciers around the world (see <http://geo.unizh.ch/wgms>). This service is also conducted under the United Nations GCOS Glacier Network (GTN-G). Braithwaite (2002) noted that there are 246 glaciers for which mass balance data are available but that most of the records are short and span recent years. Figure 2.9 shows the global distribution of 108 glaciers from the Mass Balance Bulletin 8 at the WGMS. Of these sites, 83 have reported data for 2002–3, and also included are 25 sites which have records in excess of 15 years. Most mass balance measurements are made in North America, Western Europe, and the former USSR; there are scant monitoring program in other parts of the world. The WGMS contributes to the Global Environmental Monitoring System (GEMS) of the United Nations Environment Programme and to the International Hydrological Programme of UNESCO. Thus it makes an authoritative contribution to informing of these important international policy organizations.

Glacier mass balance data are also available at several national data depositories. For example, the NSIDC maintains an active archive of several US glacier mass balance data sets. Many of such websites make the data available to the science community for routine science studies and are essential for continued climatology studies.

Hagen and Reeh (2004) suggested that with the increased accuracy of remote sensing measurements with time, ground measurement campaigns will be used more as calibration and verification data. This is a very important aspect because only by testing remote sensing estimates with independent (ground) measurements can the remote sensing be validated. A good example of this use of ground measurements can

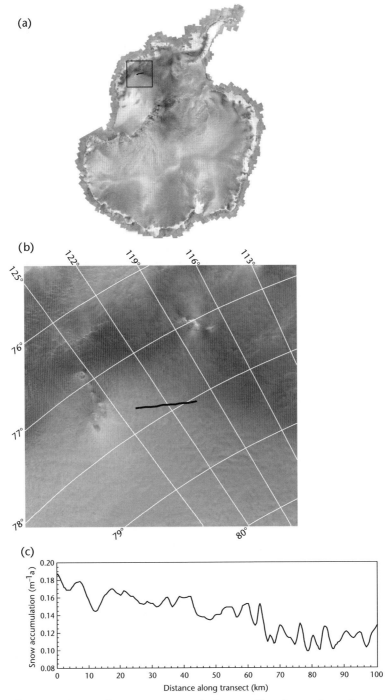

FIG. 2.8 Ground Penetrating Radar snow accumulation profile of a section of west Antarctica in 2000 from Spikes et al. (2005) obtained during ITASE. (a) The area of transect superimposed on the Radarsat map of Antarctica (see Jezek and the RAMP Product Team 2002). (b) The enlarged area of transect. (c) The variation in snow accumulation rate from 1966 to 2000 in m^{-1} a.

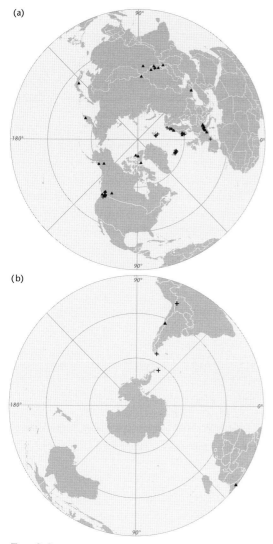

(a)

(b)

FIG. 2.9 Example of locations of 108 glaciers from the Mass Balance Bulletin 8 in the WGMS data archive where mass balance records are maintained (source: <http://www.geo.unizh.ch/wgms>) in Northern Hemisphere (a) and Southern Hemisphere (b). Triangles represent sites where data record lengths exceed 15 years or more and crosses represent locations where data records are shorter than 15 years and include 2002–3 season.

be found in the PARCA experiment special issue of the *Journal of Geophysical Research*. However, there are parts of glaciers and ice

sheets that cannot easily be measured from remote sensing (such as complex and steep terrain environments and glacier marginal areas) so *in situ* measurements will continue to be used. Furthermore, as we will see in Chapter 4, while the accurate characterization of small glaciers is important for local environmental change science, individually these ice masses make a small contribution to global environmental change science. Therefore, remote sensing approaches, which are regional to global in measurement scope, will likely remain the dominant measurement approach for the foreseeable future.

2.2.5 PERMAFROST AND SEASONALLY FROZEN GROUND

Permafrost is a layer of permanently frozen ground, that is, a layer in which the temperature has been continuously below 0°C and has been for at least some years. Seasonally frozen ground, or active layer, is usually a layer above the permafrost that freezes in winter and thaws in summer where depth of thawing from the surface is usually less than a meter or so in thickness. Permafrost is found in the Arctic and sub-arctic, in high mountain ranges, and in ice-free regions of Antarctica. Heginbottom (2002) noted that the World Data Centers for Glaciology (Boulder, Colorado, USA; Cambridge, UK; Lanzhou, China) have been archiving data on frozen ground. In 1997, GCOS and its associated organizations identified two important observational components of frozen ground that are critical in a changing global climate: the active layer (the surface layer that freezes and thaws annually) and the thermal state of the underlying permafrost. Founded in 1983, the International Permafrost Association (IPA) instigated the Global Terrestrial Network for Permafrost (GTN-P) in 1998 to organize and manage a global network of permafrost borehole temperature measurements

for detecting, monitoring, and predicting
climate change. This is a concerted approach
to draw data from several countries into a
coordinated monitoring effort. The data
monitoring network for GTN-P is still under
development and there are currently
approximately 39 sites that are active for
GTN-P in Canada, USA, Switzerland, Norway,
Sweden, Italy, Kazakhstan, and Mongolia.
Most records are archived from the late
1990s with Canada having records back to
the mid-1980s and the Asian site measurements starting in the 1970s.

Active layer measurements have been
managed effectively by the Circumpolar
Active Layer Monitoring (CALM) program.
Figure 2.10 shows the location of the
approximately 80 sites where active layer
permafrost measurements are being undertaken. In some countries, the data archive
extends back to the early 1990s but most
sites became active in the mid- to late 1990s.
The record, therefore, of ground temperatures and active layer thickness in these
regions is developing. The measurement

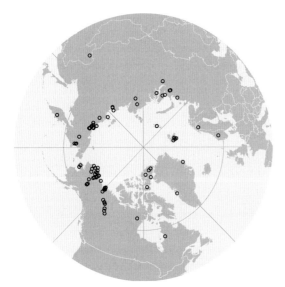

FIG. 2.10 Location of CALM monitoring sites
(from <http://www.udel.edu/Geography/calm/
index.html>).

protocol described in the CALM network's
documentation requires that measurements
of active layer data should be collected at
regular intervals from the time of snowmelt
until the annual freeze-up (Brown et al.
2000). Furthermore, studies have shown
that active layer thickness is known to vary
substantially over very short distances.
Representative measurements are required
implying that a protocol should aim to
obtain a statistical characterization of active
layer thickness in representative terrain and
vegetation types, on an interannual basis.
Measurements, traditionally, have been
undertaken at the *in situ* point scale using
small-diameter metal rods that are used to
probe the bottom of the active layer, or
through the use of thaw tubes (e.g. Nixon &
Taylor 1998) combined with measuring and
recording ground temperatures. Both
approaches can yield high-quality data, but
are necessarily restricted in their ability to
provide spatial information. In addition,
because previous measurement schemes
have not been coordinated with any long-
term vision, the CALM network has developed a structural hierarchy of sampling
design that is defined not only by scales of
permafrost/active layer variability for measurement but also by the heritage of measurement in a region (whether or not a
region can tie in to a previous field experimental design). Using a series of "tiers,"
there are five levels of sample design ranging
from Tier 1 that consists of transects across
large experimental river catchments, such as
the Mackenzie River basin, to Tier 2 which
leverages from long-term experimental
basin infrastructure, through to Tier 5 which
represents measurements made by remote
sensing instruments at fine scales (100–30 m).
A full description of the CALM sampling
approach can be found in Brown et al.
(2000). It should be noted that the method
of sampling the active layer can be very
important and impact the experimental

outcome significantly. For example, typically two approaches to sampling are used: transects and random samples. In regions dominated by patterned ground, measurements of active layer depths might seriously overestimate or underestimate the "true" variability. Also, the design should account for the presence of vegetation and its effect on the freeze–thaw dynamics. It can be demonstrated by experimentation that the most effective and economical sampling design is the systematic stratified unaligned (SSU) sampling scheme described by Thompson (1992). An attractive feature of the SSU design is its relative ease of implementation in the field and its provision of excellent areal coverage that tends to be unaffected by patterned ground features.

2.2.6 RIVER RUNOFF

Of any of the cryosphere state index variables, river runoff has, perhaps, been measured continuously for the longest time. Along with air temperature, it is possibly one of the most significant indicators of how the cryosphere has changed. It is important to remember that rivers connect different terrestrial environments and environmental changes in one environment, such as the cryosphere, can affect other environments. For example, runoff in the Indus River in Pakistan runs the entire length of Pakistan, through arid and semiarid environments, to the Arabian Sea. Its headwaters, the Upper Indus Basin, are located in the Greater Karakoram mountains where snow and glacier ice melt contribute the majority of the total annual flow (Archer 2003). Thus, the river connects a dominantly cryospheric environment with a noncryospheric environment. River flow within and from the cryosphere is an important diagnostic for climate and environmental changes, and for water resource management.

The data sources for river flow can be considered in two ways: runoff measurements made in glacierized environments (controlled by snow and glacier ice melt) and runoff measurements made in nonglacierized environments (controlled by snowmelt and/or liquid precipitation). These two groups are not mutually exclusive since river flow from glacierized basins may contribute to snowmelt and water precipitation inputs from extraglacierized areas. However, the annual hydrologic regime for each is distinct on account of different water and energy cycle controlling processes. The runoff from a glacierized basin (alpine or ice sheet) is usually restricted to warm seasons when the surface radiation budget provides sufficient energy to melt surface snow and ice. During the cold winter season, the meltwater runoff is relatively negligible because the hydraulic system is closed. In nonglacierized basins, however, runoff can be continuous throughout the year with peak flows around the spring melt event when snowpacks thaw and rapidly contribute to river flows. After the spring melt event, runoff may be contributed by rainfall or snowfall events, ground water, especially from any thawing active layer soil moisture. In the winter, runoff from large nonglacierized basins generally continues albeit at lower flows than summer discharges. This winter flow regime is the main difference between glacier-dominated and snow-dominated rivers.

Measurement of discharge has received much attention in the literature. For a full description of general river discharge measurements, several texts can be consulted (e.g. Shaw 1993; Jones 1997; Dingman 2002). Where human settlement is located nearby a river, gauging often can be successfully achieved using the standard methods in these texts, especially through the use of weirs or concrete structures.

However, in more remote locations, a major problem that has to be addressed is the fact that streamflow discharge measurements tend to yield uncertain discharge estimates because river channels may be unstabilized producing imprecise measurements (Collins 1990). Through successive seasons, the uncertainties can be reduced assuming the flow regime remains relatively constant or changes gradually.

Discharge records from 7,222 discharge gauging stations around the world (as at March 24, 2005) are archived at the Global Terrestrial Network for River discharge (GTN-R) which is part of the GTOS. The Global Runoff Data Center (GRDC) in Germany (<http://grdc.bafg.de>) serves as the digital depository of these discharge records and associated metadata. The archive includes data for low-latitude rivers and data records of seasonal or annual meltwater runoff from glacierized basins have not been identified in any systematic way as part of GTN-R. Generally, if the basins are part of a routine submission procedure to GRDC, these data are available for analysis. However, in many instances, runoff measurements from alpine glaciers and ice sheets are conducted as part of site-specific experiments that are relatively short-lived in duration. Runoff from some alpine glaciers has been measured over 30 years, especially in the European Alps (e.g. Braun et al. 2000; Collins & MacDonald 2004), Scandinavia (e.g. NVE 2002), and in selected parts of the USA, such as Alaska and Washington state (e.g. Fountain et al. 1997), and Canada (Environment Canada 1999). The data for these experiments are sometimes easily available but sometimes more difficult to secure for research studies. Runoff records for most other glacierized basins, including ice sheets, are not well integrated globally and it is perhaps an area of study that requires some effort.

Measurements of runoff in high-latitude basins that are not dominated by glacier meltwater contributions are better organized for cryospheric studies. In fact, a recent publication by Lammers et al. (2001) reported on the most comprehensive assessment to date of Arctic river runoff data based on *in situ* gauge measurements. Their motivation for the study (called the Regional, Electronic Hydrographic Data Network for the Arctic Region, or R-ArcticNET) was that according to Vörösmarty et al. (2000), the Arctic Ocean is the most "river-influenced and landlocked of all oceans" (Lammers et al. 2001) so it is of critical importance for scientists to have a consistent data set of discharge measurements in the circumpolar Arctic basin for environmental change studies. Monthly discharge data from 3,754 gauges are archived in the R-ArcticNET data base (<www.R-arcticnet.sr.unh.edu>). Figure 2.11 shows the locations of these sites. It is interesting, yet predictable, that the lower-latitude regions have denser stream gauge networks compared with higher-latitude regions. In general, the Eurasian sites have a longer heritage of runoff gauge measurements than North America and Europe although there is much variability and incompleteness in the length of records. Calculations of annual discharges for the nine largest basins in the Arctic have been used by the Arctic Monitoring and Assessment Programme (AMAP, <http://www.amap.no>), a working group of the Arctic Council which is a high-level intergovernmental forum for governments and people of the Arctic regions.

2.2.7 RIVER AND LAKE ICE BREAK-UP AND FREEZE-UP

A significant control of cryospheric hydrology, especially in the Arctic, is the timing and extent of river and lake water freeze-up and break-up. These two parameters are

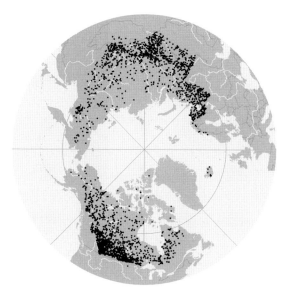

FIG. 2.11 Runoff gauging stations (black dots) in the R-ArcticNET archive (data from <http://www.r-arcticnet.sr.unh.edu>). The gauges comprising this archive are located only in river basins that flow into the Arctic Ocean. This explains why, for example, in this data set there are no gauges along the Pacific coast regions from southern Alaska to Washington state.

critical to the hydrologic cycle because they control the annual water flux magnitude and its timing to the oceans. In a recent paper, Magnusson et al. (2000) note that *"Calendar dates of freezing and thawing of lakes and rivers were being recorded by direct human observations well before scientists began to measure, manipulate, and model these freshwater ecosystems."* A data set is now available that spans the period from 1846 to 1995 for ice on lakes and rivers. The data set assembled by the North Temperate Lakes Long-Term Ecological Research program at the Center for Limnology at the University of Wisconsin-Madison was compiled with submissions from the Lake Ice Analysis Group. The resulting Global Lake and River Ice Phenology Database is maintained at the National Snow and Ice Data center and

contains freeze-up and break-up dates for 750 lakes and rivers. Freeze-up date is defined as the first date in the season on which the water body was observed to be totally ice covered. Break-up date refers to the date of the last break-up observed before the summer "open water" period begins. Some records are longer than others. For example, 429 water bodies have records longer than 19 years. A total of 287 of these are in North America and 141 in Eurasia. There are 170 locations that have records longer than 50 years with 28 longer than 100 years. These data, therefore, allow analysis of broad spatial patterns as well as long-term temporal patterns of lake ice freeze-up and break-up in the Northern Hemisphere. Using the data from the NSIDC archive, Fig. 2.12 illustrates the distribution of observing sites for both lake and river locations for the 1960–61 winter season (Benson & Magnusson 2000).

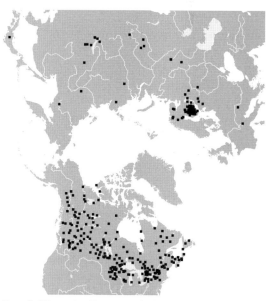

FIG. 2.12 Distribution of observation sites of river ice and lake ice freeze-up and break-up for 1960. Data are taken from the archive of Benson and Magnusson (2000) at NSIDC.

2.3   CONCLUSIONS

This chapter has examined, albeit briefly, some of the *in situ* data sets available for analyzing the state of the cryosphere. *In situ* measurements provide an accurate representation of state variables at the point or microscale and have enabled our understanding of energy and mass transfer processes at that scale. Countless research projects in geographical, earth, and engineering sciences have yielded high quality data sets. However, the combined spatial and temporal coverage of these field campaigns remains limited. The ability of these data sets to comprehensively represent regional to global cryospheric processes is uncertain at best. For example, our understanding of snowpack energy and mass physical exchange dynamics suggests that these processes can be resolved at scales of 100 m or so in the mountains and perhaps 1–2 km over more homogeneous terrain. Quantifying bulk snow properties using sparse *in situ* networks of point measurements is unlikely to yield accurate spatial representation of snow water equivalent or other bulk variables at the regional scale. Intensive field measurement programs, usually conducted for short durations, can represent processes at the local scale well. However, the link to regional and global scales is uncertain in most cases. Our ability to represent other cryospheric state variables from local to global scales of variability also suffers from a similar disconnect.

In situ measurements alone, therefore, are inadequate to quantify global cryospheric

changes. At the same time, many publicly funded operational *in situ* measurement networks are in decline and the need for understanding of the cryosphere in global environmental change context has never been greater. The fact is that *in situ* measurements, when used with other monitoring approaches, can make a significant contribution to global cryospheric understanding. For example, remote sensing is an approach that has found a strong niche in cryospheric science. Coupling *in situ* measurements with remote sensing observations, either for calibrating retrieval algorithms and models, or validating remote sensing products, is of critical importance to the credibility of applying remote sensing data to global cryospheric change. Furthermore, the use of numerical models for the simulation of cryospheric physical processes requires both remote sensing and *in situ* measurements in the model development phase (conceptual design or model formulation), the implementation phase (as forcing fields for the model or perhaps through complex data assimilation schema), and at the validation phase (to quantify accuracy and precision). A strong reason exists, therefore, for continued *in situ* cryosphere measurements. However, no longer should we expect *in situ* data alone to represent cryospheric state variables globally. They should be used within the context of their inherent spatial and temporal constraints as well as to develop and improve remote sensing and/or numerical simulation models of the cryosphere.

3

PROCESSES OF CRYOSPHERIC CHANGE

3.1 INTRODUCTION

The connecting links between the large-scale cryospheric changes discussed in Chapter 1 and the intricacies of measurement, monitoring, and modeling discussed in Chapter 2 are the concepts of mass and energy budgets. Regions of the Earth where surfaces and volumes of snow and ice are prevalent display ways of storing water and regulation of energy that are unique. By measuring the inputs, throughputs, and outputs of mass and energy into, through, and out of snow- and ice-covered areas over unit time periods, it becomes possible to account for the availability, storage, and distribution of that mass and energy. The interpretation of the unique behavior of the mass of snow and ice is the domain of the snow and ice hydrologist (e.g. Kuzmin 1961) and the interpretation of the processes associated with energy exchange are the domain of the boundary layer climatologist (e.g. Oke 1987).

Snow reflects about 80% of the sun's radiation, whereas soil and sea water absorb 80% or more. This simple fact means that the amount of land (and sea ice) covered by snow is critically important to the Earth's radiation balance and so to the global climate system. Changes in radiation balance are extreme when the snow cover is first established and when it disappears. In latitudes above 65°N, the rapid increase in solar radiation in spring coincides with the rapid melting of shallow snowpacks and this produces extreme changes in surface radiation balance over periods of as little as 10 days.

3.2 SNOW AND ICE AS ENERGY REGULATORS

The energy regulation functions (described below) profoundly alter the climate of the snow and ice, by comparison with the atmosphere, ocean, and soils which surround them (Fig. 3.1).

Snow and ice function as energy banks: they store and release energy. They store latent heat of fusion, sublimation, and crystal bonding forces. To sublimate 1 kg of snow requires the same amount of energy as raising the temperature of 10 kg of liquid water by 67°C; by comparison, to melt 1 kg of snow (already at 0°C) requires the same amount of energy as raising the temperature of 1 kg of water by 79°C.

Snow and ice also function as radiation shields: cold snow reflects most shortwave radiation and absorbs and reemits most longwave radiation. As snowmelt progresses, the snow cover reflects less shortwave radiation because of a change in its physical properties. The proportion of shortwave radiation reflected from a snow or ice cover is high compared with soil and vegetation. Bare soil and vegetation will absorb as much as eight times shortwave radiation as a fresh snow cover. Shortwave radiation that is not reflected is absorbed in the top 30 cm of the snowpack. Snow cover behaves

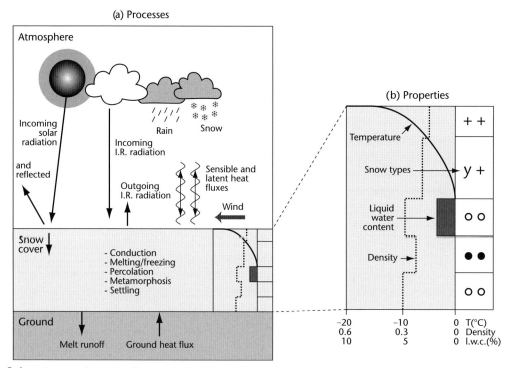

FIG. 3.1   Mass and energy fluxes controlling the energetics of a snow cover and their relation to snowpack structure properties and processes, the atmosphere, and the ground (from Pomeroy & Brun 2001).

almost as a black body. The longwave radiation is absorbed and reradiated as thermal radiation. The wavelength of emission depends on the surface temperature of the snow cover.

Snow and ice also function as insulators. They are porous media with high insulation capacity. This insulation can result in strong temperature gradients that restructure the snow composition. The thermal conductivity of a snow cover is low compared with soil surfaces and varies with the density and liquid water content of the snow cover (Table 3.1). From the table it can be calculated that a surface temperature wave of a given magnitude would penetrate 2.5 times as deeply into granite compared with fresh snow.

3.2.1   THE ENERGETICS OF THE SNOW SURFACE

The concept which is commonly used to summarize the effects of energy regulation at the surface is the energy balance, in the following form:

$$Q^* = Q_H + Q_E + Q_G \qquad (3.4)$$

where Q^* is the net all-wave radiation flux density, Q_H is the turbulent sensible heat flux density, Q_E is the turbulent latent heat flux density, and Q_G is the subsurface heat flux density all in the form of W m^{-2}. Energy used in melting snow is normally accounted separately from latent (evaporative) heat because for continuous snow

TABLE 3.1 Thermal regime (J.R. Mackay, personal communication).

	κ	ρ	c	α	$\sqrt{\alpha}$
Typical thermal regime values (CGS units)					
Gravel	0.003	2.0	0.18	0.008	0.09
Icy silt	0.006	1.6	0.31	0.012	0.11
Dry peat	0.0004	0.4	0.5	0.002	0.04
Ice	0.005	0.9	0.5	0.012	0.11
Icy peat	0.005	0.9	0.4	0.012	0.11
Snow (fresh)	0.0002	0.2	0.45	0.002	0.04
Snow (packed)	0.0005	0.3	0.45	0.004	0.06
Water	0.0013	1.00	1.00	0.0013	0.036
Granite	0.006	2.7	0.19	0.013	0.11
Wet mud	0.0006	0.3	0.45	0.004	0.06

Handy equations

Depth of freezing:

$$z = b\sqrt{t} \tag{3.1}$$

Rate of freezing:

$$b/2\sqrt{t} \tag{3.2}$$

Depth to which temperature of ground perceptibly decreases in time, t:

$$z = \sqrt{12}\ \alpha t \tag{3.3}$$

Definitions

κ thermal conductivity; the quantity of heat that flows in unit time through a unit area of a plate of unit thickness having unit temperature difference between its faces (cal (cm sec °C)$^{-1}$ or 418.4 W (m °K)$^{-1}$).

ρ density; mass per unit volume (g cm^{-3}, multiply by 1,000 for kg m^{-3}).

c volumetric specific heat or heat capacity, amount of heat necessary to change the temperature of a unit mass by one degree (cal (g °C)$^{-1}$, multiply by 4.184 for J (kg K)$^{-1}$).

α thermal diffusivity; $\alpha = k/\rho c$ determines the rise in temperature (cm^2 s^{-1}, multiply by 0.01 for m^2 s^{-1}).

b a proportionality factor which is a function of soil characteristics.

L specific latent heat of fusion of ice (80 cal (g °C)$^{-1}$; 334 kJ kg^{-1}).

t unit time.

z depth of freezing or temperature change.

cover in open environments, the proportion of total phase change energy is so much greater than evaporation (c. 90% according to Shook & Gray 1997). In most places, incoming shortwave radiation is the principal source of energy but other processes can involve energy exchanges of similar orders of magnitude. Much of the solar radiation incident on a snow surface is reflected with albedos as high as 0.9 for compact, dry, clean snow, dropping to 0.5–0.6 for wet snow, and 0.3–0.4 for porous, dirty snow. Albedo decays with time since the previous snowfall, but with very different decay rates for shallow and deep snowpacks.

Over snow surfaces exposed to the atmosphere, outgoing longwave radiation is usually larger than incoming radiation, leading to a net loss of longwave radiation from the snow cover; by contrast, under forest canopies, a downward net longwave flux can develop. The radiation balance over snow shows wide variations as a function of cloud cover, forest canopy, and topography.

The turbulent fluxes of sensible and latent heat are also important for the energy balance of a snow cover. They depend respectively on the vertical gradients of temperature and humidity in the atmosphere and turbulent transfer of heat and water vapor at the snow surface. Because gradients of meteorological variables are rarely available, bulk transfer calculations are necessary.

For discontinuous snow covers in open environments, local advection of sensible energy from bare ground to snow patches becomes an important component of the energy balance. The ground heat flux to snow on a daily basis is considered to be a small component of the energy balance. But, because it is persistent, it can have an important cumulative effect early in the melt season in retarding or accelerating the time of melt and in affecting the environment between snow patches. In locations with incompletely frozen soils, the ground heat flux is positive; however, over permafrost soils the flux is negative in late winter. This negative ground heat flux is important in delaying the snowmelt season in the North, in spite of high sensible and radiant heat fluxes.

3.2.2 THE ENERGETICS OF THE SNOWPACK

Oke (1987) has explained that the energy balance of a snowpack is complicated not only by the fact that shortwave radiation penetrates into the snowpack but also by internal water movement and phase changes. He invites comparison of a snowpack volume in terms of energy balance of (a) a frozen snowpack and (b) a melting snowpack with (c) the water balance of that same control volume (Fig. 3.2).

The energy balance of a snow volume depends upon whether it is a "cold" ($<0°C$)

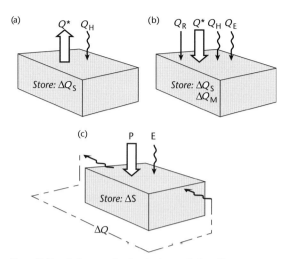

FIG. 3.2 Schematic depiction of the fluxes involved in the energy (a and B) and water balances (c) of a snowpack volume (from Oke 1987). The energy balances are for (a) "cold" or frozen pack, (b) "wet" or melting pack.

or a "wet" (0°C, often isothermal) snowpack. When considering a snow volume, eqn. 3.4 has to be replaced by

$$Q^* = Q_H + Q_E + Q_G + \Delta Q_S + \Delta Q_M \qquad (3.5)$$

where ΔQ_S is the net heat storage change and ΔQ_M is the latent heat storage change due to melting or freezing. In the case of a "cold" snowpack (Fig. 3.2a) such as is commonly found at high latitudes in winter with little or no solar input, Q_E and ΔQ_M are likely to be negligible because there is no liquid water for evaporation, little atmospheric vapor for condensation or sublimation, and both the precipitation and the contents of the snowpack all remain in the solid phase. Similarly, heat conduction within the snow will be small because of the low conductivity of snow (Table 3.1) and the lack of solar heating, so that ΔQ_S and Q_G are also negligible. The energy balance therefore reduces to that between a net radiative sink (Q^*) and a convective sensible (Q_H) heat source (Fig. 3.2a).

In the case of a "wet" snowpack during the melt period (Fig. 3.2b), the surface temperature will remain close to 0°C, but the air temperature may be above freezing. Precipitation may then be as rain and the energy balance is further complicated, as follows:

$$Q^* + Q_R = Q_H + Q_E + Q_G + \Delta Q_S + \Delta Q_M \tag{3.6}$$

where Q_R is heat supplied by rain with a temperature greater than that of the snow.

There is another factor which should be borne in mind and that is the infiltration of meltwater into soils, involving a significant energy flow. If meltwater refreezes in frozen soils, there will be a significant release of latent heat and further melting of frozen soil-water. Dry snow experiences thermal conduction, thermal convection, and wind pumping which govern the heat flux between bottom and top of the snow cover.

Temperature gradients induce water vapor gradients and consequent diffusion of water vapor from warmer parts to the colder ones. When temperature gradients exceed $10°C \cdot m^{-1}$, destructive ice crystal sublimation and recrystallization occur as water vapor moves along the gradient.

The rate of snowmelt is primarily controlled by the energy balance near the upper surface where melt normally occurs first. In temperate climates, snowpacks tend to be uniformly close to the melting temperature when melt commences at the surface. In cold climates, however, change in internal energy of the snowpack may be significant.

In an alpine, midlatitude region, the relative importance of each of the energy balance terms varies with season (Fig. 3.3). Note in particular the changing dominance of the shortwave radiation versus longwave radiation from December to April; the relative importance of the sensible heat versus latent heat and the comparatively small

importance of the rainfall energy (Brun et al. 1989). Although snow cover reduces the available energy at its surface because of its high albedo to solar radiation and high emissivity of longwave radiation, its insulative properties exert the greatest influence on soil temperature regime (Table 3.2). Snow cover acts as an insulating layer which reduces the upward flux of heat, resulting in higher ground temperatures than would occur if the ground were bare (Fig. 3.4). Judge (1973) estimated that on the average, ground surface temperatures in Canada are 3.3°C higher than air temperatures.

The exact value of this insulation effect can be calculated from eqn. 3.7 in Table 3.2. The water balance of a snow or ice volume (Fig. 3.2c) is given by

$$\Delta S = P - (E + Q) \tag{3.8}$$

where ΔS is the change in snow or ice storage, P is precipitation, E is evaporation, and Q is discharge, all expressed as millimeter per annum. Further discussion of this equation will be delayed until the second half of this chapter.

3.2.3 THE ENERGETICS OF GLACIERS

The energy balance at the surface of a glacier is the sum of the individual energy components and may be expressed as

$$Q_M = Q^* + Q_H + Q_E \tag{3.9}$$

where Q_M is energy available to melt ice. Q^* is frequently broken down into net shortwave (Q_S) and net longwave radiation (Q_L) flux densities because the relative importance of each component varies so much from place to place (Paterson 1994).

In glacier energy balance studies in continental locations, net radiation has been calculated to account for approximately 66% of ablation energy (e.g. Braithwaite 1981);

Fig. 3.3 Daily variation of energy balance components of an alpine snow cover in the French Alps during three periods. (a) mid-winter, (b) late winter, and (c) spring (from Brun et al. 1989).

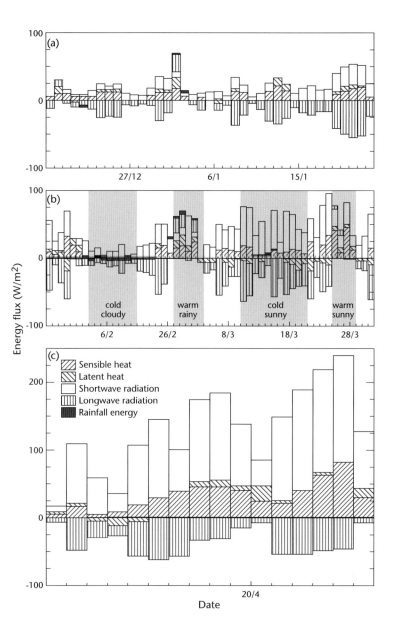

however, in more maritime climates, such as Iceland, net radiation may account for as little as 10% (e.g. Ahlmann & Thorarinsson 1938) because of the presence of warmer and moister air masses. Q_S is highest on low-latitude, high-altitude glaciers in the Andes; Q_L is highest in humid conditions and close to rock surfaces and valley sides;

Q_H and Q_E become important where warm, humid air moves over glacier surfaces such as western North America, southwest Iceland, western Norway, the Pacific coast of Chile (Laumann & Reeh 1993), and the glaciers of equatorial Africa (Hastenrath & Kruss 1992). The effects of debris cover on a glacier are interesting. Debris having

TABLE 3.2 Effect of snow cover on the mean annual ground temperature at the bottom of the active layer (modified from Kudryavtsev 1965).

Snow density	Thermal diffusivity (α)	Thickness of snow z (m)						
		0.1	0.2	0.3	0.4	0.6	0.8	1.0
0.075	0.0010	0.094	0.181	0.259	0.329	0.451	0.551	0.632
0.110	0.0015	0.081	0.155	0.224	0.288	0.400	0.491	0.572
0.150	0.0020	0.071	0.136	0.197	0.253	0.355	0.442	0.518
0.190	0.0025	0.064	0.123	0.178	0.230	0.324	0.407	0.480
0.225	0.0030	0.058	0.113	0.164	0.213	0.302	0.381	0.450
0.250	0.0035	0.054	0.105	0.153	0.198	0.282	0.357	0.425
0.300	0.0040	0.051	0.098	0.143	0.186	0.267	0.338	0.403
0.340	0.0045	0.048	0.093	0.136	0.178	0.254	0.323	0.386
0.380	0.0050	0.045	0.088	0.130	0.169	0.242	0.309	0.371
0.415	0.0055	0.043	0.081	0.124	0.161	0.232	0.297	0.356

$$T_A = T + T_0\left(1 - \frac{1}{f}\right) \tag{3.7}$$

The value of $\left(1 - \frac{1}{f}\right)$ is:

$$f = e^{0.018\, z \frac{\sqrt{\pi}}{\alpha T}}$$

T_A: mean annual ground temperature, at the bottom of the active layer at a given time (°C).
T: mean annual air temperature (MAAT) for the period.
T_0: yearly amplitude (half range) of mean monthly air temperatures (°C = T_0) for the period.
z: mean maximum snow depth (cm).
T: a period of one year, in hours (8,760).
α: thermal diffusivity ($\alpha = k/pc$) of snow (m² hr⁻¹). Refer to Table 3.1 for definitions.

a lower albedo than ice will absorb more Q_S and encourage higher ablation; but debris can also protect underlying ice from ablation by shielding it from Q_S. Protection will occur if the debris cover is thick enough to prevent heat from the surface being conducted through to the ice in the course of a daily temperature cycle. The crossover point, where ablation is at a maximum, occurs when the debris is 0.5–1 cm thick (Ostrem 1959).

The same general energy budget equation as that which is used for snow is appropriate for glaciers. Over melting glaciers, net shortwave radiation is generally the dominant term (Greuell & Smeets 2001), the latent heat flux is relatively small, and net longwave radiation and the sensible heat flux are of intermediate magnitude (Fig. 3.5). The subsurface heating and the heat flux supplied by rain is commonly negligible when averaged over a year but specific combinations of cold glaciers and intense warm rainstorms must be considered separately (Hay & Fitzharris 1988).

A part of the shortwave radiative flux will be absorbed below the surface of the ice. Water formed by melt or deposited on the glacier as rain will penetrate into the snow and/or ice. On its way down, some of the water will be retained in the pores owing to capillary forces. If the water

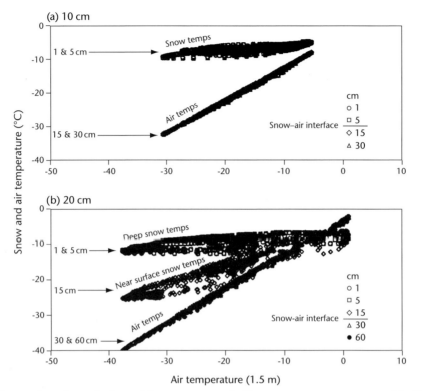

FIG. 3.4 Thermal regime of a boreal forest snowpack. Temperature was measured half-hourly at various heights above the ground in Prince Albert National Park, Saskatchewan. Temperatures measured at heights greater than the snow depth are air temperatures. (a) Early winter, 10 cm deep, 100 kg m^{-3} dense. (b) Early mid-winter, 20 cm deep, 150 kg m^{-3} dense (from Pomeroy & Brun 2001).

encounters cold snow, it will refreeze and locally increase the temperature. The water that refreezes within the snow is called internal accumulation. If the water encounters ice on its way downward it will generally from runoff, thus contributing to ablation. However, if the slope is small, part of the water will accumulate on top of the ice. There it may freeze as so-called superimposed ice if the underlying ice is cold.

3.2.4 THE ENERGETICS OF SEA ICE AND VARIOUS TERRAIN TYPES

The energy balances of snowpacks and glaciers are comparatively simple when compared with those of sea ice and variable

terrain types because of the high spatial variability of the sea ice and the terrain surfaces. As indicated in Chapter 1, there are many varieties of sea ice and each of them has different spectral albedos. Large changes can occur rapidly over the course of a summer melt season and also during the fall freeze back. Perhaps the most important aspect of the temporal behavior of the spectral albedos of sea ice is the general decrease that takes place with the onset of the melt season. In a study reported by Grenfell and Perovich (1984), shortwave radiation flux densities were measured near Point Barrow, Alaska on shorefast ice, first and second year floes and a deformed rubble zone between May 18 and June 17, 1979. This represents

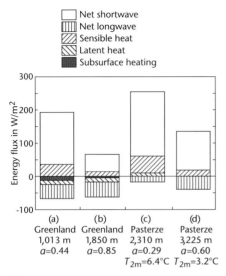

FIG. 3.5 Components of the surface energy balance derived from mid-summer balance on the western part of the Greenland Ice Sheet and on the Pasterze glacier, Austria. Average values (a) over 38 days at Camp IV just below ELA; (b) over 70 days at Carrefour in the ice sheet accumulation zone; and (c) and (d) over 46 days at the Pasterze on both sides of the ELA (from Ambach 1979; Greuell & Smeets 2001).

the transition from cold winter conditions to well-developed melt conditions. They documented three separate evolutionary sequences which are graphed separately in Figs. 3.6a, 3.6b, and 3.6c.

At the scale of the polar regions, terrain types present similar problems in terms of spatial variability. In an oft-quoted study by Weller and Wendler (1990) closed boreal forests, open woodlands, tundra, the pack ice of the Arctic Ocean, glaciers, and large ice sheets were compared in terms of their distinctive energy budgets during summer and winter (Fig. 3.7). Large differences are shown between energy balances during summer with the boreal forest and the tundra acting as major heat sources for the surrounding terrain, and the glaciers acting as major heat sinks. In winter, thin pack ice is the major source of heat energy. In the

Antarctic, summer conditions are quite different. Because there are few exposed rock surfaces and no tundra the albedo remains high everywhere and the energy budget is quite low in summer. Little melting occurs on the slopes and the plateau of the Antarctic Ice Sheet.

3.2.5 PERMAFROST

Permafrost is defined as ground that remains at or below 0°C for at least two consecutive years. This means that moisture in the form of either water or ice may or may not be present. Permafrost may therefore be unfrozen, partially frozen, or frozen depending on the state of the ice/water content.

There is a tendency to regard a frozen soil as one in which the water has been replaced by ice. In fact at most temperatures of interest, frozen soils contain ice and water. The more fine-grained a soil is, the greater is the amount of water remaining at a given negative temperature. As the water content is reduced by progressive formation of ice, the remaining water is under an increasing suction. The suction developed by freezing can be seen from the observation that a previously unfrozen clay sample, after being frozen and then thawed, shows a completely changed structure. It consists of hard almost shaly flakes. The clay flakes have consolidated because of the effective stress associated with the suction. This changed structure means that the bulk strength is lowered as the discontinuities between the flakes are planes of low strength.

Another critically important implication of suctions developed in the ground on freezing is the phenomenon of frost heave. Water is drawn toward the freezing soil and, on entering the frozen zone, becomes ice. The frozen soil then contains more ice and has a higher moisture content than

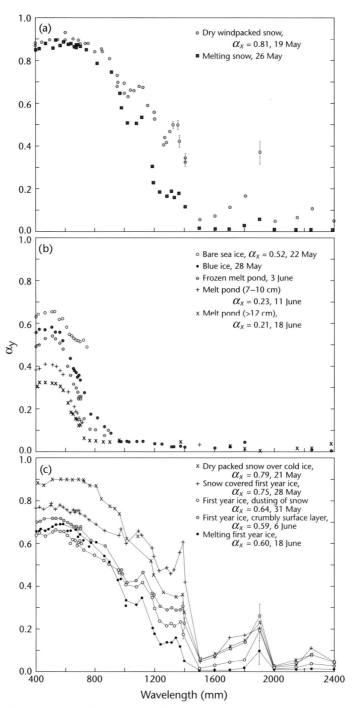

FIG. 3.6 Spectral albedos observed over snow, bare ice, melt ponds, snow covered ice, and melting first year ice (from Grenfell & Perovich 1984).

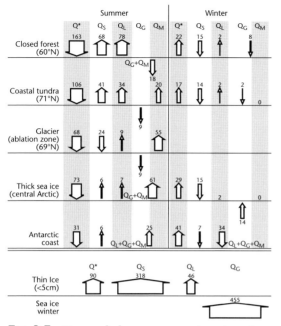

FIG. 3.7 Energy balance over various terrain types in the polar regions (from Weller & Wendler 1990).

before freezing. Consequently the volume of the soil increases and this results in frost heave. This effect is over and above the 9% expansion that occurs when water freezes. Frost heave occurs preferentially in silty soils and does not occur in coarse grained materials. Clay rich soils have such a low permeability that water migration is reduced by comparison with the silty soils. Again, the practical significance of frost heave lies in the great loss of strength that occurs when frost heaved soils thaw. Foundations for roads and pipelines in permafrost regions require large quantities of coarse grained materials to reduce the heaving during winter.

As French (1996) points out, there are three major considerations related to the water/ice content of permafrost:

1 The freezing of water in the active layer at the beginning of winter each year results

in ice lensing and ice segregation. The amount of heave will vary according to the amount and availability of moisture in the active layer, with poorly drained silty soils showing the maximum heave effects. There is also secondary heave effect as unfrozen water progressively freezes. This moisture migrates in response to a temperature gradient and causes an ice-rich zone to form in the upper few meters of permafrost (Mackay 1983).

2 Ground ice is a major component of permafrost, particularly in unconsolidated sediments (Mackay & Black 1973). If ground ice-rich permafrost thaws subsidence of the ground results. A range of processes associated with permafrost degradation are summarised under the term thermokarst.

3 The hydrological and groundwater conditions of permafrost terrain are unique (Hopkins et al. 1955). Subsurface flow is restricted to unfrozen zones called taliks and to the active layer.

There are three groups of features whose formation necessarily involve permafrost and which therefore are diagnostic of permafrost conditions: (a) Patterned ground, including ice wedge polygons (Fig. 3.8), stone polygons, sorted circles (Fig. 3.9),

FIG. 3.8 Patterned ground on Jan Mayen Island, illustrating high centered polygons, 1978 (photo by the late Alfred Jahn).

FIG. 3.9 Sorted stone circles in Hornsund Fjord, Svalbard (photo by the late Alfred Jahn).

sorted stripes, and nonsorted circles; (b) palsas; and (c) pingoes. Permafrost terrain is generally regarded as highly sensitive to thermal disturbance. Mackay (1969) has summarized some of the major processes involved (Fig. 3.10). The fundamental points made by Mackay are (a) the distinction between severe and minor disturbance and (b) the association of severe disturbances with frost susceptible soils and high ground ice content.

3.3   SNOW AND ICE RESERVOIR FUNCTIONS

Snow and ice function as storage reservoirs (Figs. 3.1 and 3.2c). They are reservoirs of water that profoundly alter hydrology. During melt, snow and ice move as meltwater in preferential pathways within the snowpack or ice mass. Runoff and streamflow generation will differ depending on whether the underlying soil is frozen or unfrozen.

Mass budgets are generally shown as

$$I - O = \Delta S \qquad (3.10)$$

where I is an input term, O is an output term, and ΔS is the change in storage term.

But what is often overlooked is that such a formulation ignores the time integration over which the storage change occurs and the relevant area.

A more satisfactory hydrologic bookkeeping equation is

$$\int_{t_1}^{t_2} P \cdot \mathrm{d}t - \left(\int_{t_1}^{t_2} Q \cdot \mathrm{d}t + \int_{t_1}^{t_2} E \cdot \mathrm{d}t \right) = \frac{\mathrm{d}S}{\mathrm{d}t} \qquad (3.11)$$

where Q is instantaneous discharge from a basin; $\int_{t_1}^{t_2} P \cdot \mathrm{d}t$ is precipitation over the basin between t_1 and t_2; $\int_{t_1}^{t_2} E \cdot \mathrm{d}t$ is evaporation and transpiration over the basin between t_1 and t_2. Selection of the time period between t_1 and t_2 affects the dimensions of the storage term; selection of the size and homogeneity of the drainage basin has implications for the resolution level of the budget.

3.3.1   MASS BUDGET FOR SNOW

If one is interested in the within-season mass budget of snow, it is common to rearrange the equation as follows

$$\Delta S_{\mathrm{Snow}} = \int_{t_1}^{t_2} P_{\mathrm{Snow}} \cdot \mathrm{d}t - \left(\int_{t_1}^{t_2} Q \cdot \mathrm{d}t + \int_{t_1}^{t_2} E \cdot \mathrm{d}t \right) \qquad (3.12)$$

and to solve for change in snow storage (S_{Snow}).

In the illustrative example (Fig. 3.11), interest attaches to the monthly and annual balance of water and ions in Jamieson Creek basin in the Coast Mountains of British Columbia (Zeman & Slaymaker 1978).

The annual mass budget under most circumstances has no net storage of snow and reduces to

$$P_{\mathrm{annual}} = Q_{\mathrm{annual}} + E_{\mathrm{annual}} \qquad (3.13)$$

However, on glaciers in the accumulation zone and in high-latitude polar regions

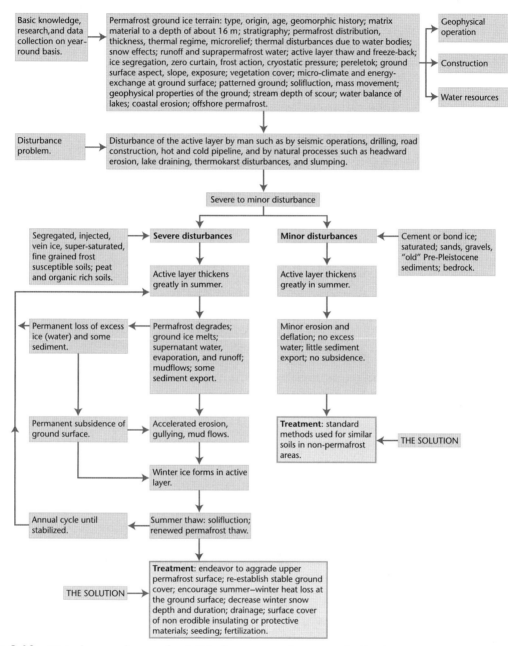

Tundra Research

| Basic knowledge, research, and data collection on year-round basis. | → | Permafrost ground ice terrain: type, origin, age, geomorphic history; matrix material to a depth of about 16 m; stratigraphy; permafrost distribution, thickness, thermal regime, microrelief; thermal disturbances due to water bodies; snow effects; runoff and suprapermafrost water; active layer thaw and freeze-back; ice segregation, zero curtain, frost action, cryostatic pressure; pereletok; ground surface aspect, slope, exposure; vegetation cover; micro-climate and energy-exchange at ground surface; patterned ground; solifluction, mass movement; geophysical properties of the ground; stream depth of scour; water balance of lakes; coastal erosion; offshore permafrost. |

Geophysical operation

Construction

Water resources

Disturbance problem. → Disturbance of the active layer by man such as by seismic operations, drilling, road construction, hot and cold pipeline, and by natural processes such as headward erosion, lake draining, thermokarst disturbances, and slumping.

Severe to minor disturbance

Severe disturbances

Minor disturbances

Segregated, injected, vein ice, super-saturated, fine grained frost susceptible soils; peat and organic rich soils.

Cement or bond ice; saturated; sands, gravels, "old" Pre-Pleistocene sediments; bedrock.

Active layer thickens greatly in summer.

Active layer thickens greatly in summer.

Permanent loss of excess ice (water) and some sediment.

Permafrost degrades; ground ice melts; supernatant water, evaporation, and runoff; mudflows; some sediment export.

Minor erosion and deflation; no excess water; little sediment export; no subsidence.

Permanent subsidence of ground surface.

Accelerated erosion, gullying, mud flows.

Treatment: standard methods used for similar soils in non-permafrost areas.

THE SOLUTION

Winter ice forms in active layer.

Annual cycle until stabilized.

Summer thaw: solifluction; renewed permafrost thaw.

THE SOLUTION → **Treatment**: endeavor to aggrade upper permafrost surface; re-establish stable ground cover; encourage summer–winter heat loss at the ground surface; decrease winter snow depth and duration; drainage; surface cover of non erodible insulating or protective materials; seeding; fertilization.

Fig. 3.10 Disturbance of permafrost (Mackay personal communication).

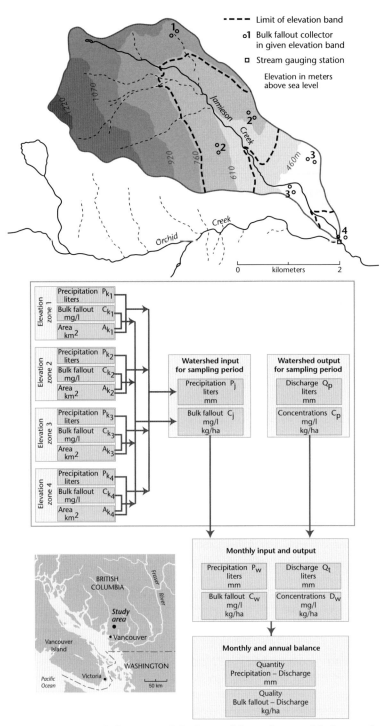

FIG. 3.11 Flow diagram for mass balance model of monthly and annual balances for snow cover in Jamieson Creek basin approximately 25 km north of Vancouver, British Columbia (from Zeman & Slaymaker 1978).

there will be a net snow storage term and eqn. 3.8 will be needed.

3.3.2 MASS BALANCE FOR GLACIER ICE

The within-season mass balance for glacier ice is normally written as

$$b = c + a \qquad (3.14)$$

where b (mass balance), c (accumulation), and a (ablation) are expressed as equivalent volume of water.

Commonly, the annual mass balance of a glacier is determined with the aid of runoff measurements (Q) from the snout of the glacier combined with the monitoring of stakes arranged at a number of elevations on the glacier, precipitation gauges (P), and a micrometeorological station to determine evaporation (E). The form of this equation is then

$$B_n = P - E - Q \qquad (3.15)$$

This is the **hydrological method**. B_n is the mass balance at the end of the balance year. This is usually subdivided into a winter balance B_w and summer balance B_s. In the example illustrated (Fig. 3.12), the net mass balance of a number of glaciers on Svalbard is compared over the period 1950–2000 (Dowdeswell et al. 1997).

3.3.3 THE MASS BALANCE OF AN ICE SHEET

The spatial scale of an ice sheet determines the need for accurate representation of spatial variability of ice mass. This problem was intractable until satellite-based remote sensing techniques became available. The form of the relevant mass balance equation is

$$B_n = M_a - M_m - M_c \pm M_b \qquad (3.16)$$

where M_a is annual surface accumulation; M_m is annual loss by glacial surface runoff; M_c is annual loss by calving of icebergs; and M_b is the annual balance at the bottom (melting or freeze-on of ice). Equation 3.16 suggests that the total mass balance can be obtained by two methods: (a) by direct measurement of the change in volume by monitoring surface elevation change and (b) by the budget method, determining each term on the right-hand side of the equation separately. Figure 3.13 shows an example of estimated snow accumulation rates derived from both historical and recent data (Bales et al. 2001).

3.3.4 MASS BALANCE OF SEA ICE

The predominant feature of the Arctic's physical environment is the presence of a sea ice cover which is perennial in the central Arctic and at least seasonal in the marginal seas (Laxon et al. 2004). Sea ice's relatively straight forward (compared to land snow cover) and rapid (compared to land ice sheets and glaciers) response to atmospheric forcing suggests that observations of sea ice cover may provide early strong evidence of global warming in the Arctic. Moreover, the sea ice cover is a spatially integrated indicator of environmental change by contrast with the spotty temperature records available for the Arctic.

On the other hand, the dynamic response of sea ice to environmental change depends on a complex interplay of mechanical and thermodynamic processes. Because of ice deformation, a typical 100 km² patch of sea ice will contain a variety of ice thicknesses. These thicknesses range from open water to very thick ice, including pressure ridges extending possibly 30 m or more below the surface. On top of this matrix there is often a relatively thin snow cover. Although thin, this snow cover can cause substantial insulation of the ice and reduce its growth rate.

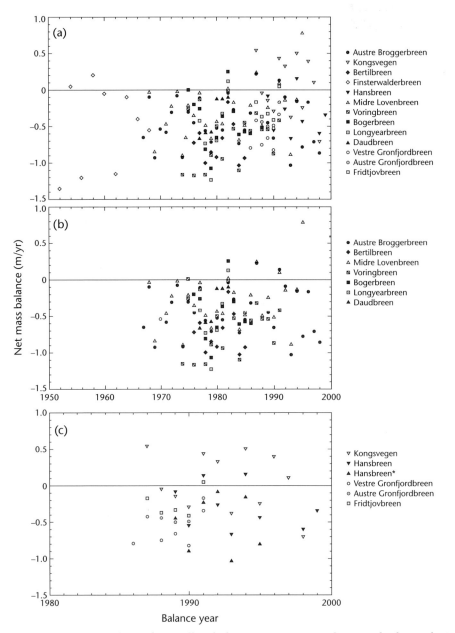

Fig. 3.12 Net annual mass balance for Svalbard glaciers. (a) For 13 glaciers, (b) for 7 glaciers with area <6 km², (c) for glaciers that calve into the sea (from Dowdeswell & Hagen 2004).

This spatial heterogeneity, especially around the edges of the sea ice, makes the mass balance of the sea ice the most complex mass balance in the cryosphere. Local growth and melt and horizontal transport and deformation alter the local mean thickness (ice volume per unit area) and involve exchanges of mass (fresh water) and energy with the atmosphere and ocean. Plate 3.1 provides an illustration based on results of

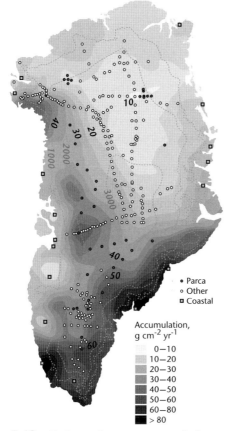

FIG. 3.13 Estimated snow accumulation rates on Greenland derived from historical and recent data after rejection of those considered dubious or those which referred to very short time periods (from Ohmura & Reeh 1991; Bales et al. 2001).

a sea ice model (Hilmer et al. 1998). In this figure, the annual mean ice transport is given by the vectors, the annual mean thickness by the solid contours, and the net freezing rate (net ice growth minus ice melt; directly proportional to the salt flux delivered to the ocean surface) by the colored shading. The general pattern is anticyclonic circulation within the Arctic basin and outflow of ice through the Fram Strait that is balanced by net ice growth over much of the basin. Mean transport pattern

leads to convergent deformation and thickening along the Canadian Arctic islands and Greenland, with divergence and correspondingly thin ice along the central Eurasian coast (Flato 2004).

3.4 SNOWFALL

Snow forms in clouds when the temperature is less than 0°C and supercooled water and cloud condensation nuclei are present. Ice crystals form around cloud condensation nuclei and grow through aggregation of small ice crystals and riming from water droplets into the snowflake form. For snowfall to occur there must be sufficient depth of cloud to permit the growth of snow crystals and sufficient moisture and aerosol nuclei to replace those removed from the cloud in falling snowflakes.

The effects of wind redistribution on the evolution of a snow cover are most obvious in open environments. Four processes are involved: (a) erosion of snow cover, (b) transport of blowing snow, (c) sublimation of blowing snow in transit, and (d) deposition of snow. Dyunin et al. (1991) noted that wind redistribution of snow is the primary process of desertification in steppe environments and they associate the phenomenon of northern desertification with suppression of vegetation in the Russian steppes and forests and the resulting increase in frequency of blowing snow.

3.4.1 INTERCEPTION BY VEGETATION

Snow interception is controlled by accumulation of falling snow in a forest canopy. The snow is subsequently affected by sublimation, melt, and unloading of snow by canopy branches and wind redistribution. Intercepted snow receives snow from snowfall, snow unloaded from upper branches, drip from melting snow on upper branches, and vapor deposition during supersaturated atmospheric conditions. Intercepted snow can sublimate to water vapor or become suspended by atmospheric turbulence, followed by further sublimation or deposition to surface snow. Deposition to the surface may also occur by melt and drip to the surface or by direct unloading from the branches. Hence, intercepted snow may reach the ground as a solid, liquid, or vapor, but not all intercepted snow eventually reaches the ground (Fig. 3.14). The interception efficiency of a canopy is the ratio of snowfall intercepted to the total snowfall and this efficiency is an integration of the collection efficiencies for individual branches. The collection efficiency of a branch is limited by three factors: (a) elastic rebound of snow crystals falling on to snow and/or branch elements; (b) branch bending under snow load; and (c) strength of the snow structure. In general, canopy scale interception efficiency declines with snowfall amounts; branch scale collection efficiency is low for low snowfall, high for medium snowfall, and low for high snowfall.

The net effect of interception processes is that under a forest canopy both snow depth and water equivalent vary.

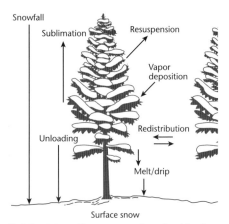

FIG. 3.14 Mass fluxes associated with the disposition of winter snowfall in a boreal forest (from Pomeroy & Schmidt 1993).

3.4.2 SNOW ACCUMULATION

The areal extent of snow varies more rapidly and more dramatically than that of any other widely distributed material on Earth. In the Northern Hemisphere, the monthly mean area covered by snow on land ranges from 5 (10^6) to 4.7 (10^7) km^2 as between the Northern hemisphere summer and winter.

Snow cover is the net accumulation of snow on the ground and is the end product of both accumulation and ablation processes. It is therefore the product of complex factors. The areal variability of snow cover is commonly considered on four spatial scales: global (10^6–10^8 km^2); regional (macroscale) with areas of up to 10^6 km^2 (10^2–10^6 km^2); local (mesoscale) from 10^{-2} to 10^2 km^2; and microscale from 10^{-6} to 10^{-2} km^2.

At global scale, snow cover duration is longest near the poles and on high mountains. Because early and late seasonal snows are ephemeral, analysts often have difficulty in deciding exactly what criteria determine the length of the winter snow cover period. This problem may be partially overcome by rejecting ephemeral occurrences. But different assumptions make it difficult to compare maps of snow cover distribution.

Factors controlling snow cover distribution and characteristics include temperature (colder temperatures associated with relatively dry and light snow), wind (redistribution as snow drifts), forest environments (openings versus within stand locations and transport of intercepted snow), physiography (elevation, slope, aspect, roughness). The average annual snow accumulation in the upper Columbia River basin (Fig. 3.15) illustrates the influence of physiography well. In the same region of western Canada, the influence of continentality and elevation on snow-load and water equivalence has been well documented (Schaerer 1970; Fig. 3.16)

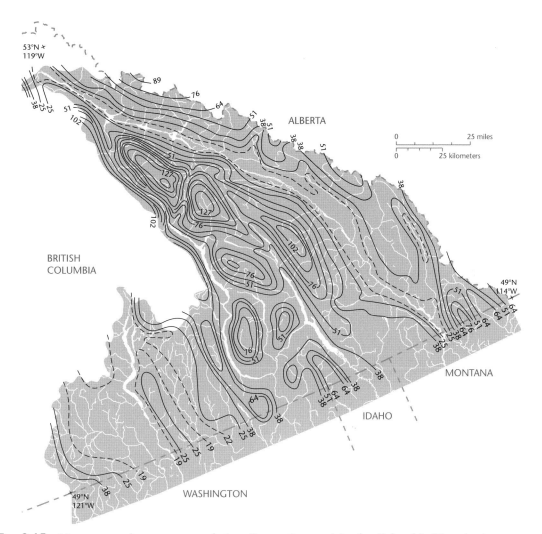

FIG. 3.15 Mean annual snow accumulation (in centimeters) in the Columbia River basin as a function of physiography (M. Church personal communication).

Trends in snow cover in the Northern Hemisphere, namely North America and Eurasia, for the period 1973–94 show a statistically significant annual decline but maximum (winter) snow cover extent shows no significant changes (Groisman et al. 1994). Brown (1997) has confirmed a century-long decrease in spring snow cover in Eurasia, but the trend in North American spring snow cover for this period was not statistically significant. Preliminary analysis of a 24-year record (1979–2003) in snow extent derived from visible and passive microwave satellite data indicates a decrease of approximately 3–5% per decade during spring and summer (Frei & Robinson 1999; Armstrong & Brodzik 2001).

Snow depth data, which are more numerous, are notoriously unreliable unless associated density values are reported. In much of Russia, for example, March is the month with maximum values of snow depth

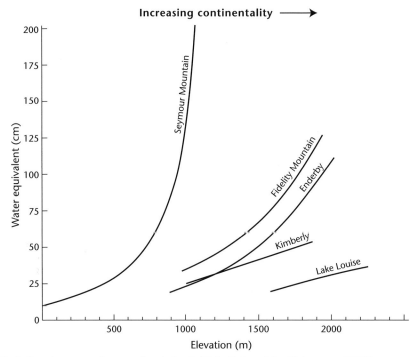

FIG. 3.16 Variation in ground snow-loads in British Columbia (Schaerer 1970).

(Kopanev 1982) but snowmelt floods occur in May over the Great Russian Plain and in June and July further to the northeast (Koren' 1991). Snow water equivalent information has been analysed for the western part of the FSU (Haggerty & Armstrong 1996) and for the western USA (Cayan 1996).

Sturm et al. (1995) have developed a seasonal snow cover classification (Table 3.3) based on snow characteristics at time of maximum snow extent in each region. Snow classes identified were tundra, taiga, alpine, prairie, maritime, and ephemeral (Fig. 3.17), as well as a mountain category which is highly variable and difficult to map at global and continental scales. These snow classes provide an initial perception of the snow properties that can be expected in the regional ecosystems of the Northern Hemisphere.

Snow occurs in many different forms that change rapidly in response to local climatic conditions. Each climatic province has a prevailing snow type, which can be characterized in terms of grain type and size, hardness, density, layering, and depth (though it must be understood that there is also great variety of snow types within each region). The kind of variation that is envisaged is well illustrated by the three major snow regions of Alaska: tundra snow on the Arctic slopes of the Brooks Range; taiga snow in the continental interior; and maritime snow in the southern coastal mountains.

Seasonal snow accumulation varies among climate provinces and through the seasons. Years of excessive snowfall result in transportation delays, large snowmelt floods, power line failure, collapse of buildings, and property loss from snow avalanches. Years of little snowfall result in groundwater depletion, reduced surface water supply, lack of soil frost protection for agriculture, and losses in the winter recreational industry.

TABLE 3.3 Description of snow classes (according to Sturm et al. 1995).

Snow cover class	Description	Depth range (cm)	Bulk density (g cm⁻³)	Number of layers
Tundra	A thin, cold, windblown snow. Maximum depth, ~75 cm. Usually found above or north of treeline. Consists of a basal layer of depth hoar overlain by multiple wind slabs. Surface zastrugi common. Melt features rare.	10–75	0.38	0–6
Taiga	A thin to moderately deep, low-density, cold snow cover. Maximum depth, 120 cm. Found in cold climates in forests where wind, initial snow density, and average winter air temperatures are all low. By late winter, consists of 50–80% depth hoar covered by low-density new snow.	30–120	0.26	>15
Alpine	An intermediate to cold, deep snow cover. Maximum depth, ~250 cm. Often alternate thick and thin layers, some wind affected. Basal depth hoar common as well as occasional wind crusts. Most new snowfalls are low density. Melt features occur but are generally insignificant.	75–250	No data	>15
Maritime	A warm, deep snow cover. Maximum depth can be in excess of 500 cm. Melt features (ice layers, percolation columns) very common. Coarse-grained snow due to wetting ubiquitous. Basal melting common.	75–500	0.35	>15
Ephemeral	A thin, extremely warm snow cover. Ranges from 0 to 50 cm. Shortly after it is deposited, it begins melting, with basal melting common. Melt features common. Often consist of a single snowfall, which melts away; then a new snow cover reforms at the next snowfall.	0–50	No data	1–3
Prairie	A thin (except in drifts), moderately cold snow cover with substantial wind drifting. Maximum depth, ~100 cm. Wind slabs and drifts common.	0–50	No data	<5
Mountain, special class	A highly variable snow cover, depending on solar radiation effects and local wind patterns. Usually deeper than associated type of snow cover from adjacent lowlands.	–	No data	Variable

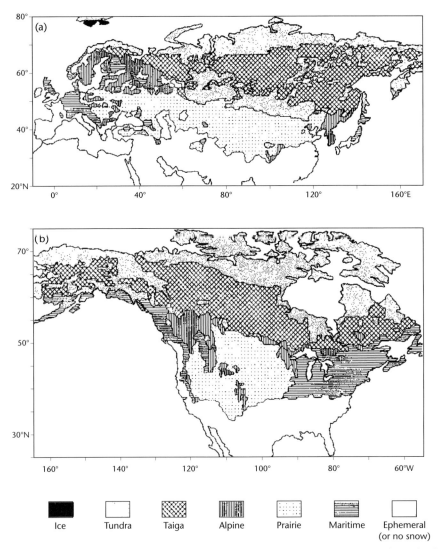

FIG. 3.17 Snow class distribution over Eurasia and North America. Classes are described in Table 3.3 (from Groisman & Davies 2001).

Processes in snow tend to be more complex than those in most other solid earth materials because snow is thermodynamically active and its phase changes constantly. Both its continuum properties (density, temperature, stress distributions) and its microscale properties (crystal morphology and crystal bonding) change in response to small changes in the environment. Snow crystal metamorphism commences as soon as a snow flake falls on the ground. The strength, deviatoric properties, and compressibility of the snow are critical in relation to snow behavior.

3.4.3 SNOW COVER STRUCTURE

Snow stratification results from successive snowfalls over the winter and processes that transform the snow cover between snowfalls (Fig. 3.18). There are two interacting processes that transform the snow

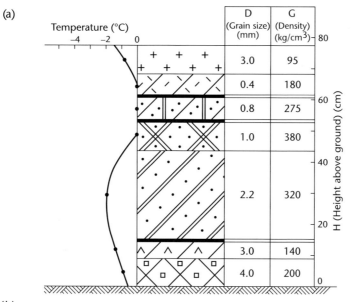

(a)

(b)

Deposited Snow

Feature	Symbol	Subclassification				
		a	b	c	d	e
Specific gravity	G					
Free water (%)	W	dry	moist	wet	very wet	slush
Grain shape	F	F1-F7 crystals	partly settled	rounded grains	grains with facets	depth hoar
Grain size (mm)	D	<0.5	0.5–1	1–2	2–4	>4
Compressive yield strength (g/cm²)	K	0–10	0–10²	10²–10³	10³–10⁴	>10⁴
Snow temperature, T (°C)	Ice layer, i			Impurities, J (%)		

Snowcover measurements

	Vertical	⊥ to inclined surfaces		
Coordinate (cm)	H	M	Inclination of surface, N (degrees)	
Total depth (cm)	HS	MS	Water equivalent of cover, HW (mm)	
			Snowcovered area / Total area Q (tenths)	
Daily new snowfall (cm)	HN	MN	Age of deposit, A (h, days, etc.)	

Snow surface conditions

Surface deposit	Surface hoar	Soft rime	Hard rime	Glazed frost
Symbol	V1	V2	V3	V4
Graphic symbol	⌐_⌐	∨	▼	∞

FIG. 3.18 (a) Use of symbol to describe a snowpack. (b) Classifications and symbols for snow measurements and surface conditions (NRC 1954).

cover: snow settling and snow metamorphism (Jones, H. G. et al. 2001). Because most of the physical properties of snow strongly depend on snow density and on the type of and size of grains comprising a snow layer, settling and metamorphism are fundamental. Snow metamorphism depends on temperature, temperature gradient, and liquid water content. The distinction between dry and wet snow is critical and the classification of snow grains under the influence of metamorphism is useful where a process-sensitive classification can be found (Table 3.4, Fig. 3.19).

TABLE 3.4 International classification of snow crystals (adapted from Colbeck et al. 1990).

Basic classification	Shape	Process classification
1. Precipitation particles	a) Columns b) Needles c) Plates d) Stellar dendrites e) Irregular crystals f) Graupel g) Hail h) Ice pellets	a–g) Cloud-derived falling snow h) Frozen rain
2. Decomposing and fragmented precipitation particles	a) Partly rounded particles b) Packed shard fragments	a) Freshly deposited snow b) Wind-packed snow
3. Rounded grains	a) Small rounded particles b) Large rounded particles c) Mixed forms	a–c) Dry equilibrium forms
4. Faceted crystals	a) Small rounded particles b) Large rounded particles c) Mixed forms	a) Solid kinetic growth form b) Early kinetic growth form c) Transitional form
5. Cup-shaped crystals and depth hoar	a) Cup crystal b) Columns of depth hoar c) Columnar crystals	Hollow kinetic crystal b) Columns of a) c) Final growth stage
6. Wet grains	a) Clustered rounded grains b) Rounded polycrystals c) Slush	a) No melt–freeze cycles b) Melt–freeze cycles c) Poorly bonded single crystals
7. Feathery crystals	a) Surface hoar crystals b) Cavity hoar	a) Kinetic growth in air b) Kinetic growth in cavities
8. Ice masses	a) Ice layer b) Ice column c) Basal ice	a) Refrozen water above less-permeable layer b) Frozen flow finger c) Frozen ponded water
9. Surface deposits and crusts	a) Rime b) Rain crust c) Sun crust, firn-spiegel d) Wind crust e) Melt–freeze crust	a) Surface accretion b) Freezing rain on snow c) Refrozen sun-melted snow d) Wind-packed snow e) Crust of melt–freeze grains

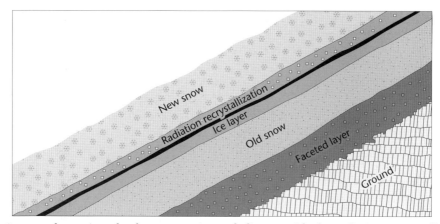

FIG. 3.19 Snowpack stratigraphy from a continental climate with the radiation recrystallized layer above an ice layer and the weak faceted layer next to the ground (from McClung & Schaerer 1993).

The size, type, and bonding of snow crystals are responsible for pore size and permeability of the snowpack. Depth hoar layers have large grains (2–10 mm) and pore sizes to match and are formed under high temperature gradient metamorphism. Rounded grains are very small (0.1–0.4 mm) and pore size is small and there are no large voids that would encourage collapse. In low wind speed environments, fresh snowfall has low hardness and a density range from 50 to 120 kg·m^{-3}. But density can change rapidly under metamorphism. Higher hardness is associated with high wind speeds, greater age, and concurrent blowing snow during deposition.

Dry snow contains relatively little liquid water, but the liquid-like layer that surrounds the snow crystals is important to snow chemistry. Dry snow metamorphism is controlled by the vertical temperature gradient that develops because of radiative cooling at the surface, warm and/or wet soils, and the low thermal conductivity of the snow cover. Temperature gradients induce water vapor pressure gradients, vapor diffusion from the warmest crystals, and consequent change in the shape and size of the crystals.

Wet snow is characterized by a significant amount of liquid water in snow and occurs primarily at mean snow temperatures at the melting point (Jones, H. G. et al. 2001).

The flow of water through snow is affected by impermeable layers, zones of preferential flow called flow fingers, and large meltwater drains. Meltwater drains are large and extend to the base of the snowpack, whereas flow fingers usually initiate and terminate at snow layer boundaries. The inhomogeneities of flow paths, water content, and water flux during melt have important implications for the chemistry and microbiology of snow covers.

Wet snow metamorphism involves the rounding of snow crystals and their enlargement at the cost of the smaller crystals. There is a consequent decrease in cohesion and hardness as liquid bonds replace solid ones between the crystals. The decrease of cohesion in wet snow induces avalanche activity and changes the ability of animals to move over the snow surface and inside the pack. Denser and more cohesive snow forms, and during active snowmelt, the density of a snowpack may change from 350 to 550 kg·m^{-3} in the course of a single day.

Basal ice forms where the meltwater flux exceeds the infiltration rate of frozen soils and there is a strong negative heat flux from the snow to the soil. Such ice layers may become quite thick (up to 70 cm) and persist after the snow cover has melted, but they are most prevalent in permafrost regions. Rain on snow events play a major role in wetting the snow surface and in forming ice crusts. A change in the incidence of rain on snow events could be an important indicator of climate change.

3.5   SNOW AVALANCHES

There are four distinct kinds of snow metamorphism:

1 Destructive metamorphism of dry snow: A few days after deposition, the crystal shape is lost. Powder snow is fine grained (0.5–1 mm) with density between 0.15 and 0.25 $g \cdot cm^{-3}$.
2 Constructive metamorphism of dry snow: Depth hoar has grain size between 2 and 8 mm and density from 0.2 to 0.3 $g \cdot cm^{-3}$.
3 Melt metamorphism: This characterizes changes produced in snow by the presence of liquid water. Large polycrystalline grains grow (c.15 mm) and "rotten snow" develops.
4 Pressure metamorphism: This is the densification of dry neve on glaciers.

The character of snow avalanching in a general sense depends on the climate: whether maritime, continental, or transitional (McClung & Schaerer 1993; Table 3.5). A maritime snow climate has relatively heavy snowfall, generates deep snow cover and experiences relatively mild temperatures. The maritime ranges of North America, for example, have average annual snowfall of 10–15 m. Avalanche formation usually takes place during or immediately following storms and failures occur in the new snow near the surface (Fig. 3.19). Warm air promotes rapid stabilization of the snow near the surface. Rain immediately following deep, new snowfall is a major cause of avalanching. Rainfall may also cause formation of ice layers, which may become future sliding layers when buried by subsequent snowfall.

A continental snow climate has relatively low snowfall, relatively shallow snow cover, and cold temperatures. A distinguishing feature of avalanches in continental climates is that they are often caused by buried structural weaknesses in older layers of the snow. Recrystallization weakens old snow in the presence of cold temperatures and high-temperature gradients. The low temperatures also allow structural weaknesses to persist.

There are, broadly speaking, two types of snow avalanches: loose snow avalanches and slab avalanches. Loose snow avalanches

TABLE 3.5   Characteristics of maritime, transitional, and continental snow climates* (from McClung & Schaerer 1993).

Snow avalanche type	Total precipitation (mm)	Air temperature (°C)	Snow depth (cm)	New snow density (kg m⁻³)
Maritime	1,280	−1.3	190	120
Transitional	850	−4.7	170	90
Continental	550	−7.3	110	70

* Mean values compiled from 15 winters of US data (Armstrong & Armstrong 1987).

start at or near the surface and only surface and near surface snow is involved. Slab avalanches are initiated at depth in the snow cover and are usually more dangerous. Dry loose snow avalanches are triggered by a local loss of cohesion due to either metamorphism or the effects of sun or rain. Wet loose avalanches are usually triggered by heavy melt due to sun or rainfall.

Dry slab avalanches are responsible for most of the damage and fatalities from avalanching. The most common trigger for natural dry slab avalanches is addition of weight by new snowfall, blowing snow, or rain. Wet slab avalanches occur by three principal mechanisms: loading by new precipitation; changes in strength of a buried weak layer due to water; and by water lubrication of a sliding surface. Slush avalanches are a class of wet slab avalanche which form under the following conditions: rapid onset of snowmelt in spring; snowpack is usually partially or totally saturated; and depth hoar is usually present at the base of the snow cover. They occur commonly in northern Scandinavia and Alaska.

Information on hazards posed by snow avalanches are important to development and winter sports (Fig. 3.20). Failure criteria for snow on steep slopes and the dynamics of avalanche motion are extremely complex phenomena which are largely modeled by empirical approximations.

3.6 Snowmelt, runoff, and streamflow generation

Snow cover provides an insulating blanket over soil and lake ice for the winter period and provides an important episodic flux of latent heat, water, and chemicals into soils and water bodies during the spring melt. In many high-latitude and high-altitude environments, snowfall accounts for over 40% of the precipitation and the release as

meltwater over a few days can be the single most important hydrological event in these environments.

In the Coast Mountains of British Columbia, the contrast in the duration of snowmelt and glacier melt-dominated runoff is illustrated from the examples of Miller Creek (glacier melt) and Central Creek (snowmelt) (Fig. 3.21).

Meltwater infiltration into unfrozen soils usually occurs in environments with deep snowpack or maritime climates. Soils frozen to depths of <15 cm behave as unfrozen soils with respect to infiltration. The proportion of snowmelt that infiltrates to unfrozen soils depends on the application rate of the snowmelt, the hydraulic conductivity of the soil layers, and the water retention characteristics of the soil. Discharge rates in excess of the saturated hydraulic conductivity will infiltrate until the soil becomes saturated and ponding of water at the base of the snowpack occurs.

Frozen soils develop where snow cover is thin or extremely cold winter temperatures prevail. Frozen soils normally have lower infiltration capacities than unfrozen soils of similar saturation level because the presence of ice reduces the effective porosity of the soil. However, frozen soils that contain large cracks or macropores accommodate infiltration of all snowmelt water.

Although modeling the physical process of melt and the movement of water through the snowpack enables an accurate estimate of snowmelt at a site, several difficulties arise in predicting snowmelt at a basin scale.

First, snowpack and meteorological conditions vary from place to place, especially in mountain environments. Second, most basins do not have a suitable network for meteorological snowpack data collection to allow physical modeling and third, longer term estimates can only be based on the assumption that past averages can be representative of future states.

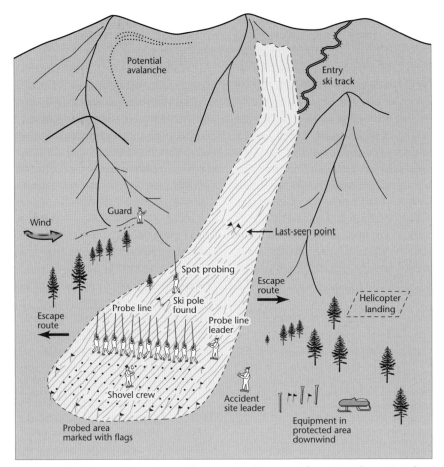

FIG. 3.20 Snow avalanche accident site with organized rescue (from McClung & Schaerer 1993).

The US Army has produced a set of basin snowmelt equations which are based on energy balance considerations, but empirically derived indices are substituted for analytical solutions. Thus they have produced empirical equations for snowmelt from (a) open areas; (b) forested areas; and (c) for melt during rain. Figure 3.22 is an example of one such setting (from Braun & Slaymaker 1981).

A cruder method (the degree day method) attempts to relate snowmelt exclusively to air temperature. The degree-day is defined as the number of degrees departure from a reference temperature (commonly 0°C) for a 24-hour period. Simplicity is both an asset and a shortcoming of this method.

A third method which is often used is a multiple regression method. Statistically derived equations relating snowmelt to all available independent variables, such as longwave and shortwave radiation, wind and temperature, wind and humidity, and rainfall and temperature, are combined to produce the best predictive equation. In general, the choice of an approach depends on the availability of data, the type of prediction needed, the generality of the model, the scale of operation, and the resources available to the investigator.

Beyond the problem of extrapolating snowmelt from sites to regions is the issue that snowmelt results from two quite

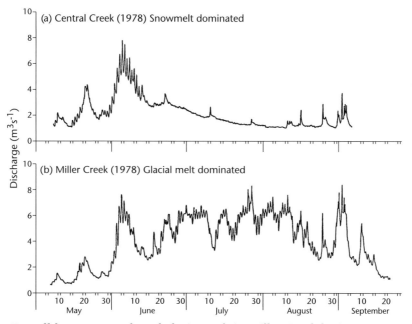

Fig. 3.21 Runoff from snowmelt and glacier melt in Miller Creek basin, Coast Mountains, British Columbia, 1978.

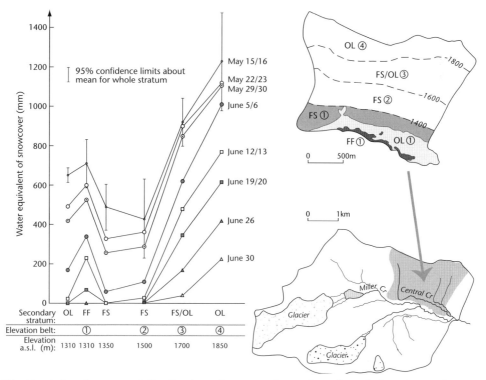

Fig. 3.22 Spatial and temporal distribution of snowpack as a function of elevation and vegetation cover in Miller Creek basin, Coast Mountains British Columbia (May to June 1978; from Braun & Slaymaker 1981).

distinct sets of processes: the snowmelt processes themselves and the controls on runoff production. Especially in mid-latitudes, the effects of the snowpack are minor. By far the most important control on the size and form of basin outflows is the interactions at the interface between the snow and the soil. This is particularly true in forests where soils are porous, usually deep, and permeable even when frozen. Price et al. (1978) presented an interesting contrast between snowmelt production at Knob Lake, Quebec, in the Canadian subarctic and at Perch Lake, Ontario. At Knob Lake, infiltration rates were essentially zero because of the presence of frozen soils even after the snowpack had completely melted. By contrast, at Perch Lake the most important controls on snowmelt runoff are the hydrologic properties of the soil. Even under low winter temperatures "honey comb" frost conditions prevail in the soil, thus allowing continuing infiltration of meltwaters. These two examples defined two end points in the generation of snowmelt runoff: no infiltration and total infiltration.

Marsh and Woo (1981) extended this discussion to High Arctic streamflow regimes by identifying that runoff in the High Arctic environment is sustained by various sources of water, including spring snowmelt, the melting of semipermanent snow banks, glaciers, and rainfall. If spring melt dominates, a simple Arctic nival regime results and if this is followed by summer glacier melt, a proglacial flow regime results. In some nonglacierized basins, however, if snowmelt is delayed until mid-summer or if semipermanent snow banks are abundant, a proglacial type of runoff pattern can be produced. The overall result is that various combinations of several sources of water will generate a suite of runoff regimes that range from the simple nival to the typical proglacial regime.

3.7 SNOW CHEMISTRY

Snowfall contains crustal elements such as Ca and Mg, from terrigenous dust, anthropogenic pollutants, weak organic acids, neutral organics from natural sources, and trace metals. These chemical species are incorporated in snow by three main processes: imprisonment during initial formation of ice crystals; capture of gases, aerosols, and larger particulates in clouds; and scavenging below the cloud layers during snowfall (Fig. 3.23).

The chemical composition of snowfall depends on factors such as the origin of the air masses that are scavenged, the altitude at which snow is deposited, and the meteorological conditions during snowfall. Chemical concentrations often decrease exponentially with time because of the progressive scavenging that occurs during a snowfall event. A scavenging coefficient Ω, is used to characterize the efficiency of precipitation in removing pollutants from the atmosphere.

Maritime air masses give snow with Na and Cl, whereas polluted air masses from industrial areas deposit acid snow with NO_3 and SO_4 prominent. In general, snowfall at higher altitudes has lower concentrations of chemical species because the depth of air available for scavenging is smaller.

The temporal and spatial variation in the chemical composition of snowfall produces a snow cover that is chemically heterogeneous. Modification of that snow cover occurs through surface exchange at the snow–atmosphere interface, surface and subsurface chemical reactions, snow grain metamorphism within the pack, and basal exchange processes at the snow–soil interface.

Leaching of snow grains by meltwater causes fractionation of solute species and the meltwater front becomes progressively

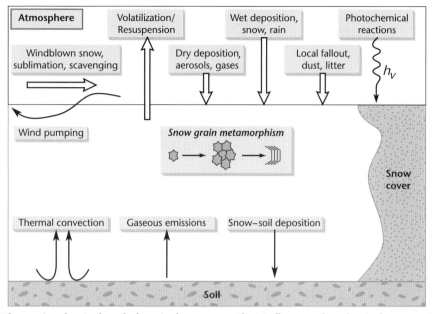

Fig. 3.23 The main physical and chemical processes that influence the physical composition of cold dry snow cover during accumulation (from Tranter & Jones 2001).

more concentrated as it moves through the snowpack. Deeper snow increases the duration of snow–meltwater interaction and gives rise to higher snowmelt concentrations. Meltwater flowing through macropores or flow fingers is more dilute than melt flowing through the matrix, because of the shorter contact time.

There is an interesting interaction between growth of algal biomass and the tendency for increasing concentration of nutrients in meltwaters. Low amounts of free water lower algal activity and nutrients may then accumulate.

3.8 SNOW ECOLOGY

Percolation of meltwaters through the snow cover causes the chemical composition of both the snow matrix and the meltwaters to change (Fig. 3.24). The concentration and distribution of solutes in the snow–meltwater system is controlled by a variety of physical and biological processes. These processes are the leaching of solute from snow grains, meltwater–particulate interactions, and microbiological activity. There are also snow–atmosphere exchanges which increase the solubility of certain species, such as SiO_2, HNO_3, and HCl, in water.

The leaching of solutes from snow crystals, known as solute scavenging, results in increasingly concentrated snowmelt as it moves through the snowpack. The result of several diurnal melt–freeze cycles is often to increase the concentration of ions in the first meltwaters issuing from the snowpack.

Chemical reactions between meltwater and inorganic–organic particles can affect the concentration of solute in meltwater. For example, snowmelt acidity may be neutralised by carbonaceous dust. In addition, organic debris, such as leaf litter, may

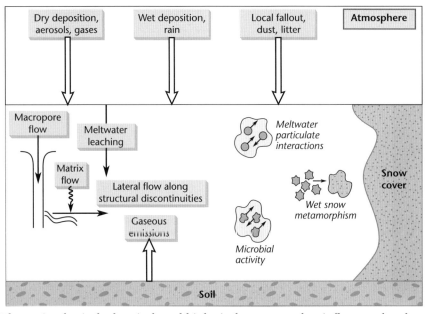

Fig. 3.24 The main physical, chemical, and biological processes that influence the chemical composition of snow cover during thaw (from Tranter & Jones 2001).

encourage ionic exchange between meltwater and itself (Walker et al. 2001).

Microbiological and invertebrate activity in snow covers is stimulated during spring melt. Photosynthesis results in an increase in algal biomass at the expense of nutrient concentrations in the meltwaters. The loss of nutrients in snowmelt waters over the whole melt season may approach 30% in some years (Jones 1991). Microorganisms such as bacteria, algae, and fungi are commonly found thriving in snow, glacial ice, lake ice, ice shelves, and sea ice. They are suitable habitats for microorganisms only when liquid water is present for at least part of the year. The sea ice of the Arctic Ocean often contains blooms of snow algae during the summer (Gradinger & Nurnberg 1996). Microbes are abundant in snow fields and in depressions and ponds found on glaciers (Wharton et al. 1985). Microorganisms play a fundamental role in the biogeochemistry of snow and ice and are closely involved in

the primary production, respiration, nutrient cycling, decomposition, metal accumulation, and food webs associated with these habitats (Hoham & Ling 2000).

The thermal properties of snow cover also allow larger organisms to survive in the relatively benign microhabitats of the subnivean space. The snow cover, in this sense, is an ecotone between two different environments: the dry, very cold, windy, and changeable atmospheric air and the humid, relatively warm, and stable air of the space underneath.

3.9 GLACIER MELT

According to Benson (1961), there are five zones on a glacier whose boundaries vary from year to year (Fig. 3.25):

1 A dry snow zone, where no melting occurs even in summer. The only dry snow zones

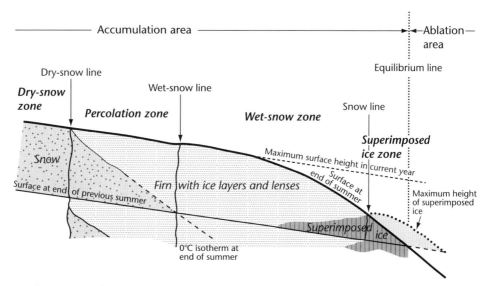

FIG. 3.25 The five zones of a glacier: dry snow, percolation, wet snow, superimposed ice, ablation zone (from Muller 1962).

occur in the interiors of Greenland and Antarctica and near the summits of the highest mountains in Alaska, Yukon, and central Asia.

2 A percolation zone which starts below the dry snow line. Water can percolate a certain distance into snow at temperatures below 0°C. Refreezing of meltwater is the most important factor in warming the snow.

3 A wet snow zone, which starts below the wet snow line.

4 A superimposed ice zone, which starts at the snow line (firn line or annual snow line). At these lower elevations, so much meltwater is produced that the ice layers merge into a continuous mass (superimposed ice). This is the boundary between firn and ice on the glacier surface at the end of the melt season.

5 All five zones may be found in parts of Greenland and Antarctica. The major Antarctic ice shelves have only dry snow and percolation zones and the entire mass loss results from calving of icebergs and a small amount of melting at the base. By contrast, the Barnes Ice Cap on Baffin Island appears to have only superimposed ice and ablation zones in most years. All of the above are "cold" glaciers in that the temperature is below the melting point.

In a "temperate" glacier, the ice is at the melting point throughout, except for a surface layer, about 10 m thick, in which the temperature is below zero for part of the year. Temperate glaciers cannot have percolation zones because in that zone, by definition, the temperature of deeper layers never reaches 0°C. Superimposed ice forms only when the firn temperature is below 0°C. On a temperate glacier, there can therefore be only a limited superimposed ice zone and the equilibrium and snow lines effectively coincide. A temperate glacier therefore has only wet snow and ablation zones (Plate 3.2; Fig. 3.26). The Salmon Glacier in the northern Coast Mountains of British Columbia and the South Cascade Glacier in the North Cascades, Washington State display the snow line and the wet snow and ablation zones quite clearly.

FIG. 3.26 Change in length of the South Cascade Glacier between 1960 and 1983. The ablation zone is clearly distinguished from the wet snow zone (photograph by Andrew G. Fountain).

3.10   FORMATION OF AN ICE COVER

Lock (1990) has a detailed discussion of the growth and decay of ice. The simplest case is that of a shallow pond of quiescent water on which a thin type of ice cover, which is usually transparent, will form. It is a continuous sheet of polycrystalline ice called skim and can be seen on pools, reservoirs, lakes, and sluggish streams (velocities <0.5 m·s^{-1}). Deeply supercooled aqueous solutions generate downward growing dendrites, commonly known as candle ice.

More dramatic changes occur when the water is not calm. Turbulent exchange interrupts the undisturbed crystal growth and multiplies the effective nucleation rate, thus creating a greater number of ice crystals. The crystals which grow under these conditions are known as frazil ice. A sintering mechanism generates first clusters and then larger aggregates of frazil crystals. They are able to stick to any submerged solid surface, and when they do so, they become anchor ice. This anchor ice, which may be as much as 1 m thick, has the ability to lift rocks and plants off the river bottom when it eventually is released. At the same time, the unattached flocs of frazil generate slush balls or slush patches. The increasingly congealed slush then forms into pancake ice (Plate 3.3) and, in turn, pancakes may collect and freeze together to produce ice floes. In the ocean, the consolidation of pancakes into floes eventually leads to a continuous sheet of primary sea ice (cf. Table 1.3).

Further growth may take two forms: secondary ice that is produced on the bottom of the ice cover either by direct freezing or by the accretion of frazil or superimposed ice that forms on top of the ice cover when it is inundated with water. The consolidation and jamming of primary river ice may cause sudden flow surges which can lead to inundation and this kind of superimposed ice is called naled. The other major source of superimposed ice is snow. If snow falls on open water, the snow crystals provide nucleation sites for the production of frazil and if the snowfall is heavy enough, it may lead to a slush cover from which primary ice may be formed. Later in the season, snow falling on discrete pieces of surface ice (skim, plates, pancakes, or floes) introduces additional mass. The additional weight of the snow causes the ice to submerge and water is added to the ice by capillarity. Subsequent freezing creates snow ice or white ice.

Bottom ice growth by addition is strongly influenced by the temperature and velocity fields below. Once a continuous sheet of ice covers a large area of sea, lake, or river, the process of secondary growth will depend on the removal of latent heat upwards through the cover; this is thermal growth or congelation.

Open fault lines commonly occur in the first-year ice and can be detected as "bright" features in radar images (Fig. 3.27).

3.11 RIVER AND LAKE ICE

Prowse (2000) has provided a useful summary of river ice ecology in four parts: (i) autumn cooling; (ii) freeze-up; (iii) main winter; and (iv) break-up (Figs. 3.28 and 3.29). Under autumn cooling, the formation of ice can reduce the amount and quality of winter habitat or can create new refugia. For example, border ice develops along the margins and offers a low velocity refuge and protection from predation. Frazil ice however can be repellent as it abrades the gills and may cause suffocation. Anchor ice may cause death of benthic inverte-brates and some fish and may also promote ice growth into the spawning areas or redds. Accumulation of anchor ice can also alter riverine habitat by modifying pool

FIG. 3.27 Prince Patrick Island Northwest Territories, Canada. The photograph was taken on July 2, 1982, during the FIREX/RADARSAT field experiment. The faults formed in June when solar heating had raised the temperature to within a few degrees of 0°C. Puddling on the ice from the melting of snow and ice was near maximum. Remnants of barchan snow dunes can be seen; most of the higher ground in the photograph is due to deeper snow cover (photograph by Arnold M. Hanson).

depth. Anchor ice commonly attaches to aquatic vegetation, boulders, and areas of gravel and coarse sand and when it is released, these materials and benthic organ-isms are carried downstream.

Under freeze-up, complete freeze-up of the water surface occurs. Loss of river dis-charge to channel storage occurs behind the advancing freeze-up front eventually leading to backwater flooding of riparian zones. During the main winter, isolated pool and ice-cavity habitats develop. These are critical for the survival of fish and aquatic plants.

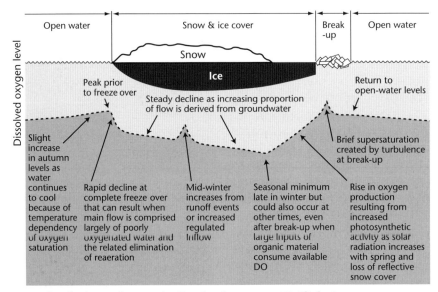

FIG. 3.28 Example of winter changes in dissolved oxygen. (1) Slight increase as water cools. (2) Peak prior to freeze over. (3) Rapid decline after complete freeze over. (4) and (6) Steady decline as groundwater contribution increases. (5) Mid-winter increase from runoff event. (7) Seasonal minimum. (8) Rise in oxygen production with increased photosynthesis. (9) Brief supersaturation with turbulence with break-up. (10) Return to open water levels (from Prowse 2000).

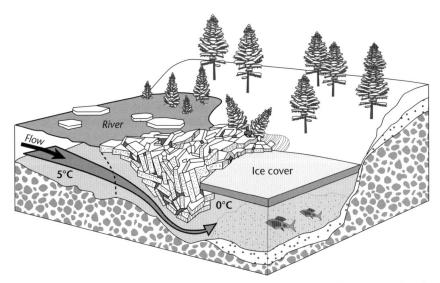

FIG. 3.29 Dynamic transition in processes and habitat created by rapidly moving break-up front of river ice (from Prowse 2000).

When the ice cover is fully developed (Plate 3.4), DO levels are significantly reduced. By the time that very low levels of DO are achieved, fish will have abandoned most reaches of the rivers and concentrated in a few preferred zones, where they remain relatively inactive for the rest of the winter. They are also vulnerable because of overcrowding and lack of oxygen for individual fish.

The presence of an ice cover has important implications for the transporting of dissolved and suspended substances. Ice also, in the form of frazil ice, is an agent of sediment transport. Its greatest transport potential is as large anchor ice deposits that detach from the bed, transporting boulders as large as 30 kg. As a result of decreased sediment transport capacity, there is a tendency for deposition of large amounts of finer sediment.

The break-up period is the most dramatic from the perspective of environmental change. Breaking fronts on large rivers may travel at 5 m·s^{-1} and water levels may rise at 1 m·min^{-1}. The issue of the role of break-up discharge in the long-term formation of northern rivers is still a topic of contention. For many northern rivers, ice break-up occurs concurrently with the spring freshet and this is often the major hydrologic event of the year for sediment transport (Church 1974).

3.12 SEDIMENT BUDGETS

The sediment mass balance equation is defined as

$$\Delta S_s = I_s - O_s \qquad (3.17)$$

where S_s is sediment stored in a drainage basin on slopes, floodplains, in channels, in lakes, and in ice; I_s is sediment produced by weathering or from dynamic sediment sources; and O_s is sediment exported from the system (Reid & Dunne 1996; Slaymaker 2004). This sediment balance must ultimately be zero. But in most real world examples it is not zero, certainly not in large drainage basins over short periods of time. The dimensions of the storage term, and whether it is positive or negative, is of importance in the assessment of sediment transport at all spatial and temporal scales.

Interpretation of the storage term from lacustrine or marine sedimentation records is at first sight rather straightforward. For example, in the case of the Black Sea sediments, large sediment stores have been identified with the period of deglaciation since the LGM (15–7 ka BP) and the last 2 ka, when the tributary basins have been extensively disturbed by human activities (Shimkus & Trimonis 1974). But when the processes of redistribution of sediment within river basins are considered, it becomes evident that storage is more complex. Sediment on its way from source to sink gets side-tracked in a number of ways. The issue is similar in principle to that which was discussed earlier with respect to snowmelt runoff pathways, but more complex because sediment transfer is an intermittent process. It is in this context that the concept of "virtual velocity" of sediment transfer is useful. Virtual velocity can be defined as the velocity of sediment transfer through a reservoir and is simply the inverse of the residence time per meter of reservoir length. Residence time per meter equals the mapped volume of sediment per meter divided by the bedload discharge rate.

The most dramatic primary sediment disturbances to visit Earth's surface within the past 2 Ma have been the recurrent, unstable Pleistocene glaciations of the Northern Hemisphere. Church and Slaymaker (1989) have depicted a relatively rapid virtual velocity for sediments in the deglaciated upland streams of British Columbia whereas sediments moving through the major river

TABLE 3.6 Fluvial sediment response to disturbance (after Church 2002).

Site/event	A_d* (km²)	t[†] (a)	t_r[‡] (a)	Intensity[§] (tonnes km⁻² a⁻¹)	Magnification[‖]
Contemporary glacial[¶]	48			400	27×
Nigardsbreen, Norway	38			1,050	66
Engabreen, Norway				1,000	70
Iceland (summary)					50–100
Alaska (summary)**	1,000				100
Pleistocene glacial					
British Columbia	10⁴	10⁴		~10²	10²

* Area of drainage basin above measurement point for sediment transport, or at area of peak response (B.C.).
[†] Response time.
[‡] Relaxation time.
[§] Sediment yield at peak response.
[‖] Intensity/weathering rate (the denominator generally being estimated).
[¶] Mainly after Hallet et al. (1996).
** From Guymon (1974).

valleys have extremely slow virtual velocities. So slow indeed that they envisage disequilibrium of the fluvial sediment mass balance over the time scale of the whole of the Holocene epoch (of order 10 ka).

Contemporary glacial sediment yield is relatively high (Plate 3.5). Glaciers are effective abraders of subjacent rock and soil materials may be frozen into the glacier and on to the glacier sole. At the ice margin, material that melts out from the ice, or is exposed from below the ice, is immediately susceptible to transport by wind and water (Church 2002). The freeze–thaw environment and absence of vegetation cover magnify the effectiveness of these processes. Consequently, it has been established (Hallet et al. 1996) that "maritime glaciation" on Neogene rocks in southeastern Alaska must yield some of the highest glacial sediment yield rates on Earth (Table 3.6). As one moves away from the glacier margin, specific yields decline. Persistently declining yields indicate either a concentration of sediment sources at the headwaters of the system or aggradation downstream (Church 2002, p. 103). Over a very long time, the distinctive spatial signature of fluvial sediment transfer following a major glacial disturbance diffuses downstream. The signature of disturbance in time is similar and has been designated "paraglacial sediment yield" (Church & Ryder 1972; Slaymaker 1987; Church & Slaymaker 1989).

4

PATTERNS OF THE CONTEMPORARY CRYOSPHERE AT LOCAL TO GLOBAL SCALES

4.1 INTRODUCTION

In this chapter, we illustrate the dramatic ways in which remote sensing and satellite imagery have revealed spatial patterns of the cryosphere. Remote sensing of cryospheric systems is a rapidly evolving field (Duguay & Pietroniro 2005). Remote sensing is an important tool for many different scientific and practical applications related to snow, permafrost, and ice cover. Unfortunately, not all the variables can be retrieved with sufficient accuracy at the spatial and temporal scales provided by current spaceborne systems. There is a need for higher spatial resolution without decreasing temporal resolution (Duguay & Pietroniro 2005). Topography and different land cover types affect retrieval of snow, permafrost, and ice variables by remote sensing partly because topographic and atmospheric processes are nonlinearly coupled and partly because of our limited understanding of the complex radiation interactions between snow, bare ground, and vegetation.

4.2 REMOTE SENSING OBSERVATIONS

In Chapter 2, we described the data sets that represent, in general, point-scale cryospheric processes. Geographical Information Systems,

as formal software (commercially available) and informal software (using linked open-source programing language routines), are effective at managing these *in situ* data sources. They can also be used to estimate, or interpolate, what processes might be found between *in situ* measurements. However, by their nature, the measurements represent cryospheric conditions at a point and unless appropriate spatial sampling designs are used, interpolation carries uncertainty with it. Remote sensing instruments have the capability to measure indirectly cryospheric variables over continuous geographical space. Remote sensing instruments do not generally measure directly cryospheric variables (snow depth, ice velocity, etc.). Rather they measure the electromagnetic energy emitted or reflected (or backscattered) from the variable of interest which then has to be converted to a geophysical quantity of interest. There are many excellent texts available on the fundamentals of theory of remote sensing and the reader should look to these for a full description of current methods (Schowengerdt 1997; Henderson & Lewis 1998; Campbell 2002; Lillesand et al. 2003; Ustin 2004). Some excellent remote sensing resources are also available on the internet (e.g. <http://www.earthobservatory.nasa.gov/> <http://www.ccrs.nrcan.gc.ca/resource/tutor/fundam/index_e.php>).

Among the principal national and international organizations that conceive, develop, launch, and maintain civilian satellite remote sensing instruments are the National Aeronautics and Space Administration (NASA), the European Space Agency (ESA), the Japan Aerospace Exploration Agency (JAXA), and the Canadian Space Agency (CSA). Many other countries such as Argentina, Brazil, India, and China also have capabilities in developing and building satellite instruments. To coordinate international civil spaceborne missions designed to observe and study planet Earth, the Committee on Earth Observation Satellites (CEOS) was created (<http://www.ceos.org>). It is composed of 23 members (most of which are space agencies) and 21 associates (associated national and international organizations). A recommendation in 1984 from the Economic Summit of Industrialized Nations Working Group on Growth, Technology, and Employment's Panel of Experts on Satellite Remote Sensing was the initial impetus for the formulation of CEOS. The group recognized the multidisciplinary nature of satellite Earth observation and the value of coordination across all proposed missions. CEOS has established a broad framework for coordinating all spaceborne Earth observation missions. Individual participating agencies make their best efforts to implement CEOS recommendations. The main goal of CEOS is to ensure that critical scientific questions relating to Earth observation and global change are covered and that satellite missions do not unnecessarily overlap each other.

The fundamental unit of electromagnetic radiation is the photon. Photons move at the speed of light as waves, analogous to the way waves propagate through the oceans. The energy of a photon determines the frequency (and wavelength) of light that is associated with it; the greater the photonic energy, the greater the frequency of light

and vice versa. The entire array of electromagnetic waves comprises the electromagnetic (EM) spectrum. The EM spectrum is divided into continuous regions to which descriptive names have been applied. Gamma rays and x-rays exist at the very energetic levels (high frequency; short wavelength) while radio waves exist at the low energy levels of the spectrum. The visible region occupies the range between 0.4 and 0.7 μm (micrometer), or its equivalents of 400–700 nm (nanometer) and the infrared (IR) region spans between 0.7 and 100 μm. The microwave region is located between 1 mm and 1 m. These three mid-range wavelength bands (visible, infrared, and microwave) are the areas of the EM spectrum commonly exploited for ground, aircraft, and satellite Earth Observation instruments. Gamma rays and x-rays do not pass through Earth's atmosphere and so cannot be used for Earth remote sensing measurements from satellite (although gamma rays are used on aircraft platforms for certain applications – see below). Also, low frequency radio waves do not, generally, have enough energy to give a strong signal when detected by satellites and so their use is also restricted to ground or aircraft studies. Visible and infrared (VIS/IR) measurements and microwave measurements have been the most exploited part of the EM spectrum for Earth observation from ground, air, and space. At these wavelengths, EM waves can pass through the atmosphere and be measured to give information about the ground (or atmospheric) surfaces below.

There are two modes of detecting EM from ground aircraft or spacecraft platforms. First, passive instruments detect natural energy that is reflected or emitted from the observed scene. They sense only radiation emitted by the object being viewed or reflected by the object from a source other than the instrument. Reflected sunlight is the most common external source of radiation sensed by passive instruments. Scientists use a variety of passive remote sensors. For example, a radiometer is an instrument that measures the intensity of electromagnetic radiation in some band of wavelengths in the spectrum. Usually a radiometer is further identified by the portion of the spectrum it covers; for example, visible, infrared, or microwave. An imaging radiometer is similar to a static radiometer except that it includes a scanning capability to provide a two-dimensional array of measurements from which an image may be produced. A spectrometer detects, measure, and analyzes the spectral content of the incident electromagnetic radiation while a spectroradiometer measures the intensity of radiation in discrete multiple and sometimes very fine wavelength bands (i.e. multispectral).

The second mode of remote sensing instrument is the active instrument that provides its own energy (electromagnetic radiation) to illuminate the target or scene that is then observed. Active instruments send a pulse of energy from the sensor to the object and then receive the radiation that is reflected or backscattered from that object. For example, a Radar (Radio Detection and Ranging) transmits radio waves or microwaves through a directional antenna and a receiver measures the time of arrival of the reflected or backscattered pulses of radiation from distant objects. Scatterometers are high frequency active microwave radars designed specifically to measure backscattered radiation. While scatterometers are generally nonimaging in character, they can be used to build maps of a required geophysical variable. Radar altimeters send out single microwave pulses of energy and measure the reflected radiation from the surface. Lidars (Light Detection and Ranging) use a laser to transmit visible light pulses the reflection from which are detected by a receiver. Distance to the object is determined by recording the

time between the transmitted and backscattered pulses and using the speed of light to calculate the distance traveled. Finally, laser altimeters use lidars to measure the height of the instrument platform above the surface. By independently knowing the height of the platform with respect to the mean Earth's surface, the topography of the underlying surface can be determined.

The following sections describe remote sensing measurements that are used to study the cryosphere. The sections follow the same order as those in Section 2.2. Tables 4.1 and 4.2 give an overview of the current polar low Earth orbiting (LEO) satellite missions that are available and being used for remote sensing of the cryosphere. In each table, the satellite names (acronyms are defined in the caption) are arranged according to imaging and nonimaging instruments, wavelength (visible-infrared, microwave – active and passive) and then by revisit time (Table 4.1) and by spatial resolution (Table 4.2). Imaging sensors are defined as those that can generate an image swath (a two-dimensional array of measurements) while nonimaging methods produce a one-dimensional array of measurements along the subsatellite path. The nonimaging indirect method refers to measurements made by instruments such as GRACE which do not measure reflected or emitted EM radiation. The repeat period in Table 4.2 is the number of days before a satellite ground track will pass through exactly the same place again. This characteristic partially controls how frequently a target location is imaged. In addition, the swath width of the imaging satellite will control how frequently a location is imaged. For all of the imaging satellites, the wider the swath, the more frequently a location is sensed. This frequency increases with latitude (i.e. poleward). The shaded cells in Tables 4.1 and 4.2 represent those satellite instruments which provide 1–2 day global coverage.

TABLE 4.1 Available satellite remote sensing instruments for cryospheric studies arranged by increasing exact repeat period.

Exact orbit repeat period	VIS/IR	Passive microwave	Active microwave
Imaging			
Less than 2 weeks	AVHRR (9d, 2,700 km)	SSMIS (*16d, 1,700 km*) *CMIS (8d, 1,700 km)* *Aquarius (7d, 323 km)*	QuikSCAT (4d, 1,800 km) *TerraSAR (11d, 100 km)* *Aquarius (7d, 323 km)*
Less than 3 weeks	*VIIRS (16d, 3,000 km)* LandsatTM (16d, 185 km) MISR (16d, 360 km) ASTER (16d, 60 km) MODIS (16d, 2330 km)	SSM/I (16d, 1,394 km) AMSR-E (16d, 2,430 km)	—
Less than 4 weeks	SPOT (26d, 60 km)	*SMOS (23d, 934 km)*	Radarsat (24d, 50–500 km)
Less than 5 weeks	AATSR (35d, 512 km)	—	ENVISAT ASAR (35d, 100–400 km)
	MERIS (35d, 1,150 km)		ERS-2 SAR (35d, 100 km)
Nonimaging			
Less than 5 weeks	VIS/IR	Active microwave ENVISAT RA-2 (35 days) ERS-2 RA (35 days) *Cryosat II (395 days with 30 day suborbit)*	Indirect GRACE (30 days)
Greater than 5 weeks	ICESat GLAS (91 days)		

Note: The text in parentheses specifies the exact repeat orbit period and the swath width measured by that instrument. The italicized entries are planned mission instruments. The shaded entries are instruments that can provide global coverage, especially at high latitudes. See text for acronym description.

The information in Table 4.2 is of interest for cryospheric studies because it details the spatial resolving power of many of the spaceborne instruments used in remote sensing. The ability of instruments to resolve EM energy at spatial scales similar to the scale of variation of the cryospheric process is key to the success of the remote sensing approach. If the spatial resolution is too great with respect to the process's natural spatial variability, then spatial averaging represents the process; local scale variability is harder to unravel. For example, the Landsat Thematic Mapper (TM) can resolve land surface processes at 30 m. For snow cover extent, assuming cloud-free views, this is well below the expected 100 m scale of variation of snow, even in mountainous terrain. However, QuikSCAT averaging cells are at 25 km suggesting that any physical snow processes resolved by the measurement will represent averages of the region rather than local variation.

Recent in-depth texts have covered key aspects of remote sensing measurement methodologies of the cryosphere. Bamber and Payne (2004) devotes a chapter to Earth observation methods of the ice sheets. Gurney et al. (1993) comprehensively described how remote sensing measurements

TABLE 4.2 Available satellite remote sensing instruments for cryospheric studies arranged by increasing spatial resolution.

Measurement spatial resolution	VIS/IR	Passive microwave	Active microwave
Imaging			
Less than 100 m	SPOT (5–10 m)	—	Radarsat (10–100 m)
	LandsatTM (15–30 m)		*TerraSAR (1m)*
	ASTER (15–90 m)		ENVISAT ASAR (30–100 m)
			ERS-2 SAR (30 m)
Less than 1 km	MODIS (250–500 m)	—	—
	VIIRS (400–800 m)		
	MISR (250 m)		
	MERIS (300–1,200 m)		
Less than 10 km	AATSR (1,000 m)	SSM/I (10–55 km)	
	AVHRR (1.1 km)	AMSR-E (5–55 km)	*Hydros (3 km)*
		SSMIS (10–55 km)	
		CMIS (10–55 km)	
Less than 100 km		*SMOS (35–50 km)*	QuikSCAT (25 km)
		Aquarius (100 km)	
			Aquarius (100 km)
Nonimaging			
Less than 100 m	ICESat GLAS (70 m)	—	—
Less than 100 km	—	ENVISAT RA-2 (16–20 km)	—
		ERS-2 RA (16–20 km)	
		Cryosat II (250 m to 15 km)	
Less than 1,000 km	—	—	GRACE (500 km)

Note: The text in parentheses specifies the exact instantaneous field of view (sometimes erroneously referred to as pixel size) or spatial resolution of the instrument. The italicized entries are planned mission instruments. The shaded entries are instruments that can provide global coverage, especially at high latitudes. See text for acronym description.

at the regional scale are used to investigate several global environmental change cryosphere variables, such as sea ice and snow. In addition, Winther and Solberg (2002) covers many aspects of remote sensing in glaciology.

4.3 LAND AND SEA SURFACE TEMPERATURE

Surface temperature of the cryosphere is sensed as part of the planetary heat budget. Ideally diurnal temperatures should be sensed using orbiting remote sensing instruments.

The National Oceanic and Atmospheric Administration's (NOAA) Advanced Very High Resolution Radiometer (AVHRR) has been used to estimate global temperatures since the mid-1960s. A method to produce a consistent record of observations for sea surface temperature estimation has been discussed by McClain et al. (1985). Radiative transfer theory is used to correct for the atmospheric effects on the observations and channel radiances (a measure of exitant radiation from the Earth's surface) are converted to temperatures measured at the surface. The resulting estimates of surface temperature have been nominally accurate to 0.3°C. A study by Key and Haefliger (1992) demonstrated effective estimation of sea ice surface temperature using AVHRR to an accuracy of 0.1 K to 0.6 K for the Northern Hemisphere. The AVHRR Pathfinder data set for sea surface temperature was also used by Comiso (2003) to reconstruct water and sea ice surface temperature records from 1981 to 2001. Thermal infrared measurements from AVHRR were used in the study. These satellite temperature data were compared with ground-based data taken during the SHEBA experiment and were shown to be consistent from 1997 to 1998. Donlon et al. (2002) have also developed a method to estimate SSTs using passive microwave instruments. Using combined passive microwave instruments, retrievals are made to an accuracy of less than 0.5 K. However, it does not estimate temperature of sea ice which it screens out prior to retrieval and so, for polar or cryospheric applications, this approach is limited.

For land surface temperature estimation, Jin (2004) has developed a method to estimate land surface skin temperature from 1981 to 1998 using polar orbiting AVHRR measurements. A current NASA global land surface temperature data set product is derived from MODIS data (<http://www.modis.gsfc.nasa.gov>). The product is based on research by Wan et al. (2004). In an effort to take a more integrated approach, several focused projects have been executed to measure, among other things, surface temperature using satellite remote sensing. For example, the International Satellite Land Surface Climatology Project (ISLSCP) sought to increase our understanding and predictive capability of climate change and how carbon, energy, and water are exchanged between the atmosphere and the terrestrial biosphere. Established in 1983 under the UNEP to promote the use of satellite data for the global land-surface data sets needed

for climate studies, ISLSCP has played a key role in addressing land-surface processes, developing climate models, experiment design and implementation, and data set development. Two research initiatives in ISLSCP have been conducted. The first aimed to develop data sets in a common form for scientists to improve our understanding of land–atmosphere energy exchanges. The second initiative built on the first effort but by increasing spatial resolution and expanding the measurements over time. In addition, the First ISLSCP Field Experiment (FIFE) was conducted in Kansas between 1981 and 1989 to test satellite remote sensing capabilities to estimate surface temperature among other variables (American Geophysical Union 1992, 1996). The Boreal Ecosystems-Atmosphere Study (BOREAS <http://www.daac.ornl.gov/BOREAS/bhs/BOREAS_Home.html>) experiment led on from FIFE except that the effort was focused on boreal forest ecosystem-atmosphere energetics (see American Geophysical Union 1997, 1999, 2001a). These integrated experiments include multiple satellite data sources, aircraft remote sensing measurements, and ground measurements together which have enhanced our understanding of the dynamics of many land surface variables, especially temperature.

4.3.1 TERRESTRIAL SNOW AND SNOW ON SEA ICE

The main observable snow variables are snowfall, snow cover extent, and snow water equivalent. The need for terrestrial snow accumulation measurements has been driven by two applications; basin-scale hydrology and regional to global scale climatology. For snow on sea ice, the need is a climatological one. Thus, the estimates required need to be at the daily time scale, especially during the snow melting period of a seasonal snowpack, and weekly to monthly for climate change studies. Two recent overviews of the mapping of snow can be found in König et al. (2001) and Hall et al. (2005a).

The estimation methodologies of snowfall are still in their research and development phase. One approach, by Skofronick-Jackson et al. (2004), used high frequency attenuation measurements from the Advanced Microwave Sounding Unit, a precursor to the Defense Meteorological Satellite Program's (DMSP) Special Sensor Microwave Imager/ Sounder, and compared them with the emission signals from a snow scattering model. The data showed consistent results between model estimates and observations. This research is still in the early stages of development.

For snow cover extent, satellite remote sensing technology has virtually revolutionized snow mapping capabilities. In the visible part of the EM spectrum, snow contrasts well with the majority of other natural surfaces (except clouds) and can be easily detected by several satellite instruments observing at these wavelengths. In 1966, NOAA began producing weekly global snow maps of the Northern Hemisphere using satellite data at a spatial resolution of 190 km. The maps are based on a visual interpretation of photographic copies of shortwave imagery by trained meteorologists. The mapping capability was recently enhanced in 1997 with improved spatial resolution and on a daily basis (Ramsay 1998). Further improvements have been made more recently. These data are available through the Satellite Services Division of the National Environmental Satellite, Data, and Information Service (NESDIS) at <http://www.ssd.noaa.gov/PS/SNOW/>. Recently, at Rutgers University, New Jersey, Robinson has made available consistent DMSP and AVHRR derived snow maps back to 1966 for climate change studies (<http://www.climate.rutgers.edu/snow-cover> and <http://www.nsidc.org>).

These data constitute the only available long-term satellite record of snow cover extent and have been analysed carefully for inconsistencies (Robinson et al. 1993; Robinson 1997, 1999). Many authors have analysed the data for regional to global climate change studies (e.g. Robinson & Frei 2000). Figure 4.1 gives an example of a daily snow map from the Rutgers data set for the northern hemisphere and Plate 4.1 is an example of change in the number of snow days over the period 1967–2005.

In 1999, global snow mapping capabilities were significantly enhanced with the launch of NASA's Earth Observing System (EOS) Moderate Resolution Imaging Spectroradiometer (MODIS) instrument aboard the Terra satellite platform (<http://www.terra.nasa.gov>). Terra is a LEO platform with a wide swath coverage (see Table 4.1) and frequent coverage at higher latitudes. With a spatial resolution of 500 m, MODIS is a 36 waveband VIS/IR spectroradiometer with excellent capabilities for the

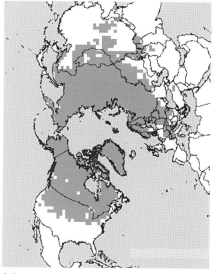

Fig. 4.1 AVHRR Snow map for December 3, 2005 produced by the Rutgers Global Snow Lab (<http://www.climate.rutgers.edu/snowcover/index.php>).

discrimination of snow from other land surface types. Since 2000, the MODIS snow and ice science team at NASA Goddard Space Flight Center has developed a set of automatic procedures to produce daily, 8-day, and monthly composite snow cover extent maps (Hall et al. 2002a). The MODIS map products provide global coverage at 500-m spatial resolution and at 0.05° and 0.25° resolution for climate-modeling grid (CMG) products. The CMG products are designed for use by climate modelers and provide a global view of the Earth's snow cover in a geographical projection with fractional snow cover reported in each cell based on the reprojected higher 500-m spatial resolution data. The automated MODIS snow-mapping algorithm uses at-satellite reflectances in MODIS bands 4 (0.545–0.565 μm) and 6 (1.628–1.652 μm) to calculate the normalized difference snow index (NDSI) (Hall et al. 2002a). Other threshold tests are also used including use of the Normalized Difference Vegetation Index (NDVI) together with the NDSI to improve snow mapping in forests (Klein et al. 1998). Because cloud cover often precludes the acquisition of snow cover from visible and near-infrared sensors, the daily maps are composited, and 8-day composite maps as well as daily maps are available. The Aqua satellite was launched in 2002 with a second MODIS instrument to enable snow cover area measurements to be extended farther into the future (Riggs & Hall 2004). The products are transferred to the National Snow and Ice Data Center (NSIDC) in Boulder, Colorado, where they are archived and distributed via the NASA EOS Data Gateway (EDG) at NSIDC. Plate 4.2 gives an example of a daily and eight-day composite snow MODIS snow map. Details of the products can be found at <http://www.modis-snow-ice.gsfc.nasa.gov>.

Fractional snow cover (FSC), the percentage of a pixel covered with snow, has been calculated using Landsat and MODIS data. Painter et al. (2003) developed an automated model that couples spectral-mixture analysis with a radiative transfer model to retrieve subpixel snow-covered area and effective grain size from Airborne Visible/Infrared Imaging Spectrometer (AVIRIS) data. Most recently, Salomonson and Appel (2004) extended the use of the NDSI to provide FSC globally with absolute errors of 0.1 or less over the whole range of FSC from 0 to 100%. Percent snow cover or fractional snow cover in each pixel (Salomonson & Appel 2004) is provided.

Landsat data have also been used for measurement of snow-covered area over drainage basins (Rango & Martinec 1979, 1982). The Landsat TM and ETM+ have been especially useful for measuring snow cover because of the shortwave infrared band – TM band 6 (1.6 μm) which allows snow/cloud discrimination. This channel is also available on the MODIS. The reflectance of snow is low and the reflectance of most clouds remains high in that part of the spectrum. Various techniques have been used to map snow cover including spectral and threshold tests.

VIS/IR sensors have an excellent capability to detect snow surfaces and provide global snow maps under day-time cloud-free sky conditions. In many cryospheric regions during winter, cloud cover can persistently obscure the land surface. To complement VIS/IR snow detection approaches, passive and active microwave measurements can detect snow surfaces under cloudy conditions and are solar illumination independent. Tait et al. (2000) demonstrated a combined use of multisatellite sensors for snow mapping. More significantly, microwave measurements can determine snow wetness and SWE (Ulaby & Stiles 1980). Since the 1970s, passive microwave instruments have been used to estimate snow extent and SWE globally on a daily basis with the launch of the Nimbus-7 Scanning Multichannel

Microwave Radiometer (SMMR) (Chang et al. 1987), the DMSP Special Sensor Microwave Imager, and continuing with the May 2002 launch of the Advanced Microwave Scanning Radiometer – EOS (AMSR-E) on the Aqua platform (Kelly et al. 2003).

SWE can be estimated from passive microware remote sensing measurements using a simple representation of a radiative transfer model that is parameterized for "average" snow grain sizes of 0.3 mm and density of 300 kg/m³ respectively. Theoretically, it is assumed that snow crystals attenuate (or scatter) naturally emitted upwelling microwave radiation from the underlying ground surface. Chang et al. (1987) demonstrated this theoretical approach with reasonable results. However, if the crystal radii and snow density differ significantly from the averages and assumptions, then SWE estimates are less accurate. Current efforts are aimed at improving the methods to estimate SWE by incorporating more dynamic parameterizations of these variables. For example, using DMSP Special Sensor Microwave Imager (SSM/I) measurements, Mognard and Josberger (2002) modeled seasonal changes in snow-grain size using a brightness temperature-gradient approach to parameterize the retrieval of snow depth in the northern Great Plains during the 1996–97 winter season. Kelly et al. (2003) developed a methodology to estimate snow grain size and density as they evolve through the season using SSM/I and simple statistical growth models. These estimated variables are then coupled with a dense media radiative transfer (DMRT) model, described in Tsang et al. (2000) to estimate SWE. Microwave emission from forest cover contributes microwave radiation from snow and can reduce the total snow signal used in the retrieval process. Thus, forest contributions need to be corrected in the retrieval process (Kurvonen & Hallikainen 1997). Furthermore, Foster et al. (1997) and more recently Goita et al. (2003)

demonstrated methods to correct for the effect of forest cover on the passive microwave retrieval of snow.

Estimates of snow accumulation by passive microwave instruments are not accurate for the detemination of SWE when snow is wet. Under such conditions the characteristic attenuation signal used for the retrieval is lost. Passive microwave instruments can be used to detect the presence of wet snow as shown by Goodison and Walker (1994) and this has shown some interesting results recently for melt–freeze cycle detection in snowfields in Alaska (Ramage & Isacks 2003). C-band synthetic aperture radar (SAR) measurements have also been used to map wet snow because wet snow absorbs C-band radar energy and returns a very low backscatter signal compared with surrounding snow-free terrain. If the snow is dry, the C-band backscatter response is from the underlying land surface with little or no attenuation by the overlying snow; for dry snow, C-band on its own does not provide a strong signal. C-band SAR systems have been the dominant frequency for ESA's Earth resources satellite 1 (ERS-1) and ERS-2 radar missions, CSA's RADARSAT mission, and Envisat Advanced Synthetic Aperture Radar (ASAR) (e.g. Koskinen et al. 1997). These SAR systems generally do not have global frequent (subweekly) coverage for which snowmelt monitoring is needed. However, Nghiem and Tsai (2001) demonstrated that NASA's SeaWinds Quick Scatterometer (QuikSCAT) can be used to map wet snow globally on a daily basis. For example, they showed rapid changes in the backscatter over the northern plains of the United States and the Canadian prairies in spring 1997. The timing of this change has been associated with the onset of flooding in the mid-western United States and southern Canada during that season. Most recently, Rawlins et al. (2005) have demonstrated the use of QuikSCAT to successfully measure snowmelt extent and timing across the pan-Arctic basin in 2000. This is an important contribution to Arctic basin hydrology.

There are two operationally available data sets of SWE: one for the globe and one for the Canadian Prairies. The global SWE data set is produced by NASA's Goddard Space Flight Center and is derived from the AMSR-E (Plate 4.3). It can be obtained through the NSIDC (<http://www.nsidc. org>). The Canadian data set can be obtained from the State of the Canadian Cryosphere (SOCC) web site (<http://www. socc.ca>). SOCC is a project of the Canadian Cryosphere Information Network (CCIN) and is a collaborative partnership between the Canadian Federal Government, the University of Waterloo, and the private sector. These data are freely available.

It should be noted that high quality robust data sets of passive microwave-derived SWE, for climate change studies, are complicated to derive because the data represent coarse spatial resolution measurements and have various uncertainties associated with their accuracy. For example, it is known that passive microwave radiation is insensitive to shallow dry snowpacks. Thus, when compared with VIS/IR products, passive microwave SWE maps underestimate global snow extent, particularly during the start of the season. In addition, passive microwave measurements at 37 GHz, the frequency that carries much of the SWE signal, tends to saturate at snow depths of approximately 50–100 cm of snow depth (perhaps 150–300 mm SWE depending on snow density). Hall et al. (2002b), Basist et al. (1996), and Armstrong and Brodzik (2002) compared snow maps derived from both MODIS and SSM/I measurements and found that the SSM/I maps showed less snow cover, than did the MODIS maps, particularly in the fall months, corroborating earlier work by researchers who compared NOAA snow maps with passive-microwave snow maps. Nevertheless, there are more

than 25 years of global passive microwave observations of the Earth and researchers are beginning to undertake longer term snow climatology studies. For example, Armstrong and Brodzik (2001) looked at the long-term record and found inconsistencies between the SMMR and SSM/I datasets. This was confirmed by Derksen et al. (2003) who analyzed the time series of SWE from both SMMR and SSM/I data over central North America for the winter season from 1978 through 1999. Recently, Armstrong et al. (2005) have produced a monthly global SWE climatology that is now available through NSIDC. This data set will be of great interest to climate scientists.

In order to characterize snow water storage with reduced uncertainty, and to measure snow accumulation at the process scale of snow variation, new and improved satellite instrument measurement techniques are required that can measure globally and at high spatial resolution. Research has shown that Ku-band radar measurements are sensitive to SWE (Papa et al. 2002) and could be developed to resolve fine spatial variations of SWE (tens of meters) through SAR technology. This could also enable more accurate retrievals of SWE globally and especially in mountain terrain where snow accumulation is significant and very important hydrologically.

Snow accumulation on sea ice has been estimated from passive microwave space-borne measurements (Markus & Cavalieri 1998). This is an important variable because sea ice controls the energy exchange processes between atmosphere and ocean water. The presence of snow cover on sea ice acts as an insulating layer to the ice and controls the ice energy budget; the presence of snow will reflect energy back to space that would otherwise impact the sea ice budget (Laxon et al. 2004). In addition, estimates of snow depth on sea ice are needed to correct measurements of ice freeboard (the amount of ice that is emergent from the ocean compared with the submerged portion) from other remote sensing instruments, such as NASA's Geoscience Laser Altimeter System (GLAS) on the Ice Cloud and Elevation Satellite (ICESat) which was launched in January 2003. While the amount of snow on sea ice is small relative to the sea ice thickness, for long-term interannual studies of sea ice it is important to account for snow thickness in the measurements of ice thickness. Data for daily polar snow depth estimates on sea ice are available in the NASA sea ice product which can be obtained from NSIDC (Cavalieri & Comiso 2004).

4.3.2 SEA ICE

Atmospheric synoptic-scale and larger atmospheric weather systems tend to drive the dynamics of sea ice extent and concentration (Laxon et al. 2004). These dynamics can be observed at the temporal scale of a few days to weeks. Hence, remote sensing observations of sea ice need to be frequent and consistent to effectively characterize sea ice variables. Polar orbiting imaging remote sensing systems on satellite platforms are ideal for monitoring sea ice routinely and frequently. Several different types of remote sensing instruments have been used to map sea ice extent, concentration, and ice type. Originally, sea ice extent was monitored using VIS/IR instruments such as the AVHRR but persistent cloud cover and continuously low sun angle for many weeks in winter reduce the capabilities of these instruments. Therefore, for VIS/IR observations, frequent coverage is essential, at least to improve the chances that cloud free locations emerge, even for a short duration. More recently, MODIS has been used to map sea ice (Hall et al. 2005b) and also to estimate the surface temperature

of the ice. These data are available through the MODIS Snow and Sea Ice Global Mapping Project (<http://modis-snow-ice.gsfc.nasa.gov>). Under cloud free conditions, high spatial resolution instruments such as TM, the Enhanced TM (ETM), or the Satellite (or Systeme) Probatoire l'Observation de la Terre (SPOT), can be used to identify and characterize in detail sea ice concentration and track individual ice floes. However, these high resolution instruments generally have infrequent revisit time periods making their applicability more suited to individual "snap shot" case studies than to routine global ice mapping studies.

Measurements from passive microwave instruments, such as the Electrically Scanning Microwave Radiometer (ESMR) with data available for the period 1972–77, SMMR (data available between 1978 and 1987), the SSM/I (data available from 1987 to the present), and the AMSR-E (data available from 2002 to the present) are independent of solar illumination and make observations at wavelengths that are generally unaffected by cloud cover. They also provide near global coverage on a daily basis. Similarly, active microwave radar instruments aboard platforms such as ERS-1 SAR (from 1992), Envisat ASAR (from 2002), and RADARSAT SAR (from 1995), are generally unaffected by clouds and solar illumination and also have finer spatial resolution compared with passive microwave observations. However, the frequency of coverage of the polar regions is less than that from the passive microwave measurements. Nevertheless, between the active and passive microwave systems, microwave instruments have provided a wealth of information about sea ice dynamics over the past 25–30 years.

Passive microwave measurements were first made by the ESMR instrument in 1972. With a single frequency radiometer measurement at 19 GHz (1.55 cm wavelength) the instrument was able to exploit the fact that sea ice emissivity at this frequency (0.80–0.98) is significantly different to open water (0.40) enabling excellent discrimination between these two surface types resulting in the effective mapping of polar sea ice extent. With the launch of the SMMR instrument in 1979 and its expanded waveband range of 0.81, 1.36, 1.66, 2.8, and 4.54 cm (or 6.63, 10.69, 18.0, 21.0, and 37.0 GHz frequencies), it became possible to map not only sea ice extent but also concentration and whether the ice was FYI or MYI (Cavalieri et al. 1984). Both SMMR and ESMR made coarse spatial resolution measurements (30–156 km). With the 1987 launch of the SSM/I and AMSR-E in 2002, measurements of sea ice concentration and type have been made for over 20 years enabling scientists to analyze interannual variations and trends in sea ice cover and type. A 17-year sea ice extent and concentration record was assembled by Cavalieri et al. (1997a, 1999). Most recently, Cavalieri et al. (2003) have combined ESMR, SMMR, and SSM/I observations of sea ice extent data by using the National Ice Center analyses to bridge the gap between ESMR and SMMR thus providing a 30 year sea ice extent data record. These data sets are available from NSIDC (<http://www.nsidc.org>; Fig. 4.2) and continue to grow with continued SSM/I measurements (Cavalieri et al. 1997b). State of the art passive microwave measurements are now made with the AMSR-E instrument which has even better spatial resolution than the SSM/I. Sea ice variables are produced using the method of Markus and Cavalieri (2000). The algorithm uses a wider range of frequencies than before plus it corrects for weather effects on sea ice concentrations through the use of an atmospheric radiative transfer model. This data product is available through NSIDC and an example is shown in Plate 4.4. With the planning of future passive microwave instruments, such as the Conical Microwave Imager System (CMIS)

on board the National Polar-orbiting Operational Environmental Satellite System (NPOESS), the ability to detect even longer trends should be assured.

At higher spatial resolutions, active microwave SAR instruments can map sea ice concentration and type in great detail. With the ERS-1, ERS-2, and RADARSAT C-band SAR instruments (operating at 5.3 GHz), and continuing with the ESA's Envisat ASAR, more than 10 years of SAR measurements have been made during which there have been many successful efforts to map sea ice routinely. The limitation to these SAR estimates is that they are generally narrow in swath character (approximately 100 km) which even at high latitude, can preclude complete polar coverage, a requisite for long-term sea ice mapping. Figure 4.3 shows the sea ice surrounding Svalbard from an Envisat ASAR image. The exception to the narrow swath mode is the Canadian RADARSAT mission which has a wide swath capability (500 km). Kwok and Baltzer (1995) have demonstrated a system using RADARSAT that can provide high spatial resolution measurements of ice growth, sea ice type, and ice movement at 3-day intervals during winter months. With such fine spatial resolution, it is also possible to track individual ice floes which if done over wide areas of the sea ice mass, can be used to identify significant ice discharge from the Arctic Ocean (Kwok et al. 2005). SAR measurements of sea ice have effectively been combined with passive microwave and VIS/IR observations in a study by Comiso and Kwok (1996). This integrated approach takes full advantage of all instrument capabilities and can provide a comprehensive approach to mapping sea ice variables. The Canadian Ice Service now routinely ingests Radarsat imagery for its operational sea ice mapping product.

The sea ice variable that has not received much attention from satellite instruments is that of sea ice thickness. Measuring a few

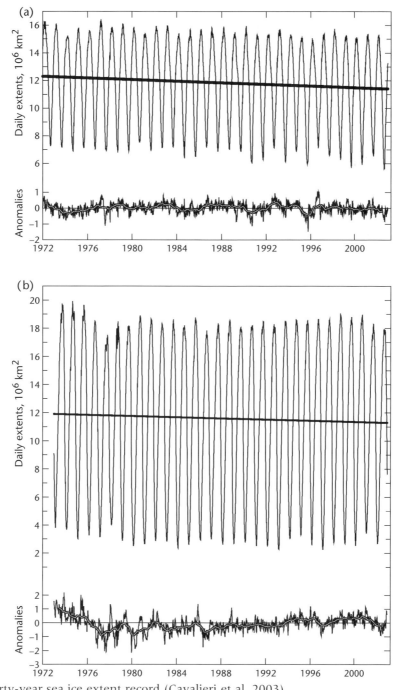

Fig. 4.2 Thirty-year sea ice extent record (Cavalieri et al. 2003).

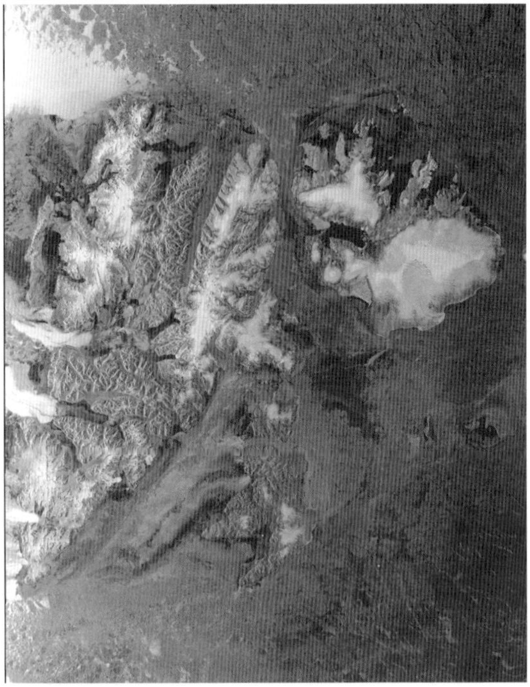

FIG. 4.3 Envisat ASAR image of the Barents Sea with first year ice surrounding the coasts of Svalbard (78°N, 20°E) acquired in image mode on December 29, 2003 (Orbit 09572). The smooth, light toned region in the upper left corner represents backscatter from open water which is a rough surface as observed by the SAR. To the right, the edge of the ice can be seen where the grey shade becomes darker. Within the ice region, the darker slithers are leads of water between the ice which give a low backscatter return. The largest of the islands is Spitsbergen. Image courtesy of ESA, <http://www.earth.esa.int/earthimages/>.

meters of ice freeboard from space, especially with seasonal snow accumulation on top, is a challenging task. An experimental approach by Laxon et al. (2003) used low spatial resolution (16–20 km) soundings from the radar altimeter on the ERS satellites to estimate sea ice freeboard. The research approach requires there to be leads in the sea ice (where open water exists) but so far, comparisons with ULS measurements have agreed to within 0.5 m of freeboard thickness. Finally, the most promising new

satellite technology for sea ice thickness mapping is NASA's GLAS instrument on ICESat. GLAS sends out a pulse of visible light and measures the return signal to obtain a precise and accurate estimate of the satellite altitude. Given an accurate Earth geoid model, it is possible to measure the local height of the target instantaneous field of view (IFOV) below. With a nominal spatial resolution of 70 m, theoretically, each IFOV has a 10 cm height accuracy which is an extraordinary accomplishment. The three drawbacks with ICESat measurements are that first, unlike microwave systems, ICESat cannot make measurements if cloud obscures the land surface. Second, the repeat orbit for ICESat is 183 days, and since GLAS is a nonimaging system (i.e. it makes discrete measurement soundings along a subsatellite ground track only) the global mapping capability is through a sampling approach rather than a full global coverage. Third, technical issues with the GLAS instrument have meant that the mission lifespan has been reconfigured to produce less continuous global sampling of the polar regions. Despite these drawbacks, ICESat has demonstrated the potential to provide the most accurate estimates of sea ice surface elevation and iceberg freeboard yet (Scambos et al. 2005; Schutz et al. 2005) and with future missions in planning, it is expected that our capabilities to map sea ice variables will only improve.

4.3.3 ICE SHEETS AND GLACIERS: ESTIMATION OF VOLUME

In Section 2.2, *in situ* measurement approaches of ice sheets and glaciers were discussed. The Greenland and Antarctic ice sheets cover approximately 1.7×10^6 km^2 and 13.7×10^6 km^2 respectively. The ice sheets also can be hostile and inaccessible and while many mass balance measurement studies are being successfully coordinated through the WGMS, *in situ* measurements are spatially limited and cannot always give the broad area context of ice sheet processes. Remote sensing, however, can provide measurements over larger areas and efforts by the geophysical research community have significantly enhanced our understanding of glacier and ice sheet processes and states through the development and application of remote sensing and modeling techniques. It should also be noted that these advances have also relied on field experiments to make *in situ* measurements of key mass balance equation processes that help in the interpretation of remote sensing measurements.

In this section, remote sensing measurement techniques to estimate ice sheet and glacier mass balance are described. Detailed summary and descriptions have been published in the past by Thomas (1993) and more recently by König et al. (2001), Bamber and Kwok (2004), Thomas (2004), and Bentley (2004). The same categorization of measurement approaches is used here as before in Section 2.2: (i) direct estimation of the glacier or ice sheet volume; (ii) measurements of specific components of mass balance. From a remote sensing standpoint, this is a convenient classification because only nonimaging systems such as satellite radar and laser altimeters, gravity measurement systems, and airborne and ground-based radio echo sounders are capable of measuring glacier volume changes. Measurements of mass balance equation components, on the other hand, specifically accumulation and ablation rate, iceberg calving, ice thickness, and ice velocity, have been measured by imaging systems to varying degrees of success (Bamber & Kwok 2004). While there is a 30-year history of satellite remote sensing measurements in this field, it is only in the last 10 years that significant advances in the quantification of ice sheet and glacier mass balance have been made. It should be noted at the outset that there are uncertainties with remote sensing measurement approaches.

For example, estimated sea level rise is on the order of 3 mm a year which is comparable to the uncertainty of mass balance estimates of Antarctica and Greenland from remote sensing (Thomas 1993). However, newer satellite instruments have significantly reduced this uncertainty and with new missions in planning, these uncertainties will be reduced further.

The remote sensing methods used to estimate mountain glacier and ice cap mass balances are very similar to those used for ice sheets. However, because glaciers and ice caps are smaller than ice sheets, higher spatial resolution remote sensing measurements are required. This can be achieved using certain satellite imaging systems but for direct volume mass balance measurements achievable by nonimaging systems, only aircraft instruments have the required precision and spatial resolution sufficient for accurate mass balance estimates. In this section, first we identify ice sheet measurement approaches and data sources and then glacier approaches to mass balance estimates.

With the launch of the European ERS-1 in 1991, ERS-2 in 1995, and Envisat in 2002, direct measurements of ice sheet surface height variations are now possible using radar altimeters (e.g. Davis 1995; Wingham et al. 1998). Bamber and Kwok (2004) give an overview of how these instruments operate. Designed for ocean surface height measurements, a radar altimeter emits a pulse of energy (usually at 13–17 GHz frequency in the Ku-band range) toward the Earth's surface and the return echo from the surface is measured by the receiver back at the altimeter. The time the pulse takes for the round trip is used to estimate the height of the instrument above the Earth. Since the platform is in a steady orbit, any variations in height are a function of variations in sea surface or land surface elevation. However, two complicating factors to the measurement approach require consideration. First, satellite RA systems cannot always track accurately the source of the return pulse signal when surface slope is steep such as over mountains or at the edge of ice sheets or coastal margins. This is because the RA maps a target as a function of the return time of the emitted pulse. A target might be closer to the RA in time yet it might not be located at the center of the emitted pulse. In these cases, the uncertainty is in the accurate identification of the geographical centre of the RA pulse return. This can occur, potentially, over 10–15% of Greenland and Antarctica. Several approaches have been proposed to correct these issues, called retracking. The second uncertainty is found when terrain is undulating. In this case the center of the measurements cannot be determined by the returned pulse signal and the RA measurement are deemed to "lose lock." The reader should consult Bamber and Kwok (2004) for a more in depth discussion of these issues.

Current theories about the state Antarctica and Greenland are based on extensive *in situ*, remote sensing observations and numerical modeling experiments. Recent research from the NASA PARCA experiment produced estimates of interior ice sheet thickening/thinning rates of the Greenland ice sheet above 2000 m above sea level (Thomas et al., 2001). Three time periods were compared with all periods consistent with one another. For each period, a different measurement approach was undertaken. For the last few decades, measurements of ice discharge from outlet glaciers with interior snow accumulation measurements showed that thinning occurred in the northwest and southeast parts of the ice sheet. Conversely, the north east and south west sectors showed long term thickening of mass. This pattern was confirmed, but with finer spatial detail, by satellite radar altimeter estimates and then by aircraft laser altimeter measurements for the 1978–88 and then 1993/94–1998/99 periods respectively (Thomas et al. 2001). In summary, it was concluded that at high elevations,

large parts of the ice sheet are in balance but with some regions exhibiting characteristic thickening and others demonstrating thinning. Coupled wth this study, Abdalati et al. (2001) demonstrated the use of extensive laser altimeter measurements to estimate the thickening/thinning of ice in the lower elevations of Greenland ice sheet. They showed that significant thinning has dominated the mass balance in these regions for the 1993/94 and 1998/99 seasons. In several cases, thinning rates of tens of cm yr^{-1} were obeserved and these locations thinning was attributed to warmer summer seasons. At several sites, however, thinning rates of 10 m yr^{-1} were measured; these rates were most likely casued by glacier dynamics and creep thinning. In a 2005 SAR interferometry study, Rignot and Kanagaratnum (2006) have shown this marginal thinning trend to continue. Finally, most recently, Zwally et al. (2005) have confirmed from satellite and aircraft altimetry measurements, that Greenland can be characterized by thinning at the margins and thickening above the ELA albeit with a small net positive balance overall (Plate 4.5).

Antarctica is a large ice sheet and, as a result of its sheer size and glaciological complexity, it has larger uncertainties associated with its current mass balance state. This is especially the case for direct measurements of mass balance components (accumulation and ablation) or for models of mass and energy balance components that are forced by highly generalized fields. To date our current knowledge about Antarctica's mass balance dynamics is based around coarse-scale numerical models that rely on *in situ* measurements and low spatial resolution radar altimeter measurements from a suite of instruments spanning the 1980s to the present (e.g. ERS-1, ERS-2, etc.). For example, Davis et al. (2005) compiled an 11 year record of height variation from ERS-1 and ERS-2 radar altimeter instruments and showed that the East Antarctic Ice Sheet is thickening at 1.8 cm yr^{-1} with an uncertainty of 0.3 cm yr^{-1}. These measurements, however, are less accurate at the margins of the ice sheet where much of the melt processes can dominate. The ablation processes of Antarctica are inextricably linked to the ice-shelf dynamics into which many of the ice streams flow. With the recent satellite observations of the disintegration of some prominent ice shelves in Antarctica (e.g. Worde and Larsen A ice shelves), much effort has been exerted to better understand the coupling between interior accumulation processes, ice discharge, and melt processes around the margins. According to many researchers and organizations, including the British Antarctic Survey, it is now understood that the ice shelves around Antarctica may act as a stabilizing agent to the interior ice mass. However, scientists are still unsure what would happen if the ice shelves disintegrate through rapid melting (perhaps from ocean surface temperature increases). It is possible that there could be a catastrophic loss of ice from the West Antarctic Ice Sheet resulting in a 5–6 m global sea level rise. Generally, the West Antarctic climate is cold and unlikely to be threatened by melting in the next 200 years although in a recent study, it has been suggested that the ice sheet may be more sensitive to climate changes than at first thought (Alley et al. 2005). The IPCC stated that *"the likelihood of a major sea level rise by the year 2100 due to a collapse of the West Antarctic ice sheet is considered low"* (Houghton et al. 2001). Nevertheless, a low probability of sea level change now must be balanced against the high cost of significant future sea level rise. To that end, there have been various efforts to improve our understanding of the state of Antarctica.

One of the many interesting aspects of the PARCA experiment was that it demonstrated the utility of the radar altimeter for ice sheet height measurements (both relative and absolute) and, at an even higher accuracy and spatial resolution, the utility

of the laser altimeter. ICESat, launched in 2003, is an innovative laser altimeter mission constructed to measure Earth's ice (land and ocean), cloud, and elevation (ocean and land) using novel laser ranging technologies. ICESat is designed to detect changes in surface elevation of the order of 1 cm yr^{-1}, in regions as small as 200 km by 200 km (Zwally et al. 2002). This level of accuracy is important because overall, a persistent elevation change of 1 cm yr^{-1} is equivalent to an ice mass imbalance of 5% and 0.4 mm yr^{-1} of sea level change. Observations of elevation changes over the entire ice sheets enable the assessment of local (individual drainage basin) to regional (major ice outlet glaciers) scale ice mass balance variations and internal ice dynamics. Work is also underway to detect and measure interannual changes in the surface mass balance enabling the evaluation of whether or not the changes are caused by recent or long-term climate variations and/or ice dynamics. Plate 4.6 shows the local to regional measurements possible from the ICESat instrument. Despite the limitations and uncertainties in the measurements, many authors (e.g. Shepherd et al. 2001) note that 20–30 cm height accuracies of ice sheet surface elevation can be obtained.

The GLAS instrument aboard ICESat was designed to measure terrestrial elevation very precisely and accurately (Abshire et al. 2005). Again, as a nonimaging system, GLAS can be used to estimate surface height variations to a 10 cm height accuracy and from within a 70 m IFOV. The big difference between GLAS and satellite RAs is that GLAS uses visible light from a laser and in doing so, can illuminate a smaller instantaneous field of view on the ground. This means that the uncertainty features found in RAs as described above are far less important for laser altimeters. However, unlike satellite microwave RA measurements, ICESat is affected by cloud cover and cannot see the ground when clouds persist.

This poses an interesting question since data acquisitions take place within the 91 day orbital period and complete ice sheet coverage of all places possible in the along track direction will be interrupted by cloud cover. However, one of the goals of ICESat is to measure accurately surface elevation changes of the ice sheets, over a 3–5 year and so the cloud obscuring problem is less significant at this timescale. Early results of studies from GLAS are in formulation but papers by, for example, Smith et al. (2005) and Nguyen and Herring (2005) show great promise that this instrument will indeed significantly improve our estimation capabilities of the ice sheet height variations through time. The GLAS instrument is also being used to test the accuracy of ice sheet digital elevation models derived from other remote sensing sources (Bamber & Gomez-Dans 2005). Data from GLAS for measuring Antarctica and Greenland surface height variations can be found at NSIDC (Zwally et al. 2003).

Finally, experimental measurements of ice sheet volume are planned using the Gravity Recovery and Climate Experiment (GRACE). GRACE precisely measures very small changes in Earth's gravitational field that result from changes in the mass of ice sheets, oceans, and ground water. It does this by measuring the distance very accurately and precisely between two GRACE satellites that are formation flying in the same orbit, some 500 km above the earth. As the GRACE-twins fly over the Earth the precise speed of each satellite and the distance between them is constantly communicated via a microwave K-band ranging instrument. The gravitational field changes beneath the satellites, as a direct result of changes in mass (topography) of the geoid, and this causes the orbital motion of each satellite to change. Changes in orbital motion cause measurable changes in the distance between the satellites and fluctuations in the Earth's gravitational field can be determined,

therefore, from these separation changes. Further details can be obtained from <http://www.csr.utexas.edu/grace/>.

4.3.4 ICE SHEETS AND GLACIERS: MASS BALANCE COMPONENTS

The second approach to glacier mass balance estimation is through the estimation of mass balance budget components, particularly mass accumulation and ablation, iceberg calving, ice thickness, and ice velocity. Starting with accumulation measurements, Winebrenner et al. (2001) demonstrated an approach, using the 6 GHz channel on SMMR, to estimate the accumulation rate on the central Greenland Ice Sheet. The measured brightness temperature is a function of emissivity and physical temperature of the snow. Assuming the temperature of the snow can be obtained independently, this leaves the emissivity, which is a function of grain size. In the accumulation zone where no melt occurs, the grain size is dominated by the age of the firn layer which can be related to the accumulation rate. The correlation of this estimation approach with *in situ* measurements is of the order of 0.8 although the residuals are large. This is due to the low spatial resolution of the measurements, which average accumulation processes over wide areas. However, the recent AMSR-E instrument has higher spatial resolution at the 6 GHz frequency and may well provide more accurate estimates of accumulation rates. Direct measurements of ice ablation are not straightforward. Most approaches have used remote sensing thermal infra-red data from AVHRR (or MODIS) in conjunction with energy balance models or positive degree-day approaches. VIS/IR data, again, are adversely affected by cloud cover in their ability to provide regular surface observations. High spatial resolution VIS/IR instruments, such as Landsat TM, can be used to identify the equilibrium line altitude, a surrogate measure of the mass balance, but again these observations are affected by the cloud problem. SAR imagery, which are unaffected by clouds, can also be used to estimate the ELA (e.g. Kelly 2002) but there is uncertainty about the exact nature of the relationship between ELA and mass balance. In addition, Abdalati and Steffen (1995) demonstrated the use of SSM/I measurements over Greenland to estimate the extent and melting of ice and it is possible that the GLAS instrument will improve these estimates in the future. Cloud-free observations of the calving of icebergs from the Greenland and Antarctic ice sheets can be achieved from VIS/IR sensors such as Landsat TM or MODIS, and RADARSAT is cloud independent. These measurements often receive news coverage especially for the larger tabular iceberg events. Converting these irregular sightings from space, to ice flux data, however, is not straightforward especially since sea ice thickness is not easily obtained from such observations (Bamber & Kwok 2004).

Ice thickness has yet to be determined directly from spaceborne instruments. However, aircraft radio echo sounders, operating at megahertz frequency ranges, can measure ice thickness because the ice is almost transparent at these frequencies. Over Greenland, measurements by Gogineni et al. (2001) have shown that the accuracy of these measurements can be made to approximately ±10 m. These measurements have been made 1993 to 2003 and formed an important part of NASA's comprehensive PARCA experiment (American Geophysical Union 2001b). Ice velocity, the last measureable balance budget component, can be measured by feature tracking and by SAR interferometry (InSAR) techniques. Feature tracking, as the name suggests, is a method that identifies features in the ice and tracks them over a period. Several authors have successfully used this

approach to estimate the velocity of ice streams and outlet glaciers (e.g. Bindschadler et al. 1996). An approach for SAR speckle noise tracking has also been developed to estimate surface velocity (Joughin 2002). Lastly, InSAR has also been used successfully to measure ice velocities over the ice sheets. Again, Joughin (2002) gives an account of the approach and combines his measurements with the speckle tracking approach. InSAR methods are complex and it is not our intention to cover them here. However, it is worth noting that while the approach can be successful, there are limiting aspects to the approach. First, temporal decorrelation of the signal from changes in surface scattering behavior makes the velocity estimates uncertain. Second, if the displacement of a target is too large from repeat passes, then the InSAR method cannot estimate the velocity. Again, there are many in-depth texts on InSAR and the reader is directed to Bamber and Kwok (2004) for further details. In summary, with both these approaches there are measurement uncertainties and there are issues concerning orbital coverage and, for the VIS/IR tracking, cloud cover issues. Nevertheless, significant progress has been made in the last 10 years to improve our estimates of ice velocity from space.

Mountain glacier and small ice caps can also be monitored using remote sensing methods. Probably the most widespread use of remote sensing has been in the mapping of glacier extent from high spatial resolution VIS/IR instruments such as Landsat TM, SPOT, and the Advanced Spaceborne Thermal Emission and Reflection Radiometer (ASTER) instrument on board NASA's Terra platform. Multispectral imagery provide good discrimination between clouds, snow, bare ice, and snow/ice-free terrain. Measurements from ASTER have been used to map global glacier and ice cap extent and the Global Land Ice Measurements from Space (GLIMS) project was set up to take advantage of the high spatial resolution of this instrument (Kiefer et al. 2000; Khromova et al. 2003; Bishop et al. 2004; Khalsa et al. 2004). Plate 4.7 shows an ASTER image (at 30–90 m spatial resolution) of part Eastern Bhutan in 2001. Progress in global mapping continues to proceed and the GLIMS system aims to map each GLIMS target at least once per year. Using automated software, changes in glacier extent will be calculated.

As noted before, to represent mass balance processes accurately, fine spatial resolution measurements are required. Coarse resolution measurements from satellite radar altimeters and passive microwave instruments, which have spatial resolutions on the order of tens of kilometers, therefore, are not capable of resolving mass balance processes except for the largest mountain glaciers or ice caps, and even then statistical validity of these measurements has to be carefully evaluated. Aircraft measurements provide a solution to the resolution issue because the spatial resolution increases as the platform altitude decreases. However, aircraft campaigns are not operationally conducted, except in a few locations such as Scandinavia. Thus, we describe potential and current efforts to collect measurements for monitoring glacier mass balance from direct estimates of volume change and then mass balance budget components.

Direct estimates of selected glacier height variations and their annual changes can be conducted using GLAS. The limiting factor in this approach is that the ICESat tracks do not take measurements all glaciers and ice caps. Studies are beginning to emerge in the application of GLAS to glacier height elevation changes (e.g. Sauber et al. 2005). These kinds of studies, which compare GLAS with field surveys and other elevation maps (conventional and satellite-derived) provide an excellent way of testing the capabilities of the GLAS instrument. Aircraft instruments,

specifically Lidar instruments can provide good estimates of glacier surface topography variations (Krabill et al. 1995). When combined with radio echo sounding measurements, from the ground or aircraft, the volume of ice can be estimated. Changes in volume from one year to the next can be used to indicate changes in mass balance of the glacier. Estimation of the mass balance budget component of accumulation cannot easily be achieved by satellite instruments. While high spatial resolution VIS/IR instruments can identify new snow extent, measurements related to accumulation are not easily obtained. Passive microwave measurements are generally too coarse to resolve accumulation processes and most SAR systems operate at wavelengths that do not interact with new snow. With respect to ablation on glaciers and ice caps, however, satellite VIS/IR and SAR measurements have been used to monitor melt zones on glaciers and to track the snow line progression through the season (Dowdeswell & Drewry 1989; Kelly 2002; Fig. 4.4). Also, several studies using SAR backscatter data have demonstrated the way that glacier snow and ice facies can be identified (e.g. Hall et al. 1995; Partington 1998) including wet snow zones, percolation zones, and firn layer zones. While these do not give rates, individually, they identify locations where ablation processes are prevalent.

Measurements of iceberg calving from tidewater glaciers are similar in approach and quality to those from ice sheets (Plate 4.8). Satellite observations of eastern Greenland's Helheim Glacier show that the position of the iceberg's calving front, or margin, has undergone rapid and dramatic change since 2001, and the glacier's flow to the sea has speeded up as well. From the 1970s until about 2001, the position of the glacier's margin changed little. But between 2001 and 2005, the margin retreated landward about 7.5 km, and its speed increased

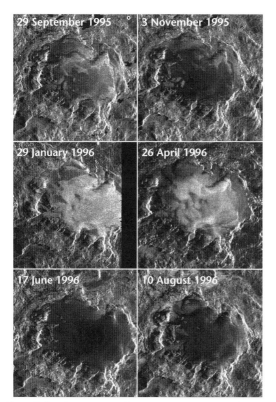

FIG. 4.4 Sequence of ERS-1 and ERS-2 SAR images of Hardangerjokulen, Norway for the winter season of 1995–96. Winter backscatters are dominated by rough surfaces which produce strong backscatter (lighter tones) while summer and fall surfaces are darker in tone, suggesting either smooth surfaces (that scatter away from the SAR receiver) or absorptive wet snow surfaces that give a weak (dark tone) response. Data courtesy of ESA.

from 8 to 11 km yr^{-1}. Between 2001 and 2003, the glacier also thinned by up to 40 m. Probably the most significant use of remote sensing for glacier studies, however, is through the use of InSAR techniques for measuring glacier velocity. For example, Ford et al. (2003) demonstrated the use of interferometric SAR analysis of the Seward Glacier in Alaska and Eldhuset et al. (2003) have demonstrated the use of InSAR for glacier motion in Svalbard. Other studies

have been limited also to specific experimental basins or glaciers but no comprehensive global mapping of glacier motion has yet been compiled. As InSAR methods become more accessible to the science community, more study sites will no doubt be measured. Of special note and of particular suitability for the InSAR approach is the estimation of ice velocity over surging glaciers. For example, a study by Luckman et al. (2002) quantified the surface flow evolution over 6 years of a glacier surge in Svalbard using InSAR from ERS-1 and ERS-2 SAR instruments. By analyzing 36 SAR images, surface velocity maps during the surge period were generated. The same measurements from ground survey would be far less comprehensive to acquire.

The kinds of remote sensing approaches demonstrated in this section illustrate the power of Earth Observation as a tool for ice sheet and glacier characterization. By integrating the different approaches, direct observations from satellite and aircraft remote sensing instruments can provide comprehensive information on glacier and ice sheet behaviors. When coupled with numerical models of ice geophysics, these measurements become even more powerful for assessing the state and changes to the world's terrestrial freshwater ice stores.

4.3.5 PERMAFROST

Mapping permafrost is not a straightforward endeavor. Remote sensing instruments are capable of sensing freeze–thaw processes only within the uppermost 5 cm of soil depth.

With the planned launch of ESA's SMOS mission in 2007, that penetration depth may increase to 10 cm or so. However, permafrost can extend tens to hundreds of meters below the surface and so remote sensing measurements only offer a small glimpse into the surface dynamics of permafrost cover. *in situ* ground measurements, however, are spatially localized but are able to reveal time varying frozen-ground dynamics on a daily basis. The spatial correlation length of permafrost variability is linked to the surface vegetation and soil type plus the volumetric water content of the soil. Figure 4.5 summarizes our current understanding of frozen ground extent for both permanent and seasonally frozen ground. This map is a hybrid of several data sources and can be obtained from NSIDC (Brown et al. 1998). The Eurasian, North America, and Greenland sections of the map was initialized by the United Nations Environmental Program/Global Resource Information Data Base (UNEP/GRID)-Arendal in Norway. The map was refined at CRREL and USGS. The data extend to 25°N, the limit of Northern Hemisphere permafrost.

As far as permafrost dynamics are concerned, large areas of the Canadian north are characterized by permafrost soils at temperatures greater than −2°C with frozen thickness less than 75 m. General circulation models predict that mean annual air temperatures may rise by between 5°C and 10°C during winter months over much of the Arctic as a response to increased atmospheric CO_2 accumulations. As a result, large regions of this permafrost may ultimately disappear under anticipated climate warming scenarios (Smith & Burgess 1999). In areas of thicker and colder permafrost, there is a higher probability that warming will result in an increase in permafrost temperature, a decrease in permafrost thickness, and a thickening of the active layer. In the Eurasian high latitude regions, a similar combination warming effect will also likely result.

Measurement of permafrost dynamics directly from Earth observing systems (aircraft or satellite) is very challenging. No operational monitor program from space or aircraft is currently underway. Since permafrost extends well below the land surface,

most remote sensing systems do not have penetration capabilities to detect permafrost conditions. VIS/IR systems are restricted to observing surface features of frozen/thawed ground or thermokarst features but the measurements need to be of a high spatial resolution, certainly less than 30 m and perhaps less than 10 m IFOV, to be able to effectively detect such features (e.g. shallow pits and depressions). The SPOT system, with its 10 m panchromatic instrument is a good candidate for this type of study. Ikonos (<http://www.spaceimaging.com>), which has not been mentioned so far, has a spatial resolution of 2 m, which can easily map thermokarst features. There are no spaceborne imaging instruments available yet that are able to determine whether the ground surface is frozen or thawed under snow and snow-free conditions. Kimball et al. (2004) demonstrated the use of the NASA scatterometer (NSCAT) to estimate the seasonal transition of the boreal forest between frozen and non-frozen conditions. This approach detected in a binary way whether the ground was frozen or unfrozen but could not give information on degree of wetness or soil moisture content (Plate 4.9). To estimate this variable with any accuracy, an active system with sensitivity to long wavelength signals is required. Two instruments, in planning and development phases at NASA and ESA could potentially provide improved information on freeze/thaw land state. The first is the Soil Moisture and Ocean Salinity (SMOS) mission under development by ESA. At 1.4 GHz (21 cm wavelength), the instrument is designed to measure soil moisture. Because of this capability, it will also be able to measure, for the first time, the detailed freeze–thaw state of the upper surface layer of soil. The second mission is NASA's Aquarius (<http://www.aquarius.nasa.gov>). It is similar to SMOS except that it has active and passive instruments operating at 1.4 GHz frequencies on board.

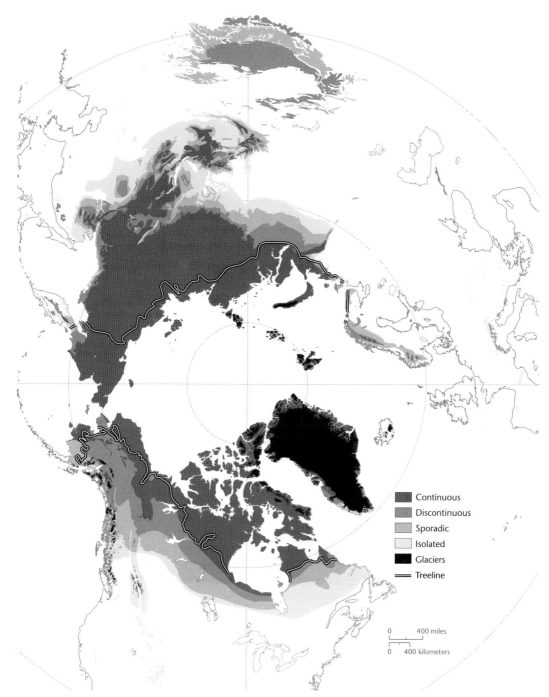

Continuous
Discontinuous
Sporadic
Isolated
Glaciers
Treeline

0   400 miles
0   400 kilometers

Fig. 4.5 Map of permafrost for the Northern Hemisphere. The circumpolar data are part of an international data set that show the distribution and properties of permafrost and ground ice in the Northern Hemisphere (20°N to 9°DN). The data classes have been grounded into spatially continuous (90–100% cover), discontinuous (50–90%), sporadic (10–50%), or isolated (<10%) permafrost zones (Heginbottom et al. 1993; Brown et al. 1997). Within each of these general classes, the original differing ground ice contents and overlying landform classes have been combined.

4.3.6 RIVER RUNOFF

Measuring cryospheric river runoff (discharge or stage) by remote sensing methods is still in its infancy and has yet to be fully developed both in technological and in scientific terms. The only potential measurements that have been demonstrated, and even then for large rivers, are from satellite radar altimeters and SARs. Birkett (1998) used NASA's TOPEX satellite RA to measure stage for both large wetlands and rivers of >1 km width for nearly 4 years of the mission life. The work was conducted on the world's largest rivers and globally important wetlands. Results from the study show that the stage estimates from the instrument can be as accurate to 11 cm height error. GLAS could potentially be used for smaller rivers since its IFOV is much smaller than those from satellite RAs. However, comprehensive results from studies have yet to appear on the use of GLAS. The instrument could potentially sense smaller rivers in the Arctic and mountains that are not measurable by RA instruments. Last, Alsdorf et al. (2001) demonstrated the use of repeat pass InSAR to estimate water surface heights in the Amazon basin. This method is attractive since the SAR is an imaging instrument and can provide broad river reach coverage, as opposed to the radar/laser altimeters which are nonimaging nadir tracking instruments. In addition, the InSAR approach can be used on smaller waters since the spatial resolution of the SAR is finer than the RA. The limitation to this method especially in mountainous terrain is that the viewing geometry of the SAR can generate uncertainties and ambiguities in the measurements which are not easily corrected. However, between these three approaches, there is great potential for estimating river stage height in remote, ungauged basins of the cryosphere.

4.3.7 RIVER AND LAKE ICE BREAK-UP AND FREEZE-UP

In the previous section on *in situ* measurements, it was noted that records of freezing/break-up of lake ice and river ice extends back to the 19th century at several locations in the northern hemisphere. Photointerpretation of aerial photography can be used to identify lake ice and note the freeze-up and break-up dates. This approach is considered an *in situ* approach in as much as it is not comprehensive over all high latitudes. Thus, satellite systems offer a more comprehensive possibility to measure these freeze-up, break-up, and ice extent and thickness. Microwave remote sensing can also be used to identify the onset of freeze-up and break-up. The backscatter properties of freshwater contrast significantly with ice and can be used to identify the change of state. Weber et al. (2003) developed a method to semi-automatically detect river ice formation. Using Radarsat-1 SAR data acquired over the Peace River in Alberta on March 7, 2001, they showed a method for classifying ice from open water (Fig. 4.6). Duguay et al. (2002) have demonstrated the use of ERS-1 and RADARSAT SAR systems in identifying lake ice. The backscatter properties of freshwater compared with contrast significantly and can be used to identify the change of state. The spatial resolution of SAR systems means that a larger variety of lake sizes can be sensed which is important for many parts of the tundra which are dominated by lakes which individually are small but collectively make up a significant fraction of the land surface. In a different study, Duguay and Lafleur (2003) demonstrated the use of combined VIS/IR and SAR measurements to estimate the thickness of lake ice, which has implications not only for the break-up timing but also for biotic organisms in those environments.

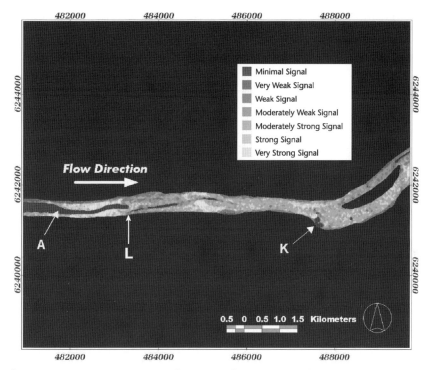

FIG. 4.6   Radarsat-1 SAR measurements of river ice formation in the Peace River, Alberta on March 7, 2001. Location A is characterized by open water while location L represents the head of the complete ice cover and location K represents localized open water in the ice (Weber et al. 2003).

Plate 4.10 shows a time series of ERS-1 SAR backscatter responses for lakes near Churchill, Manitoba during the winter of 1992–93. The colored Landsat TM subplate shows the bathymetry of the lakes in this area and the SAR is capable of identifying whether a lake is partially frozen or frozen to the base (Duguay & Lafleur 2003). Probably the most significant factor for accurate measuring freeze-up and break-up timings from satellite instruments is the frequency of sensing of the target. VIS/IR instruments can be used but cloud cover generally reduces the number of opportunities available to sense the ground. SAR systems, while operating at wavelengths that can penetrate clouds, may have low revisit frequencies unless wide swath modes are available (e.g. RADARSAT or Envisat). Passive microwave instruments can also be used and Walker and Sillis (2002) used the SSM/I 85 GHz channel to measure the spatial configuration of shallow lake freeze-up and break-up in the Mackenzie Basin. Using a combination of passive and active radar systems, the Canadian Ice Service now monitors Canadian lakes regularly. The freeze-up and break-up dates are determined to within one week accuracy and during the winter, ice coverage is measured in 10% increments. The CRYSYS website (<http://www.crysys.ca>) contains statistics for some 70 lakes in Canada which are compiled using an *in situ* observation network from the Canadian Ice Service (CIS). The future approach for lake ice monitoring will be conducted most likely with remote sensing instruments.

4.4 NUMERICAL MODELS

The previous section described many attempts by the cryospheric research community to assemble measurements of data, by *in situ* means and remote sensing observations, to better represent cryospheric processes. One way of characterizing these two approaches is that the *in situ* approach generally gives high accuracy at a localized place whilst the other generalizes, sometimes with less accuracy, but can yield internally consistent synoptic-scale estimates. In addition, with improved retrieval or remote sensing model methodologies and with continually improving instrument system technologies, remote sensing measurements will continue to improve. In spite of the uncertainties, both approaches are highly compatible, especially when models of cryospheric processes are used as a framework for investigation. A good example of this approach is the NOAA NOHRSC National Snow Analysis (NSA), which combines elements of a snowpack energy balance point process model, SNTHERM (Jordan 1991), with *in situ* measurements, aircraft and satellite observations, and downscaled analysis and forecast fields from a mesoscale, numerical weather prediction model. Details of the approach can be found at <http://www.nohrsc.nws.gov>. This integrated assimilation approach provides a physically consistent framework for integrating the wide variety of snow data that are available at often different times and works effectively when confidence in the measurement accuracy is high and representativity of the *in situ* measurements and NWP data is well understood. In the US and Europe, generally there are many high quality data sources to drive this approach.

However, in other locations, where *in situ* networks are sparse and/or have large measurements uncertainties, estimates from such an integrated approach potentially might not be significantly better than simple interpolated fields or static satellite measurements. Understanding and quantifying *in situ* and remote sensing uncertainties and errors, therefore, will improve the potential utility of measurement and observation based data in environmental change studies.

Table 4.3 gives a list of selected numerical energy and mass balance models that can be used in the study of the cryosphere to improve our understanding and predictability of cryospheric processes. The list is not exhaustive and also consists of a variety of models at different space and timescales that have been used for cryospheric studies. The reader is advised to follow the internet links to these sites to find out more about the various models and how they are implemented. EISMINT is included because although it is not a model itself, it refers to a larger intercomparison effort that compared several different mass balance models.

4.5 CONCLUSIONS: VALIDATION, COORDINATED PROJECTS, AND CLIMATE DATA RECORDS

Validation of models and remote sensing products is an important aspect of the development of the geophysical products. In the past, the emphasis was often on using *in situ* measurements to validate remote sensing or model estimates. As described earlier, more recently, *in situ* and remote sensing products are frequently being combined through the use of integrated modeling approaches. Validation of these approaches is an important activity and requires careful consideration; if before, validation consisted of comparing, independent estimates of the same geophysical variable, now this is not always the case. Thus, validation campaigns

TABLE 4.3 Selected models, with current references, for cryospheric variable modeling.

Cryosphere variable	Organization	Model	Website
	European Centre for Medium-Range Weather Forecast (ECMWF)	ERA-40, 4D-VAR, and EPS	www.ecmwf.int
	National Center for Atmospheric Research (NCAR)	MM5, WRF	www.ncar.ucar.edu
	NASA Goddard Institute for Space Studies (GISS)	Surface Temperature Analysis	www.giss.nasa.gov
Temperature	NOAA	RUC/Rapid Refresh, GFDL NCEP	www.maps.fsl.noaa.gov www.gfdl.noaa.gov www.ncep.noaa.gov
	UK Meteorological Office Hadley Centre	Numerical Weather Prediction System	www.metoffice.com/research/hadleycentre/index.html
	Canadian Centre for Climate Modeling and Analysis	AGCM, CGCM, CRCM	www.cccma.bc.ec.gc.ca/eng_index.shtml
	NASA Global Monitoring and Assimilation Office (GMAO)	GEOS5	www.gmao.gsfc.nasa.gov
Snow energy and mass balance	US Army Cold Regions Research and Engineering Laboratory (CRREL)	SNTHERM	www.crrel.usace.army.mil
Sea ice	NOAA	National Center for Environmental Prediction sea ice analysis	www.polar.ncep.noaa.gov
	Canadian Ice Service (CIS)	CIOM	www.ice.ec.gc.ca

(continued)

TABLE 4.3 (CONT'D)

Cryosphere variable	Organization	Model	Website
Ice sheets and glaciers	British Antarctic Survey	BEDMAP	www.antarctica.ac.uk/aedc/bedmap
	European Ice Sheet Modeling INiTiative (EISMINT)	various	www.homepages.vub.ac.be/~phuybrec/eismint.html
Permafrost	University of Alaska Fairbanks Geophysical Institute	GIPL 2.0	www.gi.alaska.edu
Runoff	University of Washington, US	Variable Infiltration Capacity (VIC) Macroscale Hydrologic Model	www.hydro.washington.edu/Lettenmaier/Models/VIC/VIChome.html
Integrated land surface model	NOAA NOHRSC NASA GMAO	NSA/SNODAS Land Surface Assimilation	www.nohrsc.nws.gov www.gmao.gsfc nasa.gov
	NASA/GSFC	Global Land Data Assimilation Model, Land Information System	www.ldas.gsfc.nasa.gov www.lis.gsfc.nasa.gov

Note: The list is indicative rather than comprehensive and readers should consult the literature for full details.

have become highly integrated and comprehensive (and more expensive) to execute.

Probably one of the earliest integrated field experiments was the FIFE experiment conducted in Kansas between 1981 and 1987. Although this was not a specific cryospheric environment experiment, it did pave the way for several more recent experiments. BOREAS was conducted through the 1990s and experiments such as SHEBA and CLPX were also comprehensive and fully integrated remote sensing, *in situ*, and modeling experiments. Coordination of these experiments is critical in order that they address specific research questions, and several organizations have taken up a lead role in this respect. For example, the Climate and Cryosphere project (CliC) was established by the World Climate Research Program (WCRP) in March 2000 and in 2004, the Scientific Committee on Antarctic Research (SCAR), became a cosponsor of the project. The goal of CliC is *to assess and quantify the impacts that climate variability and change have on the cryosphere, and the consequences of these impacts for the climate system* (see CliC website <http://www.clic.npolar.no>). In validation terms, CliC is currently playing an important role in the planning for the International Polar Year (IPY), which will be held between March 2007 and March 2009 (<http://www.ipy.org>). The IPY will consist of many research efforts aimed at improving our understanding of the cryosphere and this will entail many field experiments to achieve that aim. IPY will be held 50 years on from the International Geophysical Year which attracted some 80,000 scientists from around the world. It is anticipated that IPY will be as successful, because in many ways, while the science has evolved through the development of

advanced technologies and computational power, the generic issues are the same: concerted research experiments, both in the field and in the laboratory, will stimulate current efforts to advance our understanding of the cryosphere and will enable scientists to be in a position to better predict future changes. By the end of the IPY, scientists will have over 40 years of VIS/IR measurements, more than 35 years of passive microwave observations, and 15 years of SAR measurements of the cryosphere in various distributed data archives. Together with the extended *in situ* measurement record and the many integrated research experiments planned for the IPY timeframe, scientists will be further along in remote sensing product validation such that they will be in a position to develop well specified climate data records, or CDRs. In their 2004 study, the US National Academy of Sciences (NAS) defined CDRs as ". . . a time series of measurements of sufficient length, consistency, and continuity to determine climate variability and change" (NAS 2004). They subdivide CDRs into *fundamental* CDRs (FCDR), which are calibrated and quality-controlled sensor measurement data that have been improved over time, and *thematic* CDRs, which are geophysical variables derived from the FCDRs, such as sea surface temperature and cloud fraction (NAS 2004). Both fundamental and thematic CDRs can be used for significant climate and environmental change research and are how several agencies and government organizations are viewing data sources. Furthermore, with the development of new remote sensing instruments that NASA, ESA, JAXA, etc. are planning for the future, our ability to measure the cryosphere and predict its future behavior will only increase.

5

THE EVIDENCE FOR PAST CRYOSPHERIC CHANGES

5.1 INTRODUCTION

It has been suggested (Bowen 1978) that the growth in understanding of the Quaternary Period is the second of three profound developments in thinking in Earth and environmental sciences that have occurred since the 1960s. The first, global plate tectonics, revolutionized our approach to planetary dynamics over geological timescales and the third, global environmental change, is enhancing our understanding of the ways in which environmental systems are interrelated with human activity. But the second, the detailed reconstruction of Quaternary paleoenvironments has deepened our understanding of environmental change during the period of the evolution of *Homo sapiens*. Much of this understanding revolves around the frequent waxing and waning of the cryosphere; the rapid nature of the cryosphere changes that are implied and an ongoing debate about the relative importance of atmosphere, oceans, and cryosphere in triggering major changes.

These understandings are crucial to the interpretation of present and possible future changes in the following ways:

1 Environmental change is the geological norm at all spatial scales from local to global.
2 It is normal that global cooling occurs relatively slowly and that global warming is relatively rapid. Indeed, in many cases, the rate of warming has been dramatic and unpredictable in timing.
3 There is an implication that environmental systems are rarely in equilibrium with the climatic or the land use drivers that obtain at a specific geological time. It would seem that transient systems are the geological norm. Another way of stating this is that contemporary landscapes are transitional and that it is therefore important to determine the immediate precursor drivers of environmental change.

Although this growth in understanding is extremely critical to an awareness of how the geophysical Earth system can change in the comparative absence of anthropogenic activity, Steffen et al. (2004a) have warned against the use of Pleistocene analogues to interpret the Anthropocene, the contemporary epoch which is increasingly dominated by human activity and is therefore a "no analogue" situation. Historical analysis does produce contextual explanation, suggests causal explanation of global climate trends, documents a library of inadvertent experiments already performed, and provides a narrative of change (Wasson 1994).

There is a further reason for accelerating interest in Quaternary history. It is argued by some (e.g. Hewitt 2000) that the severe and rapid environmental changes imposed by alternating glacial and interglacial stages were responsible for the present genetic structure of ecosystems, including the genetic structure of human populations.

5.2 THE UNIQUENESS OF THE QUATERNARY PERIOD

During the past 2.4 Ma, the oxygen isotope record indicates over 90 stages of alternating cooler and warmer conditions. In spite of recommendations to eliminate the Quaternary Period from the geological timescale (Gradstein et al. 2005), we assume that the Quaternary Period started around 2.4 Ma ago. The implication is that there have been at least 45 warmer and 45 cooler episodes during the Quaternary (Bowen 2004). It is not possible to assume that these were glacial and interglacial stages, though the isotope record contains a temperature and an ice volume component. The isotopic record is a climatic signal of global significance. Not only have there been repeated changes but these changes have been abrupt (Broecker 1994). It is in these respects (frequent and rapid changes of mean global temperature) that the Quaternary Period is unique.

Many of the frequent and rapid Quaternary climate changes can be explained by the Croll–Milankovitch theory of orbital changes and the resulting redistribution of solar radiation around the Earth. There are three components of these orbital processes: orbital eccentricity, axial tilt (obliquity), and precession of the equinoxes (wobbling in the Earth's axis) (Fig. 5.1). The critical point to note is that each component has a predictable periodicity: approximately 100 ka for the orbital eccentricity cycle, 41 ka for the tilt cycle, and a 21 ka for the precession cycle. Although the idea that orbital variations might influence climate on Earth can be traced to the 17th century, the first detailed discussion was spelled out by Croll (1867). In 1920, Milankovitch concluded that orbital eccentricity and precession

produced effects large enough to cause ice sheets to expand and contract and that the climatic effects of obliquity were even greater than Croll had assumed. Between 1950 and the late 1960s, the theory fell into disfavor, partly because it was impossible to check its accuracy against real world data. However, loess sequences in Czechoslovakia (Kukla 1968), sea level changes (Broecker et al. 1968), and oxygen isotope ratios in marine cores (Broecker & Van Donk 1970) provided data sequences that confirmed the 100 ka cycle. Shackleton and Opdyke (1973) provided oxygen isotope and magnetic susceptibility measurements of a Pacific deep-sea core and gave the first accurate chronology of late Pleistocene climate. All frequencies corresponding to orbital forcings were documented by Hays et al. (1976).

5.3 INITIATION OF GLACIAL AGES

Although the Croll–Milankovitch theory predicts cycles of warmer and colder periods, it provides no explanation of a trigger mechanism, it cannot explain synchronous glaciation in Northern and Southern hemispheres nor can it explain the strength of the 100 ka signal (Fig. 5.2) because eccentricity should be the weakest of the orbital forcings. Another curious phenomenon is that prior to 800 ka BP and commencing around 2.4 Ma BP, the 41 ka cycle is the dominant signal in the record (Ruddiman & Raymo 1988). Prior to 2.4 Ma ago, the 21 ka cycles is a more dominant signal.

What the Vostok ice core showed was that the succession of changes through each climate cycle and termination was similar and atmospheric and climate properties oscillated between stable bounds (Fig. 5.2). Atmospheric concentrations of carbon dioxide and methane correlate well with Antarctic air temperature throughout the

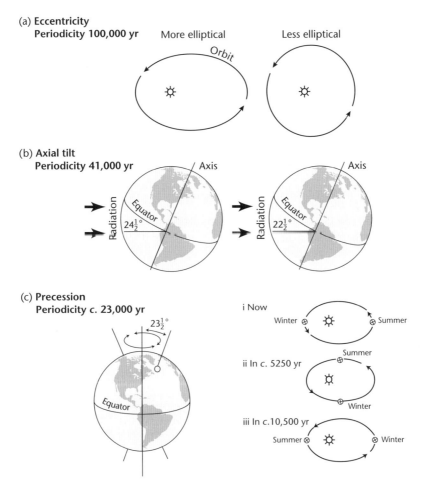

FIG. 5.1 The three components of the astronomical theory of climate change: (a) eccentricity of the earth's orbit; (b) obliquity of the ecliptic; and (c) precession of the equinoxes (Lowe & Walker 1984).

record. And, interestingly, present day concentrations of carbon dioxide and methane are without precedent during the past 420 ka (Petit et al. 1999). The interpretation of the record is that there was a similar sequence of climate forcing during each glacial termination: orbital forcing (with a possible contribution of local insolation changes) followed by two strong amplifiers, greenhouse gases acting first, then deglaciation and ice–albedo feedback. The authors also see a significant role for the Southern Ocean in regulating the long-term changes

of atmospheric CO_2 (Petit et al. 1999). The technology for retrieval of ice cores is illustrated in Plate 5.1.

A deep ice core from Dome C, Antarctic has now provided a climate record for the past 740,000 years (EPICA 2004) identifying eight glacial cycles and extending the marine isotope stages from MIS 10 (Vostok ice core) to MIS 18. The earlier period from 740 to 430 ka shows less pronounced warmth in interglacial periods but a higher proportion of each cycle was spent in the warm mode.

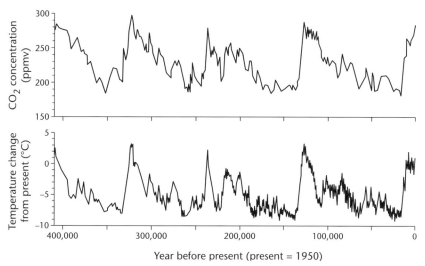

FIG. 5.2 The Vostok ice core records of changes in concentration of CO_2 and temperature (expressed as difference from present). Temperature changes are estimated from changes in deuterium concentration (from Petit et al. 1990).

A trigger mechanism for the onset of glacial conditions seems essential because orbital forcing has always existed and glaciations occur during specific periods of Earth's history. Estimated mean global temperatures declined during the Cenozoic until some critical mean temperature was achieved which allowed ice sheets to grow. But this simply poses the question why temperatures declined. There are numerous hypotheses ranging from variations in solar activity, through CO_2 (and possibly) CH_4 content of the atmosphere, the physical arrangement of the continents and oceans, the presence of large mountain ranges, and the frequency of intense volcanic activity.

Most recently, Ruddiman (2004) has claimed to recognize the effects of human activity in reversing the trends of carbon dioxide and methane concentrations around 8–5 ka BP. His hypothesis is that clearing of the land for agriculture and intensification of land use during the Holocene has so affected the climate as to delay the arrival of the next glacial episode. This is a controversial

hypothesis which requires further testing (Fig. 5.3). But it adds strength to W. Steffen's warning about a "no analogue" situation.

Most recent hypotheses about triggers of glaciation have emphasized feedback mechanisms between sea level, surface albedo, carbon dioxide, ice sheet elevation, and ocean circulation which tend to maintain the present ice house state. Denton (2000) proposed that the ocean conveyor system oscillates with a periodicity of 100 ka and therefore may act as a control on ice sheet extent over these periods. The idea is that if the thermohaline circulation is stopped during the early phase of onset of glaciation, it cannot reinstate itself. If the ocean conveyor is stopped for a long time, ice sheets can build up, dense salty deep water can only start to form when the calving of icebergs into the Arctic Ocean is reduced.

A major complication in thinking about the influence of Northern Hemisphere solar changes on the growth and decay of ice has occurred from Henderson and Slowey's (2000) calculation of the date of the last

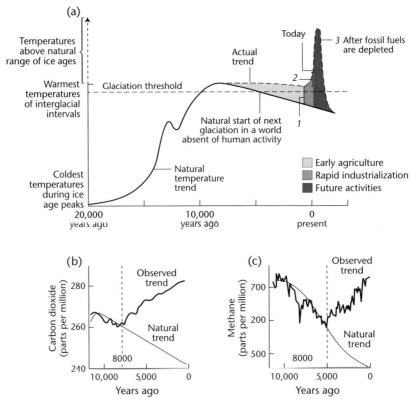

Fig. 5.3   (a) Suggested effects of (1) early agriculture (2) rapid industrialization (3) future human activities on global mean temperature; (b) reversal in CO$_2$ concentration trend around 8,000 years BP; (c) reversal in methane concentration trend around 5,000 years BP (from Ruddiman 2005).

interglacial at 135 ka BP ago. This is a date when solar insolation was maximum in the Southern Hemisphere and minimum in the Northern Hemisphere. It is speculated that feedbacks between the Pacific Ocean, insolation, and previously neglected aspects of the Antarctic Ice Sheet could perhaps be responsible. At the very least, it reminds Northern Hemisphere scientists of the comparative neglect of consideration of the role of the Southern Hemisphere. It is perhaps instructive to look briefly at the incidence of Ice Ages through geological history (Fig. 5.4).

The mean periodicity of glaciations is about 155 Ma. Although it is true to say that alternations of glacial and nonglacial stages of climate have been uniquely frequent during the Quaternary Period, it is also the case that glacial episodes have occurred with some regularity during the geological past. The world climate system appears to have switched regularly from a glacial to a nonglacial mode (or from icehouse to greenhouse state; Fischer 1981). The questions then are why the switch and why are the switches periodic. The disposition of the continents is a favored hypothesis because land at high latitude would seem to be a prerequisite for the onset of an ice age. Simulations have shown that when land masses concentrate in polar regions global mean temperature is up to 12°C lower than when they are massed around the Equator.

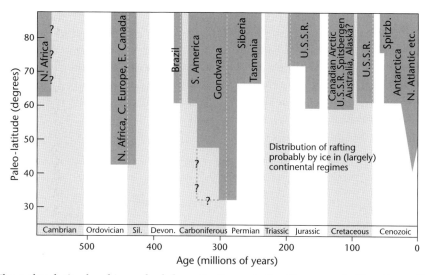

FIG. 5.4   The paleo-latitude of ice rafted deposits through the Phanerozoic Eon (from Frakes & Francis 1988).

Beaty (1978) had a particularly compelling version of this general hypothesis. He proposed that if a large landmass could be generated at a latitude high enough for snow to accumulate, an ice age would commence. Thereafter, glacial and interglacial cycles would occur according to astronomical forcing until land had moved far enough away for the polar regions to be unable to support ice sheets. On this basis, the Earth is presently locked in an icehouse state. The triggering mechanism is conceived as the increase in the Earth's albedo, owing to the chance occurrence of intense volcanism and a slight reduction in solar output. This icehouse state is self-maintaining through increased snowfall in critical areas and an averagely high albedo.

5.4   RECONSTRUCTING THE EXTENT OF GLACIAL ENVIRONMENTS

Whatever the causes of initiation of an ice age, the growth in understanding of Quaternary history over the past four decades has been exponential. The extent and thickness of former glaciers and ice sheets has most commonly been estimated by field mapping of landforms of glacial erosion and deposition. Patterns of former ice flow direction are often reconstructed through measurements of the distribution of glacial erratics and by glacial striae orientations. The principal morphological features that have been used to define the positions of former ice margins are lateral and terminal moraines, fluvioglacial outwash plains, ice marginal meltwater channels, as well as the extent of drift sheets. Within the outermost limits of a particular glaciation, the presence of linear moraines may either indicate stages in the overall retreat and thinning of the ice or they may represent periods of glacier readvance. In many instances, it is difficult to estimate the size of former ice sheets and valley glaciers since ice marginal moraines may be absent. This can arise when the ice was relatively clean or when the ice margin did not exist in a particular area for a sufficient length of time to allow the construction of moraines or when the ice calved directly into ocean or lake waters. The vertical extent of former ice sheets and glaciers is frequently aided by the distribution of periglacial landforms. In land areas located beyond or above the area directly affected by glacier ice, the boundaries between glaciated topography and the lower limit of frost shattered bedrock and patterned ground phenomena known as trim lines can be used to define the surfaces of former ice masses (Ballantyne 1990). Recognition of weathering zones is a useful method of correlating glacial limits where there is limited datable material as for example in the Torngat Mountains, Labrador (Evans & Rogerson 1986).

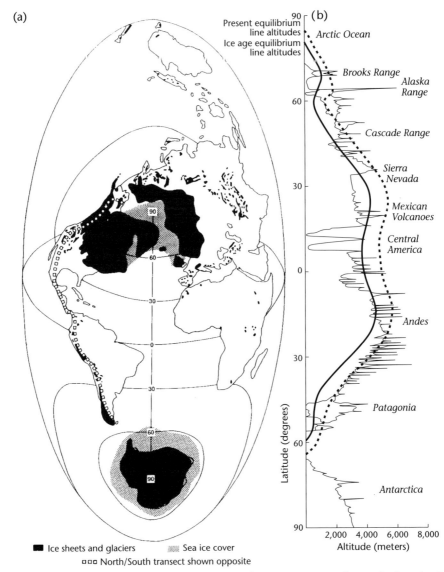

FIG. 5.5   (a) Distribution of the principal ice sheets and mountain complexes during the LGM. (b) The global lowering of ELA during LGM (from Dawson 1992).

TABLE 5.1 Relationship between volumes of last ice sheets and global sea level lowering according to the minimum and maximum ice reconstructions of Denton and Hughes (1981). The values assume isostatic equilibrium and a rock–ice ratio of four.

Ice volume	Minimum reconstruction		Maximum reconstruction	
	Total volume (10^6 km³)	Volume causing lower sea level (10^6 km³)	Total volume (10^6 km³)	Volume causing lower sea level (10^6 km³)
Ice sheets				
Laurentide	30.900	30.500	34.800	34.200
Cordilleran	0.260	0.260	1.900	1.840
Innuitian			1.130	0.983
Greenland[*,†]	2.920[*]	0.287[†]	5.590[*]	2.550[†]
Greenland[†]	0.287[†]	0.287[†]	2.950[†]	2.550[†]
Iceland	0.050	0.050	0.267	0.236
British Isles	0.801	0.773	0.801	0.773
Scandinavian	7.250	7.060	7.520	7.320
Barents-Kara	0.955	0.865	6.790	6.250
Putorana			0.581	0.581
Antarctica	37.700	9.810	37.700	9.810
East	24.200	3.330	24.200	3.330
West	13.500	6.480	13.500	6.480
Glaciers and ice caps	1.184	1.830	0.750	0.715
Totals	84.174	51.300	97.829	65.300
Sea level equivalent[§]		127 m		163 m
Eustatic sea level drop[¶]		91 m		117 m

[*] Total Late Wisconsin–Weichselian ice.
[†] Additional Late Wisconsin–Weichselian ice contributing to lower Wisconsin–Weichselian sea level.
[§] Without hydroisostatic sea floor rise, using present ocean area (361×10^6 km²).
[¶] With hydroisostatic sea floor rise.

A major difficulty in calculating the dimensions of ice sheets is that, because ice covered most of the land area, the position of the surface of former ice sheets is not defined by any morphological evidence. Even where there is ample evidence to delimit the lateral extent, ice sheet profiles are rarely available. Another difficulty is that most morphological evidence for earlier ice sheets has been removed by erosion by later ice advances. Morphological evidence for the retreat from the furthest point of ice sheet advance can usually be mapped accurately. But stratigraphic information and accurate datings are needed to interpret earlier ice sheets.

In spite of rapid progress, there are still uncertainties about precise limits of ice during the Quaternary. One noted interpretation is reproduced as Fig. 5.5. This figure should be read in association with Table 5.1 (from Dawson 1992), which indicates the minimum and maximum reconstructions of ice volumes. There is a difference of 13.65 (10^6) km³ or a 15% discrepancy between minimum and maximum ice volume reconstructed.

Complications associated with glacio-isostasy (or the depression and rebound of the Earth's crust during and following continental ice sheet cover); glacioeustasy (changes in sea level resulting from removal of water from the oceans during glacier expansion and its return by glacier melting); hydroisostasy (crustal loading of the oceans causing sea level change); and geoidal eustasy (the geoid changes position and shape responding to changing ice sheet distribution and large accumulations of sediment in deltas such as those of the Mississippi and Ganges) have to be factored into any attempted reconstruction of past ice extent (Fig. 5.6). Within areas occupied by the major Pleistocene ice sheets, isostatic uplift effects dominate (Fig. 5.7). Figure 5.8

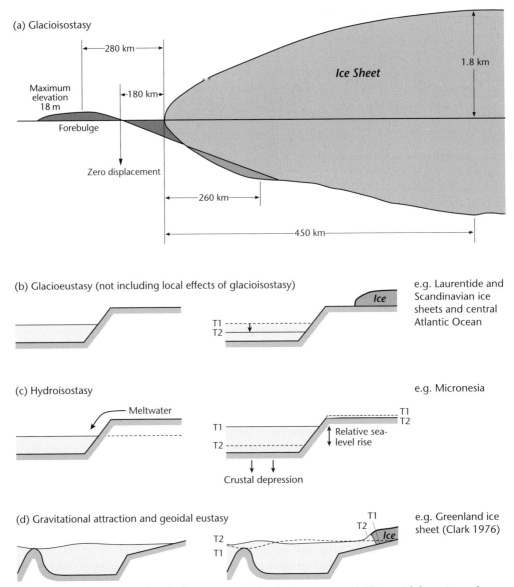

FIG. 5.6 (a) Glacioisostasy, (b) glacioeustasy, (c) hydroisostasy, and (d) geoidal eustasy (from Dawson 1992).

Fig. 5.7 Raised marine deltas along Ellesmere Island. Successively lower surfaces record postglacial emergence following deglaciation. Sea ice in the right foreground (photo by John England).

shows regional isobase maps for eastern Canada, Scandinavia, and Scotland. These maps plot the total amount of uplift minus the glacioeustatic component of sea level change and are a good indicator of the location of former ice sheet dispersal centers during glaciations. The pattern of uplift in Canada suggests at least three loading centers whereas rebound patterns in Scandinavia and Scotland are more simple and domed.

If one examines postglacial sea level curves around the coast of Britain an interesting effect becomes apparent (Fig. 5.9). In the south of the country, sea level change has been dominated by glacioeustatic submergence. Such isostatic rebound as did occur was exceeded by eustatic sea level rise, resulting in net sea level rise. In Scotland, by contrast, where the ice was thickest, the sea level curves record early emergence, while local isostatic uplift exceeded global eustatic sea level rise. A period of relative sea level rise occurred between 8 and 5 ka BP.

During the last two decades, interest in late Quaternary paleoclimate and paleoceanography has grown to the point where detailed environmental changes can be estimated (e.g. Fig. 5.10). It is now well documented that rapid changes in ocean circulation and atmospheric conditions occurred throughout the last glacial period. Well dated high resolution sediments and uplifted coral reefs have proved particularly helpful. Continental ice volume changes since the last glacial maximum (LGM) around 18 ka BP (^{14}C years) have been reconstructed from ^{18}O in benthic foraminifera (Duplessy et al. 1986) and from drowned coral reefs off Barbados (Fairbanks 1989). Maximum sea level lowering occurred around 18 ka BP (^{14}C years) at 121 ± 5 m. Sea level then rose slowly by about 20 m over the next 5 ka, followed by very rapid rise centering on 12 ka BP corresponding to a major meltwater pulse. This was followed by a slower rate of rise and then a second major meltwater pulse centered around 9.5 ka BP (^{14}C years). Almost all of these meltwater pulses were

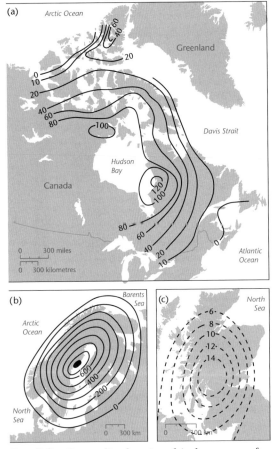

Fig. 5.8 Generalized regional isobase maps for (a) Canada, showing shoreline emergence in eastern Canada since *c.* 6,000 years BP, (b) Scandinavia showing patterns of absolute uplift during the Holocene, and (c) Scotland, showing shoreline emergence for the Main Postglacial Shoreline since *c.* 7,000–6,000 years BP (Andrews 1970; Morner 1980; Lowe & Walker 1984).

Jansen 1992; Fig. 5.11) show five major climatic episodes. (a) a Bolling-Older Dryas-Allerod interstadial from 13.2 to 11.2 ka BP (^{14}C years); (b) Younger Dryas from 11.2 to 10.2 ka BP (^{14}C years); (c) Younger Dryas II from 10.2 to 9.5 ka BP; (d) the warmer PreBoreal, Boreal, and Atlantic chronozones 9.5–5 ka BP; and (e) the cooler Sub-Boreal 5–4 ka BP.

5.5 EXTREME EVENTS

Thus the two major meltwater pulses occurred at the beginning of the Allerod and at the close of the Younger Dryas. There is a continuing discussion on the major sources of these meltwater pulses, in particular the relative importance of North America and Eurasia and also the question of the role of the Southern Hemisphere in these changes, whether causative or simply responding to changes initiated in the Northern Hemisphere.

There is currently a debate concerning the timing of climate events in each hemisphere. Opinions are divided as to whether Antarctic records lag or lead those from Greenland. Blunier et al. (1998) suggested that warm peaks in the oxygen isotope records from the Byrd and Vostok ice cores at about 36 ka and 45 ka BP occur about 1 ka earlier than the comparable records in Greenland (Fig. 5.12). However, Mazaud et al. (2000) suggest that short-term changes in Antarctic paleoclimate may be forced by alterations to North Atlantic conditions. Identification of teleconnections is a major research area.

Clark et al. (1999) have studied the interaction between ice, ocean, and atmosphere during the last deglaciation and focussed on the recognition of two periodic oceanic phenomena: Dansgaard–Oeschger (D–O) and Heinrich events (Fig. 5.11). Heinrich events are thought to be the periodic release of

discharged to the world ocean via the North Atlantic Ocean. Bard et al. (1990) calibrated the radiocarbon record by thorium/uranium dates which located the two meltwater pulses at 14 ka and 11.3 ka BP calendar years (11.8 ka and 10 ka ^{14}C years).

Reconstruction of sea surface temperatures from diatom assemblages in sediments from off the Norwegian coast (Koç Karpuz &

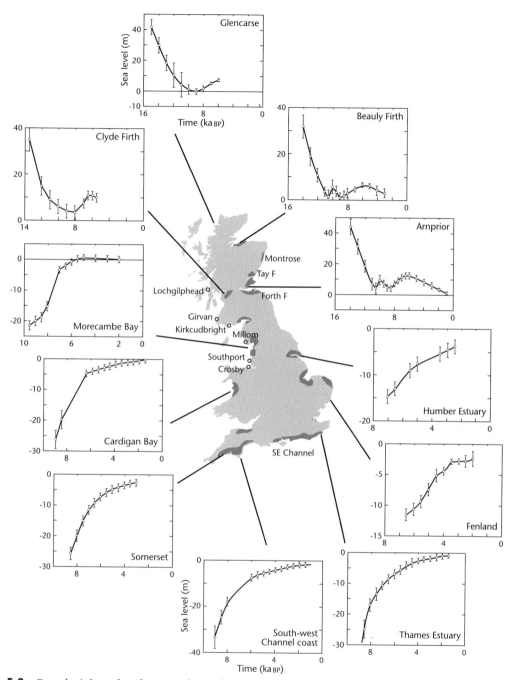

Fig. 5.9 Postglacial sea level curves from the coast of Britain demonstrating the influence of glacioisostasy in the north (emergence) and glacioeustasy in the south (submergence) (from Lambeck 1995).

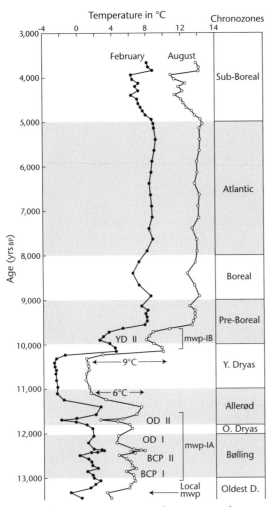

FIG. 5.10 Reconstruction of August and February SST's in the southeast Norwegian Sea over the last 13.4 ka BP (^{14}C years) (from Koç Karpuz & Jansen 1992).

icebergs from the Laurentide Ice Sheet. The events were detected from measurement of the ice rafted debris in the North Atlantic. They have a periodicity of about 7 ka and may be linked with the internal dynamics of the Laurentide Ice Sheet. There is also a decrease in salinity during periods of ice rafted debris formation. In contrast to the Heinrich events, Dansgaard–Oeschger events are millennial scale periods of cooling, probably associated with meltwater and

iceberg production (Bond et al. 1997). Although these events are distinct, they both affect ocean circulation and, because of this, climate. The meltwater released by iceberg discharges and surface runoff is thought to reduce the thermohaline driven formation of North Atlantic Deep Water, the force behind the modern ocean circulation system. The sudden loss of warm Atlantic waters that usually replace the NADW will be reflected in the cooling of the atmosphere.

It is also possible that the ocean system may oscillate freely at a 1.5 ka timescale, triggered initially by a meltwater event. Under this idea, according to Alley and Clark (1999), the Northern Hemisphere is able to cool at the same time as parts of the Southern Hemisphere are warming. This is because if melting events cause the shutdown of NADW, it may be that the ocean demand for deepwater formation must come from elsewhere, such as the Southern Hemisphere. This then may initiate warm surface waters southwards, warming the climate.

Alley and Clark (1999) also suggest that changes in the ocean circulation caused by a D–O event (and possibly an H-event) could be responsible for the Younger Dryas climate reversal around 11.2 ka BP (^{14}C years). The sudden switch off of the ocean conveyor caused SST's in the North Atlantic to become much colder resulting in colder conditions over adjacent landmasses. This had the effect of reversing the deglaciation that had been going on since the LGM, and glaciers began to readvance.

There is another process by which freshwater may be input to the North Atlantic: the discharge of proglacial lakes from North America during the last deglaciation. Lake Agassiz, for example, discharged enough water into the ocean at 8.2 ka BP (calendar years) that sea level was increased by 20 cm in a matter of days. This freshwater input

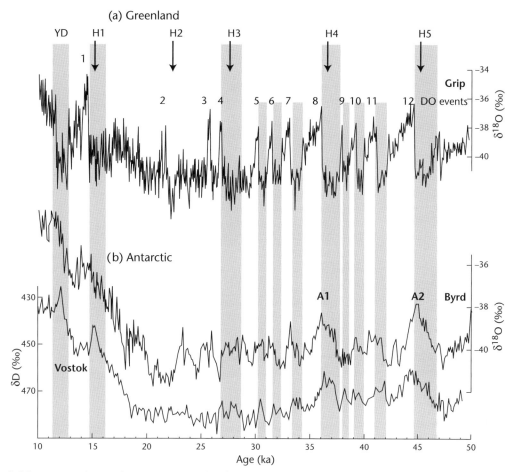

FIG. 5.11 A correlation between Greenland (GRIP) and Antarctic (Byrd & Vostok) ice core records. Heinrich and Dansgaard–Oeschger events are indicated in gray shade (from Mazaud et al. 2000).

correlates with a marked cooling event in Greenland ice cores suggesting a link between proglacial lake discharge, ocean circulation, and climate (Barber et al. 1999).

Heinrich events are indeed connected with rapid climatic variations in the North Atlantic region. The timing of events coincides closely with the pattern of climate fluctuations documented from ice cores. But the mechanism that drives them remains a matter of debate. Broecker (1994) has reviewed the possible mechanisms. It is useful at this point simply to review the relation between Heinrich events, D–O cycles,

and Bond cycles. The so-called D–O cycles were recognized from the Greenland ice core record (Dansgaard et al. 1982; Oeschger et al. 1984). These D–O cycles were characterized by abrupt jumps in temperature, dust content, ice accumulation rate, and concentration of methane. D–O events of the last glacial cycle (12 of them) have been dated around 14, 22, 25, 26, 30, 32, 34, 36, 38, 40, 41, and 44 ka (calendar years) BP.

Heinrich events are recognized from ocean bottom sediments rather than from ice cores. The sediments are characterized by a relatively high proportion of detrital

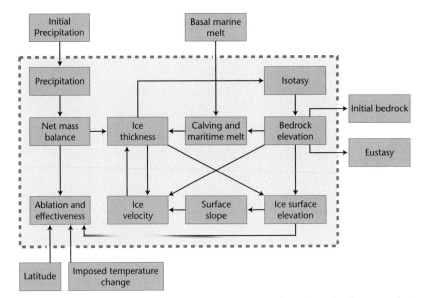

FIG. 5.12   Flow chart to show the interaction of parameters that describe large-scale ice sheet behavior. Note the feedback process focusing on isostasy (from Siegert 2001).

limestone, the clay sized minerals much older than those in adjacent glacial sediments and by a dearth of foraminiferal shells. This last characteristic is an indication of a dramatic decrease in marine productivity. Finally, these Heinrich event sediments thin by more than an order of magnitude from the Labrador Sea to the European end of iceberg route around 46°N. These sediments have been attributed to deposition from the bottom of iceberg armadas resulting from the catastrophic break up of the margins of the Laurentide Ice Sheet. Heinrich events of the last glacial cycle (5 of them) have been dated around 15, 23, 27, 36, and 46 ka (calendar years) BP.

There is a link between Heinrich events and D–O cycles. Bond et al. (1993) showed that the ice armadas were launched during the most intense cold phase of a package of several D–O cycles, and that each Heinrich event was followed by a prominent warming, which then initiated a new package of D–O cycles. This bundling of progressively colder D–O cycles into subsets is now called a Bond cycle.

The last of the D–O cold events is the so-called Younger Dryas and it is suggested by Broecker (1994) and Alley and Clark (1999) that it is closely associated with another Heinrich event. The abrupt end of the YD, around 11.5 ka BP, ushered in the Holocene, a period of relatively stable warm climate which continues today.

5.6   ICE SHEET MODELING

Ice sheet modelers ask the question "How can ancient ice sheet shape, size, and dynamics be reconstructed?" The principle behind numerical ice sheet modeling is that an ice sheet can be divided into a number of ice columns. Each of these columns represents a cell in the model's 2-D grid. Ice sheet models are usually arranged in a loop that begins by applying a series of algorithms, determining the flow of ice, mass balance, and interactions with the Earth in each cell. The loop is completed by a final continuity equation used on the full grid to calculate the interaction and flow of ice between cells.

One loop of the model is called an iteration. The accuracy of the model is dependent on the width of the grid cells and the time period which a single iteration represents. For continental scale ice sheet models, where the time-dependent change in ice sheet behavior is calculated over several thousand years, the grid cell width is usually between 5 and 25 km, and the iteration time step is between 1 and 10 years.

Most ice sheet models are centered around the continuity equation for ice (Mahaffy 1976) where the time-dependent change in ice thickness is equated to the specific net mass budget of a cell:

$$\frac{dH}{dt} = b_s(x, t) - \Delta \cdot F(u, H) \qquad (5.1)$$

where $F(u, H)$ is the net flux of ice from a grid cell in $m^2 \cdot a^{-1}$. The depth averaged ice velocity (u $m \cdot s^{-1}$) is calculated by the sum of depth averaged internal ice deformation and basal motion. The specific mass budget term (b_s) is a function of a number of processes including ice sheet surface mass balance, iceberg calving, and ice shelf basal melting. The flux between grid cells is determined by the ice thickness and the vertically averaged ice velocity of the grid cell in question and also of neighboring grid cells. Numerical methods that can be used to solve this partial differential equation are described in, for example, Press et al. (1989).

Numerical ice sheet models which detail the growth and decay of the late Quaternary Ice Sheet require the following inputs and boundary conditions. First, the subglacial topography on which the ice sheet has formed. As the ice grows, this topography will be altered by the action of glacioisostasy. Second, a time and spatially dependent function of climate over a glacial cycle is required to formulate rates of accumulation and ablation of ice as well as air temperature. Third, the time-dependent change in sea level has to be calculated. Parameters describing rates of iceberg calving and ice shelf melting

are also required. The relation between the parameters of a modeled ice sheet system can be summarized in a flow diagram (Fig. 5.12).

5.6.1 THE ANTARCTIC ICE SHEET

Antarctica is made up of two main grounded ice sheets: the West Antarctic and the East Antarctic ice sheets, a relatively smaller ice cap across the Antarctic Peninsula and dozens of glaciers located in mountainous regions at the margins of the continental landmass. The present volume of ice is about $3 \cdot 10^7$ km^3, of which 83% is in the East Antarctic Ice Sheet. The maximum surface elevation is just over 4,000 m in East Antarctica. Maximum ice thickness varies between 2,800 and 4,500 m. Ice is drained from the domes of the ice sheet by fast flowing rivers of ice known as ice streams, where the dominant method of flow is by sliding and/or basal sediment deformation. These issue grounded ice into numerous floating ice shelves that surround the grounded ice continent. Icebergs, usually of tabular form, are calved out from the marine margin of ice shelves, where ice thickness is usually around 250–300 m. This process is the most important mechanism by which ice is lost from the ice sheet, accounting for 75–85% of ice loss (Paterson 1994).

There is controversy over the stability of the Antarctic Ice Sheet during the past 30 Ma. The two dominant positions are (a) that the ice sheet has remained relatively stable over this time period and (b) that there have been a number of phases of growth and decay (Miller & Mabin 1998). In the first case, it is argued that the ice sheet was initiated by the breaking away of the Antarctic continent from South America, at which point the system of circumpolar currents and winds isolated the weather systems over the Antarctic continent from the rest of the planet. Land based evidence of uneroded ash beds 15 Ma old would support this position. On the other hand, marine based diatoms from the Transantarctic Mountains suggest fjord or lacustrine environments associated with unstable ice sheet behavior.

There is also controversy over the details of changes in the Antarctic Ice Sheet from the LGM to present. We will therefore focus on the results of the application of an ice sheet model by Huybrechts (1990, 1992), a model which is well respected by the scientific community. The model starts from a modern type ice sheet established by modeling contemporary environmental conditions. In order to reproduce the LGM ice extent, the ice sheet is subjected to shifts in imposed accumulation rates, surface temperature, and sea level fall. When a temperature reduction of 10°C and a 50–60% reduction of Holocene surface snow accumulation rates are imposed, relatively little ice sheet change occurs. But lowering the sea level by 130 m causes grounding of the ice shelves and expansion of ice sheet limits.

Ice sheet response times show that ice sheet stabilization takes thousands of years. Therefore ice sheet responses will lag climate changes. Not only so, but the actual ice sheet probably never reaches a steady state. It becomes doubly important to input to the model time dependent variations in environmental conditions (Fig. 5.13). The following scenario for ice sheet history results. Ice sheet growth and decay are affected mostly by variations in global sea level. At 30 ka the ice sheet expanded quickly in response to global sea level fall until the LGM. LGM ice expansion involved the migration of the grounding line to the continental shelf break in most places. The major present day ice sheet domes and divides were maintained throughout the glacial cycle. The modeled ice sheet extent at the LGM is consistent with that determined by the CLIMAP group (1981) but ice thickness is much less. The latter point is important as it leads to an estimate of sea

level reduction of only 50% of that predicted by CLIMAP.

Deglaciation probably occurred at around 12 ka BP in response to sea level rise. Ice decay was probably completed by 6 ka BP, although a final phase of ice decay after this time has been suggested. If this latter point is correct, there are implications for the present stability of the West Antarctic Ice Sheet.

The West Antarctic Ice Sheet is a marine ice sheet with part of its ice being grounded on land below sea level, and part in the form of floating extensions called ice shelves. Based on evidence from lake sediments of a Pleistocene warm period in the Antarctic, Mercer (1968) concluded that the West Antarctic Ice Sheet may have completely disappeared at least once and his findings led to concern that global warming might cause the ice sheet to collapse. In contrast, other major ice sheets are either largely grounded above sea level and are gradually ablating (Greenland) or have no clear history of major rapid ice mass changes in the recent geological past (Oppenheimer 1998).

The question of stability of the West Antarctic Ice Sheet is important because its turnover time is so long (on the order of 10 ka) that secular changes in the mass balance could scarcely make a major contribution to sea level rise during the next few centuries. A better understanding of bedrock and grounding line morphology, ice velocity fields, and improved models of the transition zone where the ice begins to float will be needed before conclusive results are achieved. In a model by Alley and MacAyeal (1993), the collapse of the West Antarctic Ice Sheet was predicted over the next few centuries independent of anthropogenic climate change.

Oppenheimer (1998) lists three scenarios that are plausible under the IPCC IS92 greenhouse gas emission projections: gradual dynamic response; no dynamic response;

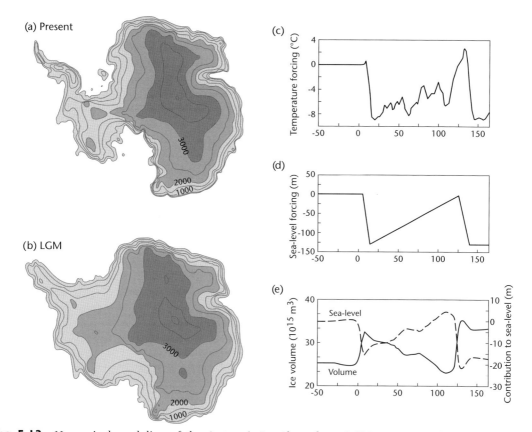

FIG. 5.13 Numerical modeling of the Antarctic Ice Sheet from LGM to present: (a) Present day ice sheet; (b) LGM Ice Sheet; (c) temperature change used to force the model; (d) sea level change used to force the model; and (e) time-dependent change in ice sheet volume and contribution to global sea level (Huybrechts 1990, 1992).

and very rapid dynamic response. He judges that the first scenario is the most likely. Under this scenario, total rates of sea level rise beyond year 2100 would be double or triple the median values projected by the IPCC for the 21st century. Global average rates of rise would exceed the recent historical values by about an order of magnitude and be comparable to today's effective rates in subsiding deltaic areas such as Bangladesh. "Such rates of change pose a serious challenge to societal adaptive capability and could prove devastating to coastal ecosystems" (Oppenheimer 1998, pp. 330–1).

5.6.2 GREENLAND

Present volume of the Greenland Ice Sheet is about 2.8 km³ and has 10% of the Earth's glacierized surface area (1.75 (10⁷) km²). The ice thickness at the center of the ice sheet is over 3,200 m but nunataks are widely exposed above the ice surface around the edge of the subcontinent. The modern climate setting of Greenland is very cold and dry in the north, and warmer and moist conditions prevail in the southeast. The climate is likely to have been significantly different during the last glacial. It is the lower elevations of Greenland that experience the greatest ice sheet response while the higher elevations show little change. This is especially true in the south where changes to the ELA and variability in the accumulation of snow influence ice sheet mass balance (McConnell et al. 2000).

Although the flow of ice within the central regions of the Greenland Ice Sheet is only a few meters per annum, the velocity of ice in the outlet glaciers is hundreds to thousands of meters per annum. The fastest known outlet glacier, Jakobshaven Glacier, has a surface velocity of between 1 and 7 km·a^{-1} (Iken et al. 1993). Since 1996, there has been widespread glacier acceleration south of 66°N and even up to 70°N by 2005 (Rignot & Kanagaratnam 2006). Accelerated ice discharge in the west and particularly in the east of Greenland has doubled the ice sheet mass deficit in the last decade from 90 to 220 km^3 a^{-1}.

Only southern Greenland experienced a significant growth in ice extent and thickness, with expansion on to the continental shelf (Fig. 5.14). Evidence for an enlarged Greenland Ice Sheet at the LGM comes from geological data at the margins of the island and glaciological modeling studies of the ice sheet. Uplifted domes across the northern, western, and eastern edges of the island, documented from marine terraces and trimlines, demonstrate the manner of ice sheet advance in those areas. The Greenland Ice Sheet flowed across the rim of mountains into fjords and across the continental shelf in several places. Interior ice sheet growth was limited to the south of the ice sheet. The position of the southern ice sheet margin is controlled by the ELA. Air temperature depression of as much as 23°C would have caused the ELA to lower and the ice sheet to grow. Ice thickness in the center of the ice sheet probably only fluctuated by a few tens of meters. During the last glacial cycle, the largest fluctuation in the size of the ice sheet was prior to 100 ka ago

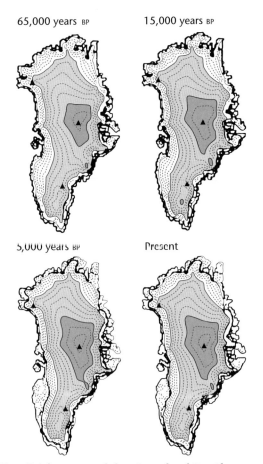

FIG. 5.14 Maps of the Greenland Ice Sheet surface at several oxygen isotope stages during the last 65 ka. The date of the modeled ice sheet and the oxygen isotope stage are indicated for each panel. Triangles refer to the locations of Camp Century, Dye 3, and Summit ice core sites. Contours at 200 m intervals (from Letreguilly et al. 1991).

when two warm periods caused the southern end of the ice sheet to decay.

The Greenland Ice Sheet was connected to the Innuitian Ice Sheet in the north by an ice stream within the Nares Strait. During ice sheet decay, ice core records show that climate over Greenland was highly variable. This may be due to complex ice–ocean–atmosphere interactions between large decaying ice sheets as well as the effects of

Dansgaard–Oeschger and Heinrich events that moved through the Davis Strait.

5.6.3 NORTH AMERICA: INNUITIAN, LAURENTIDE, AND CORDILLERAN ICE SHEETS

At the LGM, the Laurentide Ice Sheet was the second largest in the world (1.6 (10^7) km²). In its coalescence with the Cordilleran and Innuitian ice sheets the ice cover over North America was more than one third of the Earth's ice cover (Plate 5.2). All that remains today are the discrete ice caps of Alaska, British Columbia, and Yukon and Nunavut territories.

The Cordilleran Ice Sheet at the LGM extended from the Aleutian Islands through the Alaska Ranges and the Coast Mountains of British Columbia as far as Olympia in Washington State and eastwards as far as the Rocky Mountain Foothills in Alberta. There was also an independent ice cap over the Brooks Range in Alaska. The Laurentide Ice Sheet at the LGM covered the Northwest and Nunavut territories, all of the Canadian provinces except British Columbia and extended as far south as Des Moines, Iowa. At this latitude, around 41° 30′N, the ice sheet encountered unfrozen sediments and these basal deforming sediments would have resulted in rapid ice sheet flow and the formation of ice lobes. These lobes, typically represented by the James, Des Moines, and Michigan lobes are the southernmost evidences of continental ice sheets during the Quaternary Period.

Dyke and Prest (1987) and Dyke et al. (2002) have reconstructed the Laurentide, Cordilleran, and Innuitian ice sheets for 18 ka BP. At that date, the Laurentide Ice Sheet had three areas of inception: the Keewatin, Labrador, and Baffin sectors; the Cordilleran had three areas of inception, the Alaska, Coast Mountains, and Rocky Mountain sectors and the Innuitian Ice Sheet was probably initiated on Ellesmere Island (Plate 5.3, Fig. 5.15).

The southern margin of the Laurentide and Cordilleran ice sheets was an ablation zone. During the main phase of deglaciation after 13 ka BP, the surface meltwater issued by the ice sheet ran off to the proglacial zone where some of it was routed through the Mississippi. By 11 ka BP, further deglaciation resulted in the formation of the larger Post Algonquin Lake, which overflowed into the Champlain Sea, and the reversal of drainage of an expanding Lake Agassiz into the Great Lakes. During its 4 ka history, Lake Agassiz played a significant role in the deglaciation of the Laurentide Ice Sheet. At around 8.2 ka BP, ice sheet stagnation was accompanied by catastrophic drainage of Lakes Ojibway and Agassiz into the Hudson Bay. Up to 1.5 (10^5) km³ of water are estimated to have been released and this sudden influx of fresh water into the North Atlantic affected the salinity of surface waters, the formation of NADW, and climate.

The LGM ice extent in North America is extremely well known from geomorphological evidence. Terminal moraines and proglacial lacustrine features have been extensively mapped. They show that the margin of the ice sheet was characterized by a series of lobe advances and a multitude of proglacial lakes. One example of the careful mapping of the Cordilleran Ice Sheet margin is provided by Kovanen and Slaymaker (2004a) (Fig. 5.16). They mapped large-scale glacial imprints of the Okanogan Lobe which diverted meltwater and floodwater along the ice front contributing to the Channeled Scabland features during the late Wisconsinan. The glacial imprints suggest a record of surge behavior of the former Okanogan Lobe based on a comparison with other glacial landsystems. Lidar data have raised new possibilities for mapping relict shorelines and ice-flow patterns and some of that potential has been realized in the interpretation of the northern Puget Lowland, Washington State (Kovanen & Slaymaker 2004b).

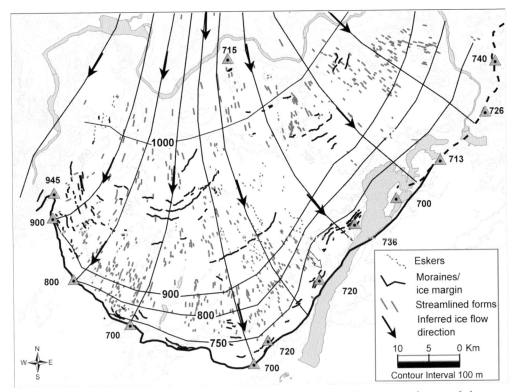

FIG. 5.15 Configuration of the ice flow system and ice surface in the terminal area of the Okanogan Lobe at the LGM (Kovanen & Slaymaker 2004a).

The periodic issuing of large volumes of icebergs from the Hudson Strait into the North Atlantic is recorded in anomalous layers of ice rafted debris on the ocean floor. These so-called Heinrich layers have a periodicity of about 7 ka. There have been several explanations for the production of Heinrich events. The most widely accepted is that the Laurentide Ice Sheet underwent surge type oscillations, termed the binge-purge model. During the binge phase, the ice sheet builds up over a frozen but warming base until the temperature reaches the pressure melting point. Once this is attained, the dynamics of the ice sheet are altered and the purge begins, involving an increase in ice velocity and the transfer of ice from the central regions to the iceberg calving front. The purge ends when so much ice has been lost that the ice sheet regains its frozen base.

The decay of ice in North America resulted in over 60 m of global sea level rise. Most of this was routed into the Atlantic either as icebergs or runoff. Paleoclimatic records from the Greenland Ice Sheet appear to indicate that air temperatures were highly changeable over the North Atlantic during deglaciation with frequent D–O events. This has led some to link climate variability with the production of glacier meltwater from North America.

5.6.4 BRITISH ISLES, SCANDINAVIAN, AND BARENTS ICE SHEETS

The majority of ice that remains within the Eurasian Arctic is found in Svalbard,

PLATE 1.1 The Mustagata ice dome (7546 m) viewed from the west near Shubashi Pass (4070 m), East Pamir (photograph by J.-P. Peulvast).

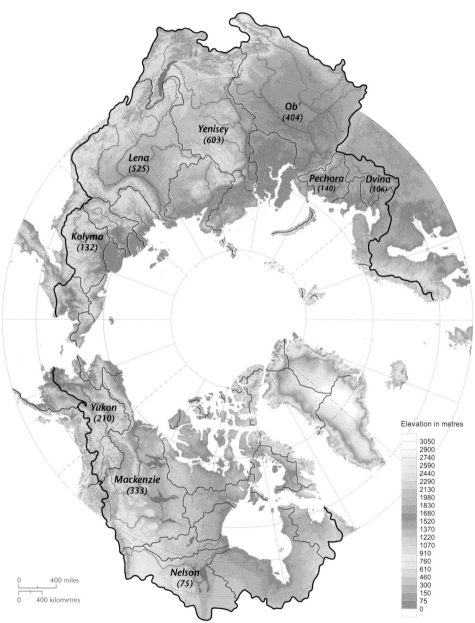

Elevation in metres

3050
2900
2740
2590
2440
2290
2130
1980
1830
1680
1520
1370
1220
1070
910
760
610
460
300
150
75
0

PLATE 1.2 River basins and runoff contributions (in km³ a⁻¹) to the Arctic Ocean (Woo 1996).

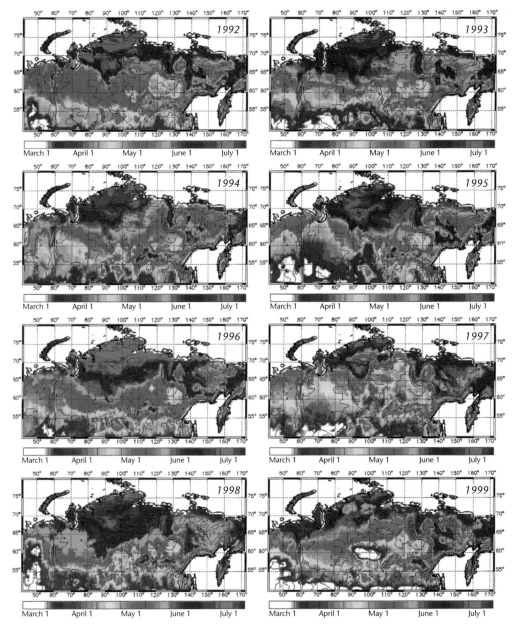

PLATE 1.3 Maps depicting the onset of thawing from ERS-1/2 scatterometer data (1992–99) over Siberia (Duguay et al. 2005).

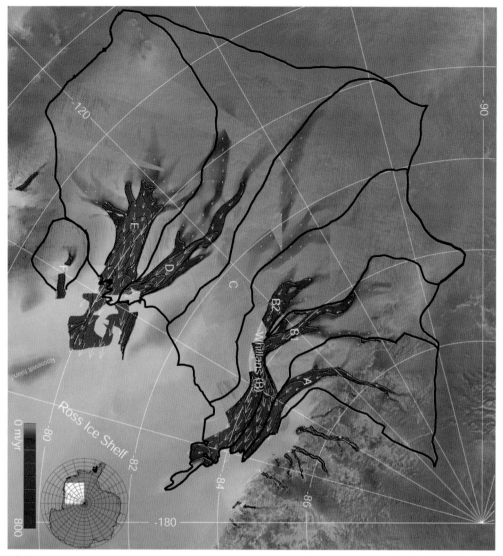

PLATE 1.4 Map of speed of surface movement in a sector of Antarctica (indicated by the white box on the inset map) from InSAR data. Flow velocity is contoured with thin black lines at 100 m per year intervals. White arrows show velocity vectors in fast moving areas. Catchment boundaries for individual ice streams are plotted with thick black lines (Bentley 2004). Copyright 2002, American Association for the Advancement of Science.

PLATE 1.5 Bowman Glacier in the Queen Maud Mountains part of the Trans-Antarctic Mountains at the south end of the Ross Ice Shelf (photograph by John C. Behrendt).

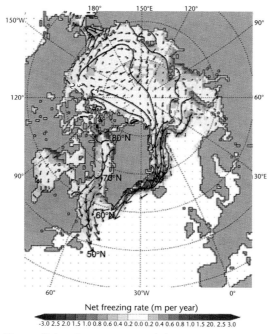

Net freezing rate (m per year)

-3.0 2.5 2.0 1.5 1.0 0.8 0.6 0.4 0.2 0.0 0.2 0.4 0.6 0.8 1.0 1.5 20. 2.5 3.0

PLATE 3.1 Components of the Arctic sea ice balance. Vectors show mean annual ice transport; contours show mean annual ice thickness, and shading indicates net freezing rate (from Hilmer et al. 1998).

PLATE 3.2 Salmon Glacier near Stewart B.C. Note the extensive ablation zone below the snow line (photograph by Richard Hartmier).

PLATE 3.3 Young pancake ice forming in the Ross Sea in February (from Shipp et al. 1999).

PLATE 3.4 Snow and ice cover before ice break-up on Canning River in the Brooks Range, Alaska, May, 1970 (photograph by Michael Church).

PLATE 3.5 Upper Iskut River, northwestern British Columbia along the Alaska panhandle boundary is fed by many active glaciers. The intensely braided river habit indicates high sediment yield (photograph by Gary Fiegehen).

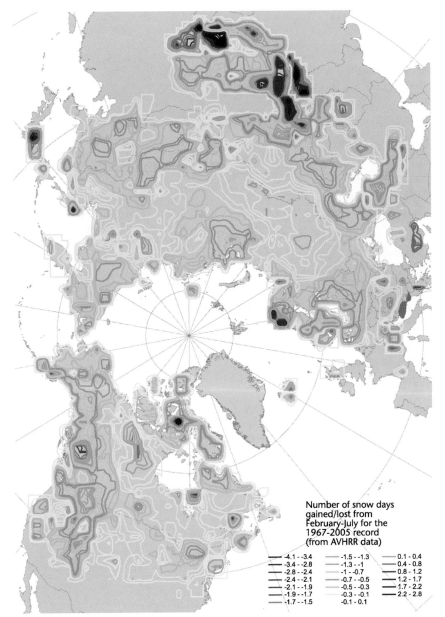

Number of snow days
gained/lost from
February-July for the
1967-2005 record
(from AVHRR data)

-4.1 - -3.4	-1.5 - -1.3	0.1 - 0.4
-3.4 - -2.8	-1.3 - -1	0.4 - 0.8
-2.8 - -2.4	-1 - -0.7	0.8 - 1.2
-2.4 - -2.1	-0.7 - -0.5	1.2 - 1.7
-2.1 - -1.9	-0.5 - -0.3	1.7 - 2.2
-1.9 - -1.7	-0.3 - -0.1	2.2 - 2.8
-1.7 - -1.5	-0.1 - 0.1	

PLATE 4.1 Change in number of snow days between February and July from 1967 to 2005 (from AVHRR data).

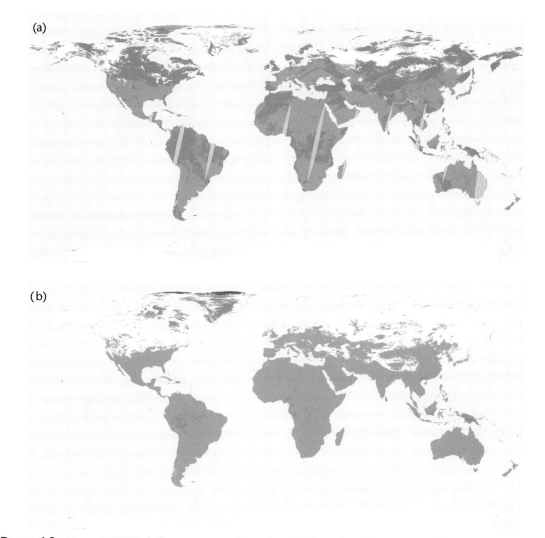

PLATE 4.2 Terra MODIS daily snow map from the CMG product (MOD10C1) for April 1, 2002 (a) and the 8-day composite product (MOD10C2) for March 6–13, 2002 (b). Data are projected to the 0.05° (or 5 km) grid. White represents snow and permanent ice, pink represents cloud cover and green represents land without cloud or snow. The black slashes in the upper panel are interswath regions that are not imaged by MODIS in a day. At this time of year, very high latitudes in the northern hemisphere are in darkness for a significant number of hours so the MODIS cannot detect snow in these regions. Data available from Hall et al. (2003).

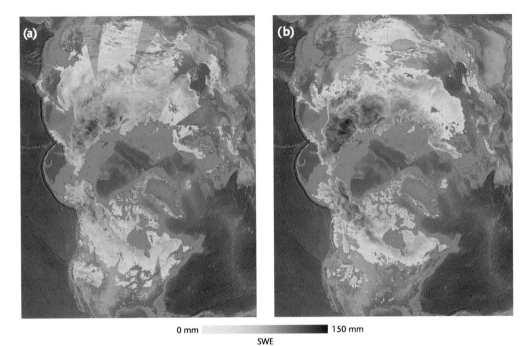

0 mm [gradient] 150 mm

SWE

PLATE 4.3 AMSR-E daily (a) SWE map for January 19, 2005 and monthly (b) SWE map for April 2005 for the northern hemisphere.

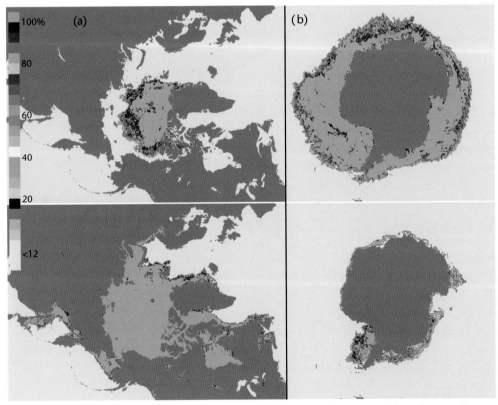

PLATE 4.4 AMSR-E sea ice concentration estimates for the Arctic Ocean (a) and around Antarctica (b) from the NASA Team 2 algorithm (Cavalieri et al. 1997b, updated 2004). The dates for each panel are: upper left is March 1, 2005; upper right is September 5, 2005; lower left is March 1, 2005; and the lower right is September 5, 2005. Figures courtesy of D. Cavalieri and A. Ivanoff.

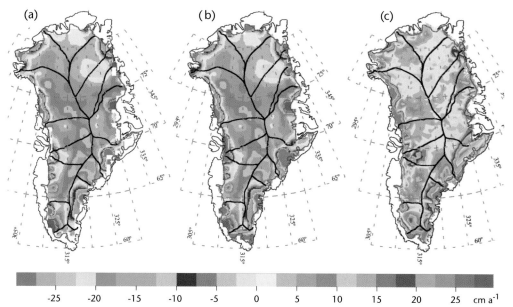

PLATE 4.5 Comparison of the distribution of Greenland Ice Sheet elevation changes for three different data sources: (a) satellite radar altimetry measurements only from the ERS-1 and 2 instruments; (b) corrected ERS altimetry estimates; and (c) interpolated airborne laser altimeter measurements from the Airborne Topographic Mapper collected in 1993 and 1999 (from Zwally et al. 2005, Figure 5).

(a)

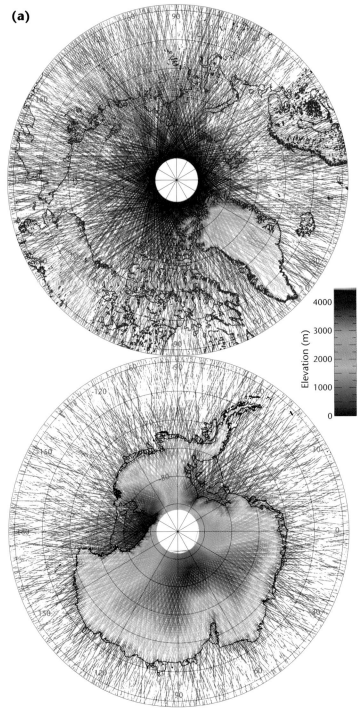

PLATE 4.6 (a) ICESat laser altimeter measurements of ocean and land/ice surface elevation between September 25 and November 19, 2003 (ICESat observing period 2A) relating to the Northern Hemisphere fall and Southern Hemisphere spring. Gaps in the data indicate the presence of clouds (courtesy of C.A. Shuman and V.P. Suchdeo, NASA/GSFC).

PLATE 4.6 (b) ICESat elevation measurements over the Petermann Glacier during observing period 2A. The colored lines represent the ICESat elevation measurements and coincident MODIS imagery is used as the background (image courtesy of C.A.Shuman, NASA/GSFC).

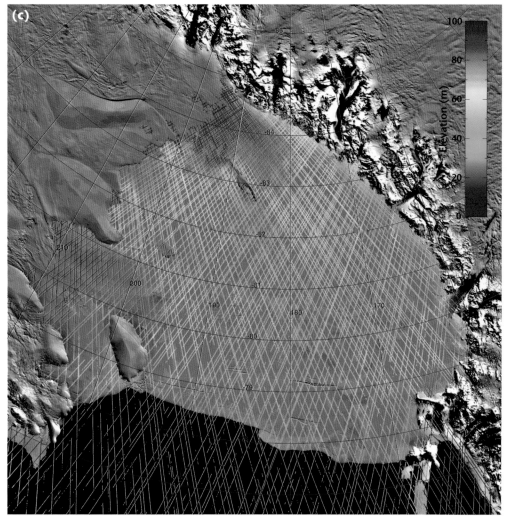

PLATE 4.6 (c) ICESat elevation measurements over the Ross ice shelf during the observing period 2A. The colored lines represent the ICESat elevation measurements and coincident MODIS imagery is used as the background (image courtesy of C.A. Shuman, NASA/GSFC).

PLATE 4.7 ASTER image of the Bhutan Himalayas acquired on November 20, 2001 and centered at 28.3°N, 90.1°E covering an area of 32.3 × 46.7 km (image courtesy of the NASA Earth Observatory and Jeffrey Kargel, USGS/NASA JPL/AGU).

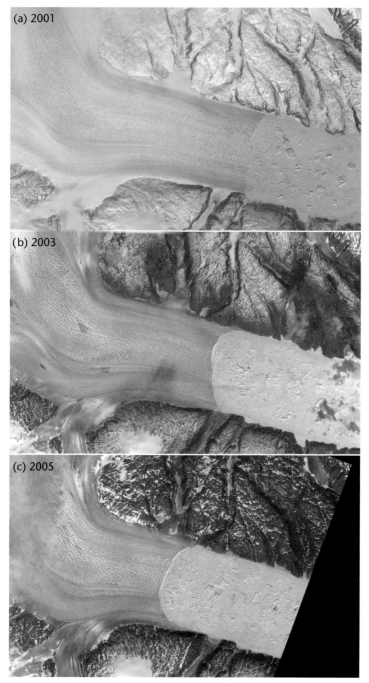

PLATE 4.8 Retreat of the Helheim Glacier, Greenland. Images from the Advanced Spaceborne Thermal Emission and Reflection Radiometer (ASTER) on NASA's Terra satellite show the Helheim glacier in May 2001 (a), July 2003 (b), and June 2005 (c). The glacier occupies the left part of the images, while large and small icebergs pack the narrow fjord in the right part of the images. Bare ground appears brown or tan, while vegetation appears in shades of red (Howat et al. 2005).

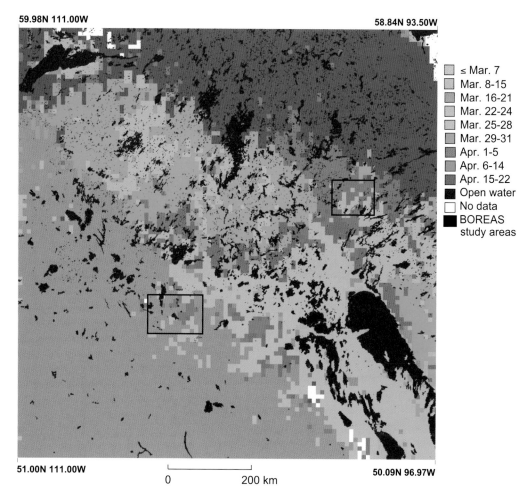

PLATE 4.9 NSCAT map showing the timing of the 1997 initial spring thaw in a region of Manitoba and Saskatchewan, Canada (Kimball et al. 2004).

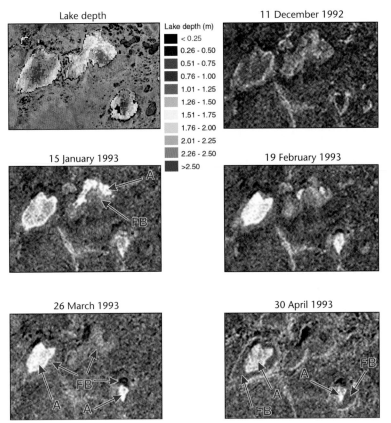

PLATE 4.10 Multisensor remote sensing of lake ice near Churchill, Manitoba. The colored figure shows Landsat TM estimates of lake bathymetry while the grey scale images show ERS-1 SAR imagery used to detect lake water freeze-up. The letters FB and A denote ice that was frozen to the bed and afloat respectively (Duguay and Lafleur 2003).

PLATE 5.1 Scientist inspecting an ice core that has been recovered from a depth of 90 m at Siple Dome, Antarctica, 1999. The auger used to recover the core is on the sled beside the core. In the background is the support tower for a drill that can continuously recover cores to a depth of 1,000 m (photo by Kendrick Taylor).

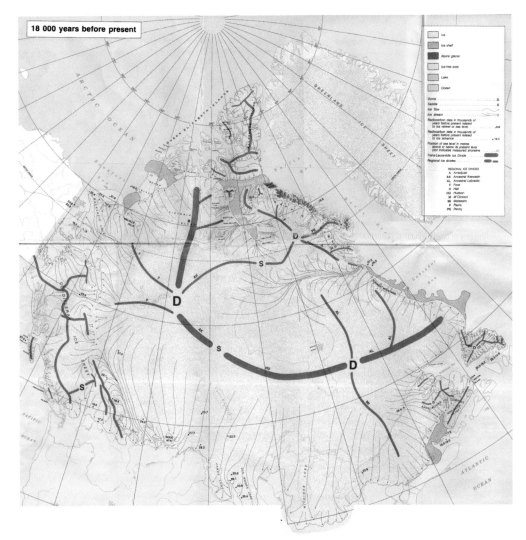

PLATE 5.2 The late Wisconsinan Glacier complex (Prest 1983; Fulton 1989).

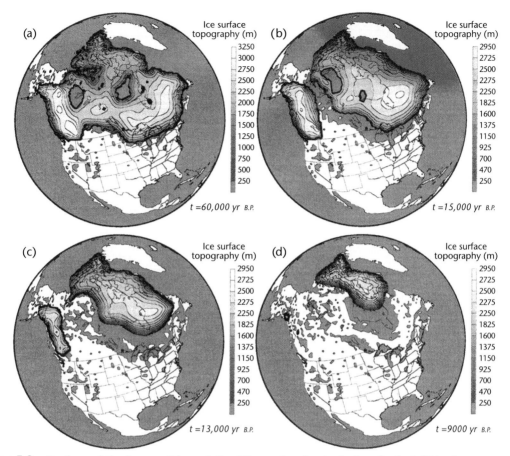

PLATE 5.3 Surface of the Laurentide and Cordilleran ice sheets during the last 60 ka from numerical modeling. Ice sheet surface at (a) 60 ka; (b) 15 ka; (c) 13 ka; and (d) 9 ka BP (from Marshall & Clarke 1999).

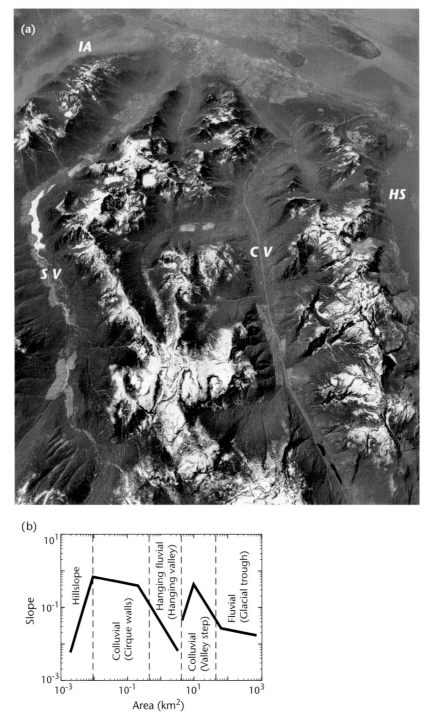

PLATE 6.1 Process domains in cirque landscapes (a) oblique photograph looking south toward city of Vancouver and (b) process domains defined by slope-area relations in the Capilano Valley (CV) (from Brardinoni and Hassan 2006).

PLATE 6.2 False color images of Neoglacial moraines and glacier recession Miller Creek, Coast Mountains.

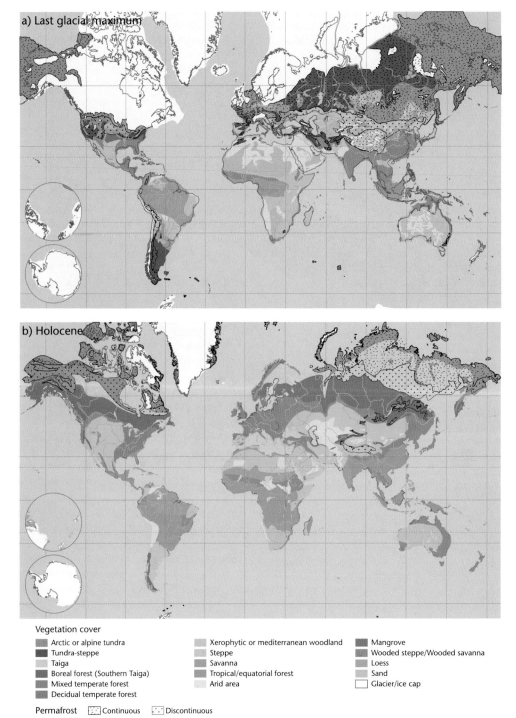

Vegetation cover

- Arctic or alpine tundra
- Tundra-steppe
- Taiga
- Boreal forest (Southern Taiga)
- Mixed temperate forest
- Decidual temperate forest

- Xerophytic or mediterranean woodland
- Steppe
- Savanna
- Tropical/equatorial forest
- Arid area

- Mangrove
- Wooded steppe/Wooded savanna
- Loess
- Sand
- Glacier/ice cap

Permafrost ▨ Continuous ⣿ Discontinuous

PLATE 7.1 Ice and biome distribution (a) at 6,000 BP; (b) at 21,000 BP (from Alverson et al. 2001).

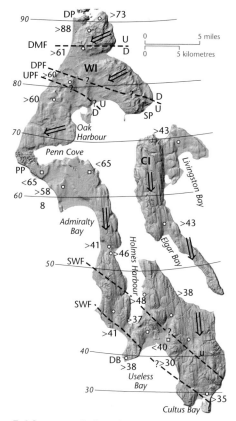

Fig. 5.16 Raised shorelines, Quaternary fault zones, and inferred ice flow direction on Whidbey and Camano islands, Puget Sound from lidar data (Kovanen & Slaymaker 2004b). Raised shorelines are indicated by open circles and the marine limit by open squares; dashed lines are Quaternary fault lines.

Franz Josef Land, Novaya Zemlya, and Severnaya Zemlya, with a grand total of about 9.2 (10^4) km². Our understanding of the LGM Eurasian High Arctic Ice Sheet has improved recently due to two field programs PONAM and QUEEN and the use of numerical modeling. A significant grounded ice sheet must have existed across the Barents Sea at the LGM because of the evidence of sediments in trough mouth fans which were transported to the shelf edge by ice streams (Fig. 5.17).

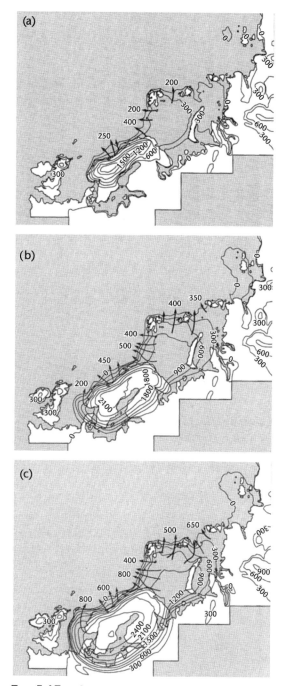

Fig. 5.17 The Eurasian High Arctic Ice Sheet "maximum sized" from 25 to 15 ka BP Surface elevation (in m.a.s.l.). (a) 25 ka; (b) 20 ka; and (c) 15 ka BP. Surface elevation contours at 300 m intervals. Maximum ice stream velocity (ma^{-1}) indicated (from Siegert et al. 1999).

Moraine features near the northern coast of Russia show that the LGM Ice Sheet did not advance significantly southwards into central Russia. Lacustrine sediments and permafrost sequences demonstrate a lack of ice across the Taymyr and Yamal peninsulas, and suggest that ice thickness across the Kara Sea would have been less than 0.5 km. The LGM Ice Sheet was over 1 km thick across the Barents Sea and had fast-flowing ice streams to the west and north. The decay of ice was by iceberg calving over the marine sections of the ice sheet and by surface melting in the south, forming massive proglacial lakes across the northern Russian mainland (Siegert, 2001).

At the LGM, the maximum phase of glaciation with the ice sheet centered over the Gulf of Bothnia. At 20 ka BP, the whole of Fenoscandia, the Baltic, and part of Denmark and the northern European mainland was ice covered. The northern margin was connected with the Eurasian High Arctic Ice Sheet over the Barents Sea. The position of the southern and eastern margins was controlled by the ELA. Thus it was an ablating margin. At the western margin, outlet glaciers and ice streams calved into the Norwegian Sea. A large ice stream occupied the Norwegian Channel. The pattern of ice sheet decay after the LGM is well known from the extensive proglacial geomorphology deposited across the mainland as the ice margin retreated. The Norwegian Channel probably became deactivated prior to 15 ka. From 20 ka to 11 ka BP, the southern ice sheet margin retreated northwards to a position halfway across the Baltic. During this retreat, the southern ablating margin released large volumes of meltwater. During the final stage of deglaciation, this proglacial lake spread across most of the 1,500 km ablating margin.

Considerable debate has surrounded the question of the coalescence of the British and Scandinavian ice sheets during the last glaciation. Maximum reconstructions show confluent Scandinavian and British ice converging over the northern North Sea (Boulton et al. 1985). The occurrence of Scandinavian erratics in British tills and offshore moraine belts lends support to this idea. But a number of recent studies suggest that coalescence occurred much earlier than the LGM, between 130 and 200 ka BP and that a large area of dry continental shelf existed between the Scandinavian and British ice sheets at the LGM (Evans et al. 2005). This is supported by ice sheet profile reconstructions (Nesje et al. 1988) and ice marginal positions identified by Andersen (1979) and many others.

5.6.5 THE PATAGONIAN AND NEW ZEALAND ICE CAPS

Many other land based regions experienced late Quaternary glaciation. Two such examples (Patagonia and the Southern Alps of New Zealand) are illustrated in Fig. 5.18. Sugden et al. (2005) have provided a succinct overview of glacier fluctuations in southernmost part of South America during the last glacial–interglacial transition. Glaciers reached or approached their LGM on two or more occasions at 25–23 ka (calendar) and there was a third less-extensive advance at 17.5 ka. Deglaciation occurred in two steps at 17.5 ka and at 11.4 ka. This structure is the same as that recognized in the northern hemisphere and suggests that at orbital timescales, the "northern" signal dominates any southern hemisphere influence. Since the signal counteracts the southern hemisphere insolation signal, it implies a forcing of the entire global climatic and oceanic system from the north.

On the other hand, during the period of transition (17.5–11.5 ka BP), at a millennial scale, glacier fluctuations mirror an antiphase "southern" climatic signal. There is a glacier advance coincident with the Antarctic Cold

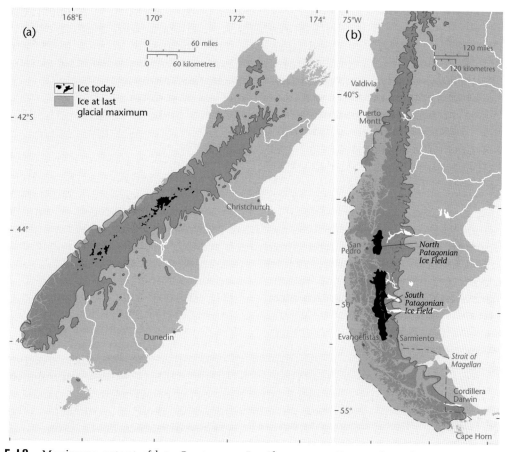

FIG. 5.18   Maximum extent of late Quaternary Ice Sheet across Patagonia and LGM ice cover on South Island, NZ from geological evidence (from Hollin & Shilling 1981; Broecker & Denton 1990; Hulton et al. 1994).

Reversal at 15.3–12.2 ka. Furthermore, deglaciation begins in the middle of the Younger Dryas. The implication, according to Sugden et al. (2005) is that during the last glacial–interglacial transition, Patagonia was under the influence of sea surface temperatures, sea ice, and southern westerlies responding to conditions in the "southern" Antarctic domain. The interesting suggestion is that this asynchrony is evident during deglaciation because of the heightened sensitivity to changes in the oceanic thermohaline circulation, perhaps related to the bipolar seesaw.

Many authors have suggested that southern hemisphere events lead those from the North by 1.5–3 ka. But Carter and Gammon (2004), reporting on data from Ocean Drilling Program Site 1119, south of Banks Peninsula on New Zealand's South Island, have concluded that a clear relation has existed between millennial climate variations on the Antarctic ice cap, in the glaciers of the mid-latitude Southern Alps of New Zealand, and in the oceanic oxygen isotope record for at least the past 0.37 Ma. They also propose that the record may serve as a proxy for both mid-latitude and Antarctic polar plateau air temperatures as far back as the late Pliocene (3.9 Ma BP). The Ocean Drilling Program Site 1119, from which these conclusions derive, is ideally located to intercept discharges of sediment from the Southern Alps. Throughout late Pliocene and Pleistocene climate cycling, southern hemisphere atmospheric dynamics were tightly correlated across a latitudinal range of at least 45–80°S. The suggestion is that lack of correlation with Greenland and North Atlantic records implies that the North Atlantic signal is of regional significance only.

Presently, in the Southern Alps, there are some 3,000 glaciers individually larger than 0.01 km², with a total area of only 1,159 km² and a combined ice volume of 53 km³ (Clare et al. 2002). The maximum ice sheet extent on the South Island of New Zealand has been mapped from moraine sequences (Broecker & Denton 1990) and the ice sheet had a maximum extent of 40,000 km² between 22 ka and 14 ka BP. Porter (1975) identified a maximum ELA depression of about 875 m across the Southern Alps.

5.7   Nonglacial Quaternary environments

Reconstructing the limits of past ice sheet cover has been enormously helped by the development of new dating methods and enhanced methods of submarine exploration. The record of ice retreat from LGM positions during the last 20 ka has become particularly well documented. This last deglaciation was forced by a rapid increase of temperature. The mean Holocene temperature is about 10–13°C higher than the mean temperature during full glacial conditions. Rapid changes involving temperature increases of as much as 10°C over a century or so (D–O events) followed by slow cooling over a millennium or so was characteristic of the period 20–10 ka, but around 10 ka BP, the temperature increased quickly from the

Younger Dryas episode and stabilized at the Holocene temperature level (±2°C).

The findings of the CLIMAP program (CLIMAP 1981) concluded that glacial cooling in the Tropics was only 1°C, ±1°C. In the last two decades, conflicting evidence has been found. Strontium–calcium ratios in corals suggest a 5°C lowering (Guilderson et al. 1994). The descent of snow lines on tropical mountains (Rind & Peteet 1985) points to a similar cooling. The combination of the lowered snow lines and the large glacial to Holocene oxygen 18 ratio found by Thompson et al. (1989) in an ice core at 6 k m.a.s.l. in the Andes seem to demand significant temperature lowering. On the other hand, foraminifera speciation (CLIMAP 1981), oxygen isotope data (Broecker 1986), and alkenone results (Sikes & Keigwin 1994) suggest a cooling of no more than 3°C. Broecker (1996) argues persuasively not only that CLIMAP (1981) did underestimate the cooling but also that there must have been strong regional variations with maximal cooling in tropical mountains (perhaps as much as 8°C) and minimal cooling (perhaps 2°C only) in the central Pacific, Indian, and Atlantic Oceans.

The Holocene has been marked by 3 relatively cool periods (c. 8 ka, 3 ka, and 0.2 ka BP) which are referred to as Neoglacials, and the most recent of which is known as the Little Ice Age. A longer so-called Hypsithermal episode between the first and second Neoglacial experienced the warmest temperatures during the Holocene. At the same time as the process of deglaciation from the LGM was occurring, regions underlain by permafrost were also becoming less extensive. French (1996) estimates that at the LGM as much as 40% of the Earth's surface may have been affected by permafrost whereas at the present time, less than 20% can be so characterized. However, it is important to recognize that the evolution of permafrost in response to climatic warming and cooling is relatively slow. During the late Quaternary, changes in permafrost thickness have taken place at a considerably slower rate than the growth and decay of ice sheets and sea level changes. For example, extensive areas of relict permafrost occur across the sea floor of the Arctic continental shelf and in northeastern Siberia, permafrost has been continuously present since the early Quaternary. The maximum thickness of permafrost in Siberia bears no direct relation to present climatic conditions but is the cumulative effect of uninterrupted periglacial conditions. Although a thickness of 1,600 m has been claimed, Duchkov and Balobayev (2001) have challenged this estimate. For further discussion consult Serreze and Berry (2005). As noted earlier, the High Arctic Eurasian Ice Sheet did not extend to eastern Siberia. The regional variations in permafrost evolution in Russia during the late Quaternary show interesting trends (Fig. 5.19). European Russia was characterized by rapid fluctuations in permafrost growth and degradation. By contrast, eastern Siberia has experienced only moderate permafrost degradation. It is proposed by Velitchko (1984) that anticyclones

developed over (a) the Eurasian Ice Sheet, (b) eastern Siberia, and (c) the Tibetan Plateau. In addition, the Arctic Ocean sea ice cover, the increased continentality in Arctic Russia due to sea level lowering, and changes in atmospheric and oceanic circulation over the North Atlantic reinforced the aridity of the region.

During the LGM, continuous permafrost existed almost as far south as 50°N and areas of discontinuous permafrost may have occurred as much as 400–600 km south of the ice sheet margin. Baulin and Danilova (1984) estimate that perhaps 700 m of permafrost growth may have occurred in Siberia during the Late Valdai Glaciation (LGM). They also consider it possible to relate the spacing of ice wedge polygon cracks and mean and annual ground and air temperatures to the Late Valdai.

5.7.1 Late Quaternary permafrost in North America and Europe

Most of the late Quaternary relict permafrost phenomena that have been identified in North America were produced during the Late Wisconsinan (LGM); the situation is similar

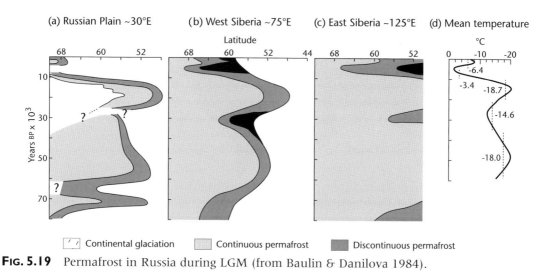

FIG. 5.19 Permafrost in Russia during LGM (from Baulin & Danilova 1984).

in Europe (where the Late Weichselian is equivalent to the LGM) except for the fact that stratigraphical studies of ice wedge pseudomorphs in the Belgian loess have allowed some earlier permafrost evolution chronology going back to isotope stage 5b (Pissart 1987).

During the Late Wisconsin, a permafrost environment existed south of the Laurentide and Cordilleran ice sheets as well as near mountain glaciers and small ice cap complexes. Unglaciated Alaska and Yukon Territory were totally underlain by permafrost (Fig. 5.20b). Whereas in Europe the

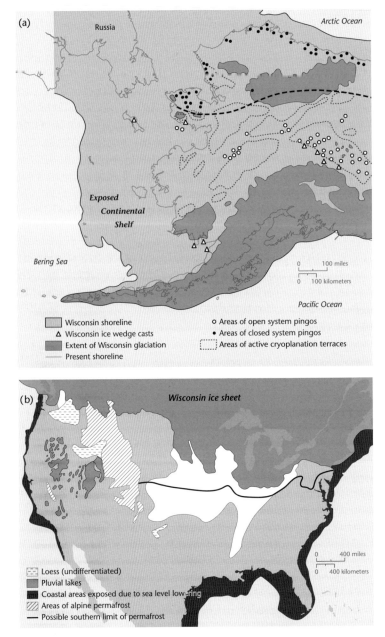

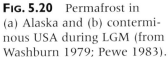

FIG. 5.20 Permafrost in (a) Alaska and (b) conterminous USA during LGM (from Washburn 1979; Pewe 1983).

periglacial zone extended up to 600 km south of the Scandinavian Ice Sheet limit, in the United States the permafrost limit was only of order 100 km south of the Laurentide Ice Sheet limit (cf. Fig. 5.20a and b).

As can be seen from Fig. 5.20, the extent of permafrost (both continental and alpine), loess, and pluvial lakes in the conterminous United States at the LGM has been mapped by Washburn (1979) and Pewe (1983). In Alaska and Yukon, glacio-eustatic sea level lowering produced a large unglaciated region that extended from northeastern Siberia to the margins of the Cordilleran and Laurentide ice sheets. The presence of ice wedges has been investigated by Pewe (1975), but no clearly dated fossil pingos or cryoplanation terraces have been associated with the LGM. By contrast, in Europe, fossil ice wedge structures, pingos, and cryoturbation structures are widespread.

Whereas mean annual temperature lowering during the LGM in North America has been estimated at no more than 11°C, in Europe, especially southern France and northern Spain, estimates of 17–20°C have been made. Both continents were under enhanced zonal wind flow and the easterly winds that were generated south of the ice sheet margins, acting in conjunction with katabatic winds and low air temperatures led to a negative heat budget at the ground and the consequent aggradation of permafrost. The differences in the degree of temperature drop and the areal extent of permafrost south of the ice sheet limits has not been satisfactorily explained but may well have to do with the fact that the ice sheets themselves extended to more southerly latitudes in North America than in Europe.

In the Mackenzie Delta region, Yukon Territory, northern Canada, there is a regional thaw unconformity which has been dated to about 8 ka radiocarbon years BP (or 9 ka calendar years BP) (Burn 1997). This unconformity, which was generated by active layer depths considerably greater than those of today, has been identified by truncated ice wedges (Mackay 1975) and variations in the cryostructure of the ground ice (Murton & French 1994). This early Holocene climatic optimum was marked by the development of many thermokarst lakes as well as the maximum development of the active layer. In the Tuktoyaktuk Peninsula, this early Holocene active layer was 2.5 times as deep as the present active layer. Up to two thirds of the deterioration in growing season conditions which have led to the southward movement of treeline since the early Holocene may be due to active coastal recession. This means that unknown proportions of the reduction in active layer thickness have been caused by climate deterioration, coastal recession, and treeline migration.

5.7.2 TREELINE VARIATIONS

Fluctuations of three major biogeographical boundaries have been studied in considerable detail using macrofossils (Bradley 1999): the arctic treeline; the alpine treeline; and the lower treeline of semiarid and arid regions. Arctic treeline fluctuations have been studied in Alaska, Canada, and Russia (Tikhomirov 1961; McCulloch & Hopkins 1966; Ritchie 1987). Figure 5.21 shows that the northern treeline in Canada was 250 km north of the present treeline during the Hypsithermal episode (6–3.5 ka BP). Unfortunately, a number of factors make the relating of paleoforest limits to past climates difficult. First, northward migration of treeline during climatic amelioration is more rapid than southward treeline migration. This is because trees, once established may survive periods of poor climate and the treeline will recede slowly. Second, this process of unequal rates of treeline migration differs from one location to another. Third, some trees, such as black spruce (*Picea mariana*) will grow in both upright and prostrate forms in response to changing snow depth.

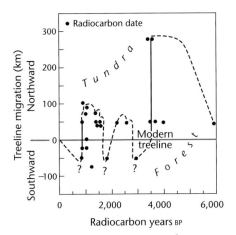

FIG. 5.21 A reconstruction of Holocene tree-line fluctuations in southwestern Keewatin, Northwest Territories, Canada. Treeline position is based on radiocarbon dated tree macrofossils *in situ* north of the present treeline, and on dates on buried forest and tundra soils north and south of the modern treeline (Sorenson & Knox 1974).

Alpine treeline fluctuation interpretation also suffers from some complications. First, the highest tree may not have been found. In such a case the paleotreeline evidence gives only a minimum paleotemperature change. Second, the present altitude of the treeline is often not precisely determined and may not be in equilibrium with present climate. Third, there is an unknown lag time between the time of maximum temperature and the time of maximum height. Fourth, regional isostatic uplift in some areas makes early Holocene treeline records difficult to interpret.

In spite of these complications, there is clear evidence that alpine treelines were higher by at least 200 m during the Hypsithermal in Europe and North America. In the Urals and Scandinavia, trees grew 60–80 m higher than at present from ninth to 13th century AD but many died in the late 13th and early 14th century, the early 1500s, and in the early and late 19th century.

Lower treeline fluctuations are of less direct relevance when considering the cryosphere except in so far as they reflect the importance of moisture for tree growth, which in turn relates to snow and glacier melt conditions in mountainous regions. Heavy reliance is placed on the evidence from fossil middens of the packrat (genus *Neotoma*) from caves throughout the southwestern United States. Lower treeline elevations were as much as 1,000 m below modern limits (Thompson 1990) over periods ranging from >40 ka to 12 ka BP.

5.7.3 CLIMATIC SNOWLINE

The snow line is the boundary between the seasonal snowpack which melts at the end of the winter season and the permanent snow which remains year round. From year to year, the actual snow line will vary in elevation, depending on the particular weather conditions during the accumulation and ablation seasons. On a glacier, this would be equivalent to the firn line or the equilibrium line altitude. If observations are made over a period of time, the average elevation of the snow line is apparent and this enables the regional or climatic snow line to be identified (Ostrem 1974). Observations of modern glaciers indicate that the ELA approximates the height at which the accumulation area of the glacier occupies about 60% of its total area. To estimate paleo-snowlines, the common practice is to reconstruct the paleo-ELA by mapping the former glacier area and then determining where the ELA would have been, based on an accumulation area ratio (AAR). This generally corresponds to the uppermost limit of lateral moraines. Mapping the difference between modern and paleo-ELA's can provide useful insights into past climatic conditions (Fig. 5.22).

A similar index is provided by glaciation levels or glaciation thresholds that define

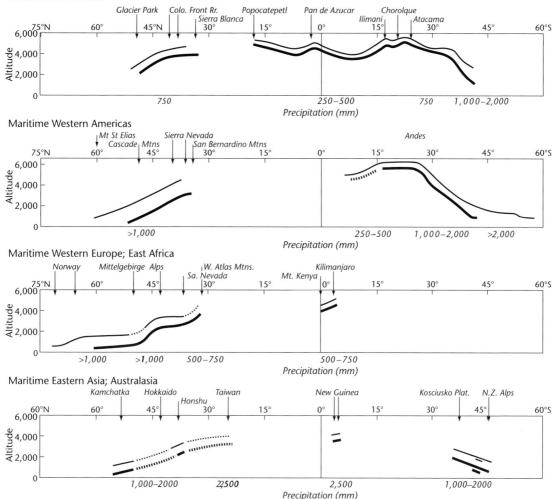

FIG. 5.22 Variations in climatic snow line and glacial age snow line (from Flint 1957).

the lowest limit at which glaciers or permanent ice fields can develop. This is usually determined by identifying the highest unglacierized and the lowest glacierized mountain summits in a region and averaging the two elevations. Both snow lines and glaciation thresholds can be used in paleoclimatic reconstructions if similar features can be mapped for periods in the past. Such threshold only provide information about extreme glacial periods (Fig. 5.23).

Results from such studies are often equivocal because:

1 Present day snow lines have not been adequately studied in relation to present climate. Climatic controls on modern snow line elevations cannot be assumed to be the same in all areas.

2 Atmospheric lapse rates have not been well documented in mountain regions and are problematic in paleotemperature reconstructions.

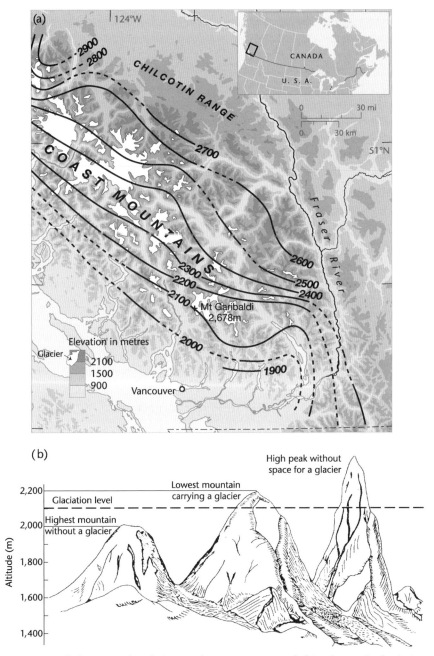

Fig. 5.23 (a) Regional glaciation levels in southwestern BC and (b) schematic depiction of "glaciation level" (from Ostrem 1966; Kovanen & Slaymaker 2005).

3 Paleosnowline reconstructions are often based on features of varying age.

4 Not all paleosnowlines are defined in the same way. A common method is to estimate the average elevation of cirque floors occupied by glaciers during a particular glacial period.

Alternatively, paleosnowline may be located at the median altitude between the terminal moraine of a given advance and the highest point on the cirque headwall. In both approaches, orientation of the cirque is an important factor and cirque populations of differing orientation will give different answers. Also, the use of cirque floor elevation without knowledge of the age of glacial deposits may provide a mixed population of snow line estimates.

However, well-dated deposits and careful selection of sites can yield useful information. An interesting example is provided by Dahl and Nesje (1996) from southern Norway. They estimated changes in the size of Hardangerjokulen from glaciofluvial and glaciolacustrine sediments, relying on the fact that as the outlet glacier advanced and retreated certain thresholds were passed, which were recorded as diagnostic signatures in downstream sediments. This enabled the ice cap area to be estimated and from that the paleo-ELA could be calculated (Fig. 5.24).

Changes in the upper limit of pine (*Pinus sylvestris* L.) provided an assessment of variations in summer temperature, and from these estimates the ELA could be obtained using a lapse rate of $0.6°C \cdot (100 \text{ m})^{-1}$. They then used a well-established relation between winter accumulation and temperature at the equilibrium line of Norwegian glaciers to reconstruct the change in winter accumulation at the ELA (assuming this relation has remained constant) (Fig. 5.24). The central lesson of this study is that it is dangerous to rely on only one method. A similar conclusion could be made from a recent study of paleoclimate change on

Mount Baker in Washington State (Kovanen & Slaymaker 2005). They compared the results of applying two different methods of ELA determination and discussed the paleoclimatic implications in relation to six radiocarbon ^{14}C dated logs in drift that were buried during an advance of the Deming Glacier, possibly during the Younger Dryas. A valuable feature of their discussion is the careful enumeration of the simplifying assumptions associated with the use of ELA estimates (Fig. 5.25):

1 that the ELA is a response to a range of atmospheric lapse rates;

2 that glaciers are in equilibrium with the local climate when moraines are constructed;

3 that estimates of past snow lines are based on interpolation between glacierized and nonglacierized landforms;

4 that local anomalies of glacier geometry, aspect, and continentality can be ignored;

5 that the ratio between the altitudinal ablation and accumulation gradients remains constant; and

6 that albedo feedbacks and step-changes in climate are ignored. A recent global assessment of tropical snow line changes at the LGM provides information on glacier equilibrium line altitudes for over 350 glacier valley localities (Mark et al. 2005, Table 5.2). They conclude that the largest share of variance in their results derives from basin morphometry (catchment size, glacier slope, etc.). Glacier aspect is also important but varies greatly by region. The observed change in ELA between the LGM and today is smallest in the Himalayas, relatively small in the southern central Andes and East Africa, and largest in the northern Andes, Mexico, and Papua New Guinea.

5.7.4 GLACIER FLUCTUATIONS

Glacier fluctuations result from changes in the mass balance of glaciers. A glacial

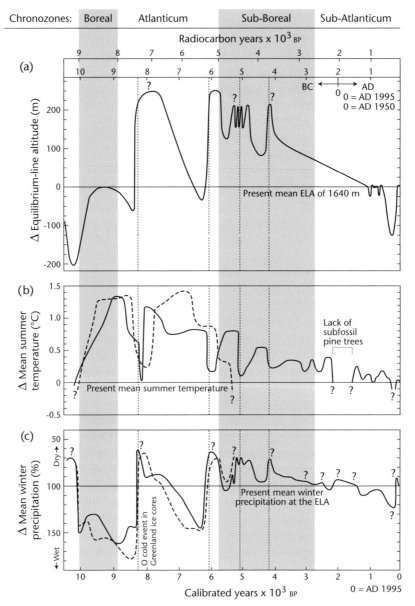

FIG. 5.24 (a) Variations in ELA on Hardangerjøkulen, south-central Norway; (b) summer temperature based on subfossil wood (*Pinus sylvestris* L.) found above modern treeline; and (c) winter precipitation derived from modern relations between accumulation and temperature at the ELA. Values are expressed relative to modern conditions. Ages are given in calibrated and uncalibrated radiocarbon years. Summer temperatures were above present levels for most of the Holocene; glacier advances were associated with periods of higher winter accumulation (from Dahl & Nesje 1996).

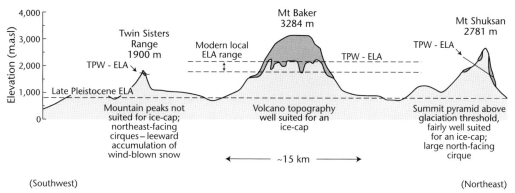

FIG. 5.25 Schematic diagram showing the difference between the TP-ELA at Mount Baker and other volcanoes and the TPW-ELA at cirque glaciers, the altitudinal range of the modern local ELA and the maximum ELA depression during the LGM (Fraser Glaciation = Late Wisconsinan) (from Kovanen & Slaymaker 2005).

advance corresponds to a positive mass balance brought about by a climatic fluctuation which favors accumulation over ablation. Glacier front positions will lag behind climatic fluctuations. Different glaciers have different response times to mass balance variations (Oerlemans 1989). There is some evidence that ice recession responds more rapidly to a climatic fluctuation than ice advance.

Many Arctic glaciers are still advancing today in response to cooler conditions of the last century, at the end of the Little Ice Age, whereas smaller midlatitude alpine glaciers over the same interval have advanced, receded, and subsequently readvanced in response to cooler conditions over the last two or three decades. The largest glacier systems respond to low frequency climatic

TABLE 5.2 Summary of reconstructed changes in ELA (ΔELA) and temperature (ΔT) for the key regions in the statistical analyses (from Mark et al. 2005).

Key region	No. of localities	Min ΔELA (m)	Max ΔELA (m)	Mean ΔELA (m)	Range ΔELA (m)	Min ΔT (°C)	Max ΔT (°C)	Mean ΔT (°C)	St. dev. ΔT (°C)
Bogotá	23	920	1,426	1,217	456	6.5	9.0	7.9	0.83
Central Andes	44	260	1,403	804	1,143	2.2	12.4	7.2	2.76
Cordillera Blanca	*21*	*470*	*1,058*	*753*	*588*	*4.2*	*6.7*	*6.1*	*0.73*
Himalaya	13	104	1,044	513	940	7.3	12.1	7.5	1.99
Iztaccíhuatl	20	820	1,250	1,030	430	6.0	8.2	7.1	0.67
Mt Jaya	12	800	1,210	945	410	9.4	11.2	10.0	0.53
PNG	51	877	1,273	1,040	397	11.2	14.6	12.8	1.13
Rwenzori	73	503	1,387	1,043	884	3.1	7.9	5.9	1.17
Total	236	104	1,426	978	1,322	2.6	14.6	8.37	3.04

The Central Andes region includes the localities of the Cordillera Blanca, which also comprise a separate key region used in the analyses. The 21 localities of the Cordillera Blanca are not counted twice toward the total, and likewise are italicized in the table.

fluctuations whereas the smaller systems respond to higher frequency fluctuations. In a recent examination of 169 glacier records (Oerlemans 2005), it is concluded that moderate global warming started in the middle of the 19th century; that the warming in the first half of the 20th century amounted to 0.5 K and that this warming was notably coherent over the globe. The warming signals from glaciers at low and high elevations appear to be very similar.

5.7.5 PARAGLACIATION

The question may be legitimately asked "What was happening to the proglacial fluvial and aeolian systems during the LGM and the deglaciation that followed?"

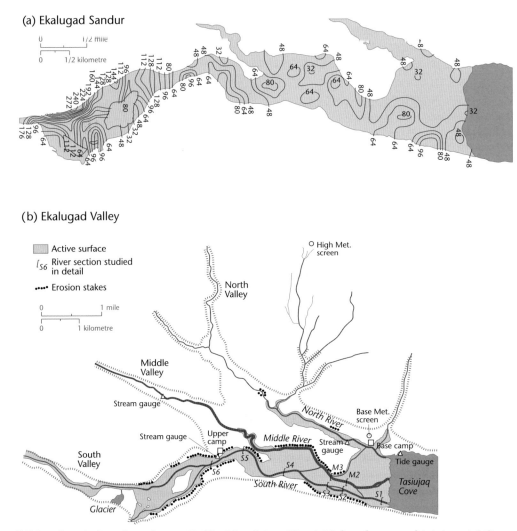

FIG. 5.26 The Ekalugad sandur on Baffin Island (see Fig. 6.10 for photograph) (a) variability of the coarse grain size on the sandur surface and (b) the topographic setting of Ekalugad Valley (from Church 1972).

Much of this will be discussed in Chapter 6, but for the moment it will suffice to say that ice sheet, permafrost, snow line, and glacier advances and retreats greatly influenced the distribution of sediments, both on land and in the oceans. Massive outwash plains, observable in Iceland, Alaska, and Baffin Island, for example (Fig. 5.26) are major evidences of the way in which glacial advances and retreats make sediment available for reworking by meltwater streams. Similarly, the great sheets of loess that are associated with the edges of the great ice sheets demonstrate the unusual availability of sediments to wind action during the LGM.

The original recognition by Ryder (1971a) of a so-called paraglacial environment derives from her observations of the regionally extensive but presently inactive alluvial fans in the interior of British Columbia. In a comparative study of the interior of British Columbia and Baffin Island (Church & Ryder 1972), the term paraglacial was defined as an environment in which nonglacial processes are entirely conditioned by glaciation. This term summarizes the landscape effects of the high degree of mobilization of sediment systems by glaciers and ice sheets during the Pleistocene (Fig. 5.27).

Benn and Evans (1998) have proposed that the term paraglacial period is perhaps more appropriate than the term paraglacial environment because the processes themselves are not unique to such environments. The paraglacial period is characterized by high rates of sediment delivery from slopes and into fluvial and aeolian systems. This period of rapid response is triggered by the instability of unconsolidated glacigenic sediments and oversteepened rock slopes

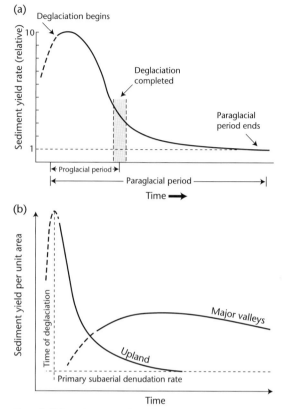

FIG. 5.27 The paraglacial environment (from Church & Ryder 1972; Church & Slaymaker 1989).

which have been debuttressed (Bovis 1990; Holm et al. 2004). Church and Slaymaker (1989) suggested that the landscape of British Columbia remains a prisoner of its history in that it is still responding, in the pattern of its fluvial sediment yield, to the LGM, which in this region occurred about 14.5 ka BP. Chapter 6 attempts to place the question of the paraglacial period into the broader context of landscapes in transition under the influence of cryospheric processes.

6

THE TRANSIENCE OF THE CRYOSPHERE AND TRANSITIONAL LANDSCAPES

6.1 INTRODUCTION

The transience of the cryosphere has been amply demonstrated in Chapters 1, 3, and 5; first, through fears of the imminent disappearance of, for example, tropical glaciers and polar sea ice (Chapter 1); second, through consideration of the processes of snow, ice, and permafrost formation and the dependence of their mass and energy balances on variable climate and human activities (Chapter 3); and third, through the Quaternary record of changing ice sheet distribution. The variable sensitivity of the components of the cryosphere makes an accurate description and analysis of this transience in space and over time exceedingly complex.

One template for the interpretation of this cryospheric transience, with the understanding that landscapes are rarely in equilibrium with their climatic and anthropogenic environment, is to focus on temporal transitions and to make liberal use of the ergodic hypothesis (Church & Ryder 1972; Knox 1984; Clague 1986; Church 2002). In Chapter 5, we focused on transitions from cryospheric to noncryospheric environments; in Chapter 6 we will focus on the nature of transitions from glacial and periglacial through paraglacial to postglacial landscapes and in the final chapter we will consider the transitions from Pleistocene through Holocene to Anthropocene environments.

6.1.1 THE LANDSCAPE AS PALIMPSEST

All areas formerly covered by ice sheets have retained some distinctive imprint of that component of the cryosphere, whether in erosional or depositional form. We will consider a selection of large-, medium-, and local-scale features, but for a fuller discussion the reader is referred to the definitive treatise by Benn and Evans (1998), especially Chapters 9 and 10. Glacial landscapes commonly consist of superimposed depositional systems of different ages resting on older erosional surfaces. Such landscapes have been compared to a palimpsest, or a parchment which was reused by monastic scholars when writing materials were in short supply (Chorley et al. 1984). Older text was erased to make room for other writing, but the erasing was not totally effective and the older inscriptions were still partially visible. This is a useful metaphor for landscapes in which the products of older processes are partially visible below more recent forms. In glaciated terrain, large-scale erosional forms are generally the cumulative effect of multiple glacial cycles whereas the sedimentary record typically records only the most recent glaciation. This is particularly true for continental interiors where general erosion and reworking of sediments during glacial advances ensures that older glacigenic sequences rarely survive within younger glacial limits.

Glacimarine sequences have greater preservation potential, particularly in subsiding basins such as the central graben of the North Sea where >1 km of Quaternary sediments are preserved.

6.2 GLACIAL LANDSCAPES: MACRO-SCALE

Glacial landscapes show systematic spatial organization and suggest climatic, geological, topographical, and glaciological controls. Evans (2003) has elaborated extensively on the concept of glacial landsystems. The concept of landsystems was developed in Australia by agricultural land surveyors (Christian 1957) and geomorphologists (Mabbutt 1968) in attempting to see the relations between geology, surface relief, and climate. Even when applied at a general level, the landsystems approach can show striking patterns of glacial erosion and deposition. For example, Fig. 6.1 shows large-scale glacial geomorphological zones in North America (Eyles et al. 1983): (1) Shield terrain, which is bedrock with little drift cover; (2) Sedimentary lowlands of (a) subglacial lodgement till plains and (b) supraglacial moraine complexes; and (3) Glaciated valley terrain. Although this is a unifying concept and is broadly informative, the dynamism of the fluctuating boundaries of glacial, periglacial, proglacial, and paraglacial zones is not captured.

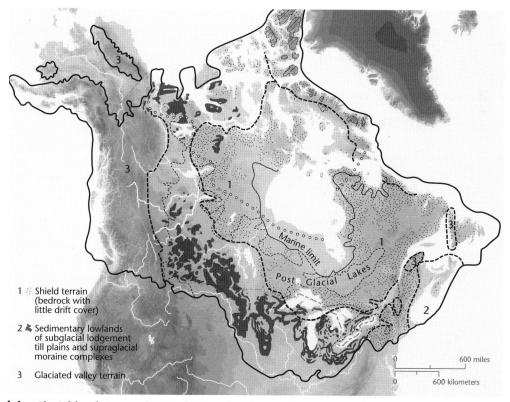

FIG. 6.1 Glacial landsystems in North America. Present ice cover is shown only for the Arctic. Southern limit of continuous permafrost is shown by open circles. Maximum outermost limit of glaciation shown by black lines (after Eyles et al. 1983).

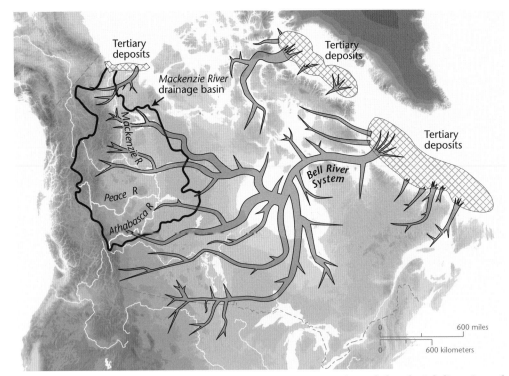

FIG. 6.2 Reconstruction of the preglacial Bell River system, Canada and the glacial diversion of the headwaters to form the modern Mackenzie River drainage basin (after Duk-Rodkin & Hughes 1994).

One of the most spectacular effects of glacial processes at continental scale is that of the diversion of major river systems. The reconstructed Tertiary Period west to east flow direction of the Bell River system in Canada (Fig. 6.2) was diverted to a NNW flow direction in the basin of the Mackenzie River by glacial/ice sheet action during the Quaternary (DuK-Rodkin & Hughes 1994). The southwestern corner of the Mackenzie drainage basin also appears to have been diverted by the Fraser River as a result of glacial deposition during the Quaternary combined with continuing tectonic uplift (Tempelman-Kluit 1980).

Another spectacular example of macro-scale glacial erosion is the landscape of the Finger Lakes region of central New York State. The 11 finger lakes have been deeply scoured (to a maximum of 306 m below sea level) and then infilled with sediment to a maximum thickness of 270 m. The timing of this erosion was between 14.4 and 13.9 ka BP (^{14}C years), coincident with Heinrich event H-1 in the North Atlantic Ocean. This observation lends support to the idea of large-scale ice sheet instability around 14 ka BP (Mullins & Hinchley 1989).

6.2.1 Cirque landscapes

Cirques are hollows formed at glacier sources in mountains and partly enclosed by steep, arcuate headwalls (Evans 2004). Because of their assumed association with small glaciers, cirques have been used as indicators of long-term average paleoclimatic conditions. Two principal cirque attributes have been used, namely orientation and altitude. There is a tendency for cirques to be preferentially oriented. In the midlatitudes of the northern hemisphere, cirques face between north and east; in the midlatitudes of the southern hemisphere, they face between northeast and southeast.

In the circumpolar regions the dominant orientations are to the north and south. Cirque orientations for given regions are normally portrayed as cumulative vector diagrams (Fig. 6.3). Evans (1977) proposed

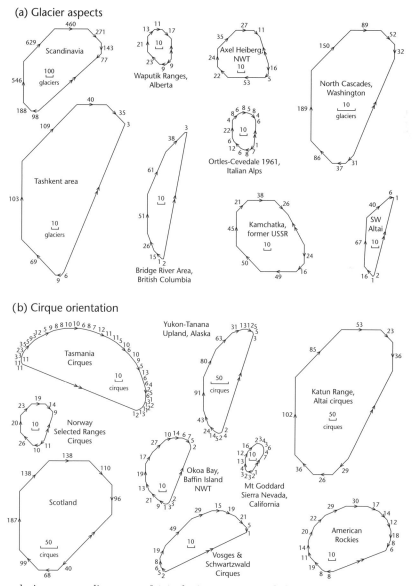

FIG. 6.3 Cumulative vector diagrams of (a) glacier aspects and (b) cirque orientations. All glacier vector diagrams are at the same scale except for the North Cascades (half scale) and Scandinavia (one-tenth scale). Cirque vector diagrams are at the same scale except for Scotland and Yukon-Tanana (half scale) and Katun (one-fifth scale). Resultant vectors are indicated by double arrows (after Evans 1977).

TABLE 6.1 Definition of cirque orientation and asymmetry categories (after Evans 1977).

Vector strength (%)	Cirque asymmetry
100	Extreme
80	Strong
60	Marked
40	Weak
20	Symmetric

a set of asymmetry classes based on vector strength, a statistic which is equivalent to the length of the resultant vector divided by the total length of all vectors expressed as a percentage (Table 6.1). The tendency for cirques in midlatitudes both north and south to face eastwards is a reflection of the preferred accumulation of snow and glacier growth in leeside hollows in the westerly wind belts. The tendency for polar cirques to face north or south results from reduced ablation and preferential glacier survival where direct solar radiation is lowest (Evans 2006).

Regional trends in cirque elevation have also been interpreted in terms of paleoclimate. Paleoclimatic reconstructions based on cirque floor altitudes must be treated with caution. At best they can only offer information on average conditions during the Pleistocene as cirques have been initiated and expanded several times under varying climates. Also cirque floor altitudes may have been affected by regional uplift or subsidence; finally, the altitude of a cirque basin does not necessarily bear a close correspondence to the equilibrium line altitude of the glacier which formed it (Richardson & Holmlund 1996).

6.2.2 FJORD AND STRANDFLAT LANDSCAPES

The most spectacular examples of glacial erosion are troughs and fjords carved by ice flow through major rock channels (Fig. 6.4). The largest troughs on Earth, the Thiel and Lambert troughs, are presently occupied by outlet glaciers of the Antarctic Ice Sheet and are approximately 1,000 km long and more than 50 km wide and up to 3.4 km deep. Very long and deep open fjords (>100 km long and up to 3 km deep) are found on Ellesmere and Axel Heiberg Islands, and along the coasts of Greenland and Norway. They are also well developed along the glaciated coasts of British Columbia, Chile, Iceland, New Zealand, and Svalbard.

In contrast with river valleys, troughs and fjords commonly have over-deepenings on their floors. Such over-deepened basins commonly form where ice discharges are relatively high such as at the junction of tributaries or at narrowings in the valley profile. Over-deepened sections commonly terminate at a sill or a threshold.

The influence of lithology and of preexisting topography is a matter of some debate. Trough and fjord alignments have been linked to bedrock lineaments such as faults and intrusions; the rectilinear pattern of some fjord systems has been linked to intersecting lines of fracture at a regional scale (e.g. British Columbia's Coast Mountains). Benn and Evans (1998) wisely conclude that "it is most likely that a continuum of landforms exists, ranging from tectonically controlled grabens through glacially modified river and fault systems to entirely glacially eroded troughs and fjords" (p. 355).

Assessing the rate of glacial erosion is difficult because of the unknown depth of preglacial valleys. In Norway, a paleic or preglacial surface has been defined and by subtracting that surface from present-day topography an estimate of glacial erosion can be defined. Nesje et al. (1992) calculated that the rate of erosion of the Sognefjord by ice must have been between 102 and 330 cm per 1,000 years (or 1,020 and 3,300 Bubnovs). One Bubnov is the rate of erosion equivalent to a loss of 1 mm in 1,000 years (von Bubnov 1963). Nesje and

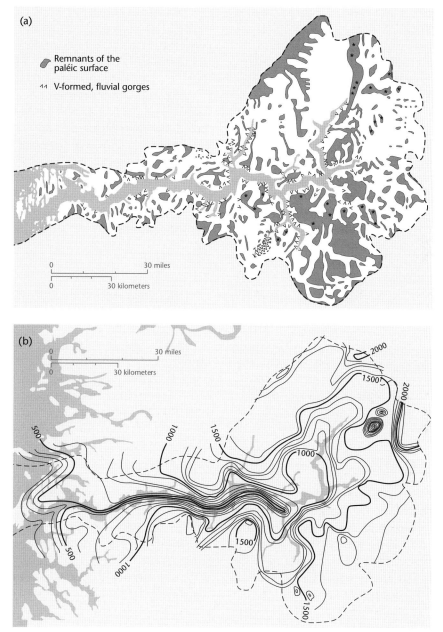

FIG. 6.4 The erosion of Sognefjord, west-central Norway. (a) the Sognefjord drainage basin, showing remnants of the paleic surface, V-shaped valleys and gorges (V symbol) and Tertiary soil remnants (stars); (b) reconstruction of the paleic surface based on interpolation of paleic surface remnants (after Nesje & Whillans 1994).

Whillans (1994) have defined four phases of evolution of that fjord: (1) Deep chemical weathering during the Mesozoic and early Tertiary; (2) Tertiary uplift, which was greater toward the west coast; (3a) Interglacials and interstadials when valley floors were filled with debris; and (3b) repeated glaciation producing overdeepening of preexisting valleys.

Perhaps the most intriguing and probably the least satisfactorily explained landscape element of at least partial glacial origin is the strandflat. The strandflat is an extensive, undulating rock platform located close to sea level around the coasts of high-latitude landmasses, including Norway, Greenland, Svalbard, Iceland, British Columbia, and Antarctica. Strandflats are up to 50 km wide, cut across geological structures and usually end abruptly at an inland cliff line or break of slope. When viewed parallel to the coast, they are remarkably horizontal but they most commonly possess very small seaward slopes, in part owing to postglacial glacio-isostatic adjustment. In some locations, several strandflats may occur at various heights above and/or below sea level.

Strandflat formation has been explained by four main mechanisms acting in combination: (a) frost action combined with active sea ice rafting; (b) marine erosion; (c) subaerial erosion; and (d) subglacial erosion.

Narrow strandflats with seaward tilts can be explained by frost action and marine erosion, and should therefore be called marine platforms (e.g. John & Sugden 1971) but the wider, often horizontal platforms of glaciated coasts present more profound problems of interpretation (Fig. 6.5). Porter (1989) has argued that strandflats formed during "average" glacial conditions that prevailed throughout most of the Quaternary, when relative sea level along many glaciated coasts was similar to that of today and coastal zones were subject to repeated alternations between periglacial, marine, and subglacial erosive processes during glacier advance and retreat cycles.

This is an important argument based on the idea that for most of the Quaternary, glacier extent has been intermediate between the full glacial maxima and interglacial minima experienced during the Holocene and that such intermediate conditions are the most relevant to long-term analyses of environmental change and landscape evolution.

6.3   Periglacial landscapes: macro-scale

Periglacial landscapes are defined variously as landscapes under the influence of frequent freeze–thaw cycles or as landscapes underlain by permafrost. Troll (1948) and many others have used the former definition which leads to the fact that 35% of the Earth's continental surface is periglacial. Most freeze–thaw cycles occur in areas of low annual temperature range, which are dominated by diurnal fluctuations. These conditions are met in subpolar oceanic locations (e.g. Jan Mayen or South Georgia) and in intertropical high mountain environments (e.g. the Andes and the East African mountains). The most extensive periglacial areas, however, are found in northern Eurasia and North America and these incorporate both tundra and boreal forest landscapes. At the LGM, the periglacial realm was expanded by about 50% (Fig. 6.6). Poser believed that the presence of three fossil forms, namely loess wedges, involutions, and climatically controlled asymmetric valleys were criteria for the presence of permafrost. His map defines four climatic zones in Europe beyond the ice sheet at the LGM: They were: (1) Continental permafrost forest climate east of the Alps; (2) Maritime forest climate without permafrost

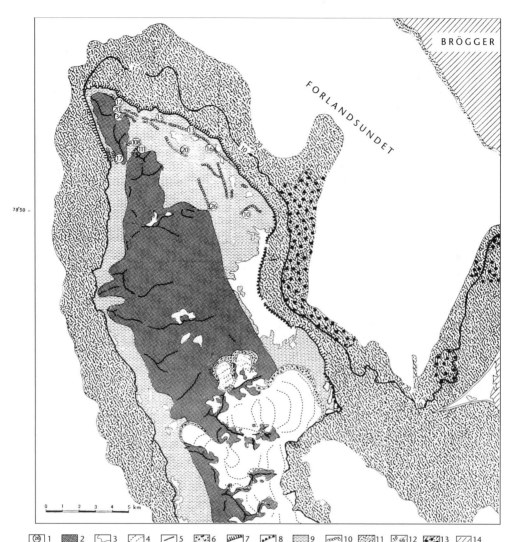

Fig. 6.5 Strandflats and associated features on northern Prince Charles Island, Svalbard. 1. Altitude of coastal features in meter; 2. mountains; 3. lagoon; 4. glacier; 5. mountain ridge; 6. moraine; 7. cliff; 8. coastal pebble ridge; 9. emerged strandflat; 10. raised pebble ridge; 11. submarine strandflat; 12. submarine contour in meter; 13. submarine strandflat with pebble ridges; 14. emerged area not shown in detail (after Guilcher et al. 1986).

west of the Alps and covering much of France; (3) Maritime tundra climate without permafrost; and (4) Permafrost tundra climate, a thin strip of land stretching from the British Isles to central Europe. As Worsley (2004) points out, it seems that the LGM occurred late in the Last Glacial stage. This implies that for much of the glacial stage, those areas which were to become glaciated were cold but nonglacial, that is, periglacial, which means that many glaciated landscapes carry a partial periglacial imprint.

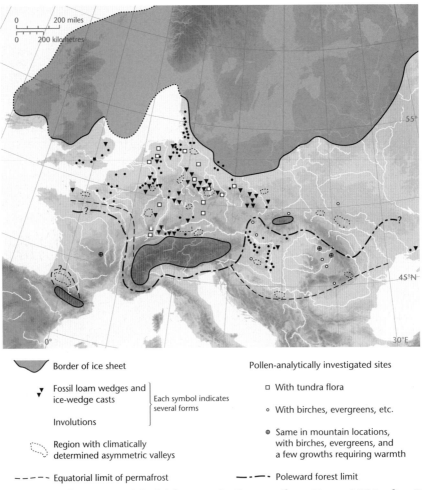

FIG. 6.6 Distribution of Würm periglacial features in Europe (from Evans 1994, after Poser 1948).

6.4 PARAGLACIAL LANDSCAPES: MACRO-SCALE

The paraglacial concept was formalized by Church and Ryder (1972) after Ryder (1971a) had introduced the term to describe alluvial fans in the interior of British Columbia (Fig. 6.7). They defined paraglacial as nonglacial processes that are directly conditioned by glaciation and added that "it refers both to proglacial processes and to those occurring around and within the margins of a former glacier that are the direct result of the former presence of ice" (Church & Ryder 1972, p. 3060). They reasoned that,

although fluvial sediment transport rates were likely to be greatest immediately after deglaciation, fluvial reworking of glacigenic sediments was likely to continue as long as such sediment was accessible to rivers. They identified three aspects of the influence of paraglacial sediment supply on fluvial transport: (a) the dominant component of reworked sediment may shift from till to secondary sources such as alluvial fans and valley fills; (b) regional uplift will condition the timing of changes in the balance between fluvial deposition and erosion such that the cascade of sediment evacuation can

Fig. 6.7 Paraglacial landforms in central Fraser River basin (air photo 1087–46 BC by the Province of B.C.).

be interrupted by sediment deposition; and (c) consequently the total period of paraglacial effect is prolonged beyond the initial period of reworking of glacigenic sediments. Church and Slaymaker (1989) emphasized the generality of this definition, specifying that it is applicable to all periods of glacier retreat and that a paraglacial period is not restricted to the closing phases of glaciation but may extend well into the ensuing nonglacial interval. The essence of the concept is that recently deglaciated terrain is often initially in an unstable or metastable state and is thus vulnerable to rapid modification by subaerial agents.

Effectively then the paraglacial period is the period of readjustment from a glacial to a nonglacial condition. Different elements of paraglacial systems adjust at different rates, from steep, sediment mantled hillslopes (a few centuries) to large fluvial systems (>10 ka). The latter implies that fluvial systems in glaciated landscapes are still relaxing from the previous glacial period. "Fluvial sediment yield in British Columbia at all scales above 1 km^2 remains a consequence of the extraordinary glacial events of the Quaternary Period – the natural landscape of British Columbia is imprisoned in its history" (Church & Slaymaker 1989, p. 453).

Further, data presented in Church et al. (1999) confirm that at the scale of Canada there is no equilibrium between hydro-climate, denudation rate, and sediment yield.

Ballantyne (2002) has identified four trends in research into the paraglacial concept since 1985: (a) extension of application of the concept to different geomorphic contexts; (b) research on present-day paraglacial processes and land systems; (c) use of the paraglacial concept as a framework for research in contrasting deglaciation environments; and (d) a growing awareness of the paleo environmental significance of paraglacial facies in Quaternary stratigraphy.

As a result of Ballantyne's slight redefinition of the paraglacial concept, namely "non-glacial Earth surface processes, sediment accumulations, landforms, land systems, and landscapes that are directly conditioned by glaciation and deglaciation," the following environments are now being actively reinterpreted with the aid of the paraglacial concept: (a) alluvial fan, debris cone, and valley fill deposits (Church & Ryder 1972); (b) rock slopes (Wyrwoll 1977); (c) sediment-mantled slopes (Ballantyne & Benn 1994); (d) glacier forefields (Matthews 1992); (e) glacilacustrine systems (Leonard 1985); and (f) coastal systems (Forbes & Syvitski 1994; Table 6.2). Ballantyne (2002) has proposed a family of exhaustion-curves (plots of sediment availability within a paraglacial system against time since deglaciation) which describe the state of primary paraglacial landscape adjustment at any time after deglaciation (Fig. 6.8). The curves are based on the rates of operation of some primary paraglacial systems, in particular rock slopes, drift mantle slopes, glacier forelands, and large alluvial fans (Table 6.2).

6.5   GLACIAL LANDSCAPES: MEDIUM-SCALE

Little thought has been given to the process of modification of glacial landscapes by fluvial and mass movement because, at the macro-scale, the glacial imprint is still so obvious in the landscape. However, if we examine the medium and local-scale elements of the glacial landscape, it is apparent that nonglacial processes are actively modifying the morphology. The interesting questions are: how rapidly is this process of readjustment proceeding and what are the metrics that can provide consistent answers?

The literature on process domains provides one avenue for exploring this question. Much of this literature concerns nonglaciated landscapes and attempts to characterize the preferred scales of operation of mass movement and fluvial processes within maturely dissected landscapes (Dietrich et al. 1992; Montgomery 2001).

6.5.1   THE TRANSITION FROM GLACIAL TO FLUVIAL DOMINANCE

Recent work by Brardinoni and Hassan (2006) has examined process domains in the cirque landscape of the southern Coast Mountains, British Columbia (Plate 6.1) in an attempt to judge the extent of postglacial recovery. Ice flows have introduced strong anisotropies into the landscape, notably cirques, a hierarchical network of U-shaped valleys, valley steps, and oversteepened valley walls. The premise of the research is that glaciated environments are transient and contain finger prints of glacial erosion that are progressively erased by postglacial processes (Evans & Slaymaker 2004). Results show a more complex geomorphic picture than that described for unglaciated orogens (Stock & Dietrich 2003). Specifically, Quaternary climate changes have left a landscape where process-specific topographic signatures rarely match the domains of currently active geomorphic process. Longitudinal profiles of channels periodically scoured by debris flows remain largely controlled by the inherited glacial topography. Relict glacial cirques and hanging valleys enclose typical

TABLE 6.2 Rate of operation of some primary paraglacial systems (after Ballantyne 2002).

System	Sources	Rate of change* (yr)	Half life[†] (yr)	Duration[‡] (yr)
Paraglacial modification of rock slopes				
Rock-slope failure	Cruden and Hu (1993)	1.8×10^{-4}	3,850	25,000
Rock-mass creep	Tabor (1971), Beget (1994), Bovis (1990), Blair (1994)	$>1.5 \times 10^{-2}$	>45	>300
		$<2.3 \times 10^{-3}$	<300	<2,000
Talus accumulation	Luckman and Fiske (1995),	1.9×10^{-4}	3,610	24,000
	Hinchliffe and Ballantyne (1999)	7.7×10^{-4}	900	6,000
Paraglacial modification of drift-mantled slopes				
Formation of mature gully systems	Ballantyne and Benn (1994)	$2.3–9.2 \times 10^{-2}$	8–3	50–200
Formation and stabilization of small debris	Ballantyne and Benn (1994)	$1.5–9.2 \times 10^{-2}$	8–45	50–300
cones	Ballantyne (1995),	$2.3–9.2 \times 10^{-2}$	8–30	50–200
	Harrison and Winchester (1997)[§]	3.1×10^{-1}	2	15
Paraglacial modification of glacier forelands				
Reduction of moraine gradients	Welch (1970), Sharp (1984)	$2.3–4.6 \times 10^{-1}$	1–3	10–20
Frost-sorting of glacier forelands	Ballantyne and Matthews (1982, 1983)	$0.8–1.3 \times 10^{-1}$	5–9	35–60
Infiltration of fines from near-surface till	Boulton and Dent (1974)	$0.7–1.4 \times 10^{-1}$	5–10	35–70
Accumulation of large alluvial fans				
Formation and stabilization of large	Ryder (1971a), Lian and Hickin	1.9×10^{-3}	365	2,400
alluvial fans	(1996), Beaudoin and King (1994),	$>7.7 \times 10^{-4}$	<900	<6,000
	Friele et al. (1999), and others			

* Rate of change represents rate of decline in sediment availability.

[†] half life represents the time period over which 50% of "available sediment" is exhausted.

[‡] Duration represents the length of the period of paraglacial sediment transfer, which is assumed to end less than 1% of initial "available sediment" remains available for transport.

[§] Probably an underestimate of period of sediment reworking.

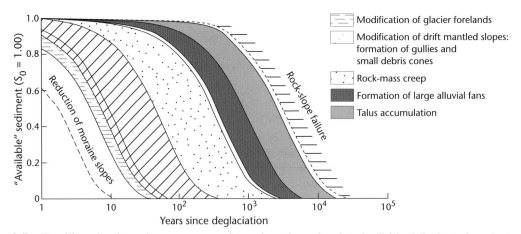

FIG. 6.8 Families of exhaustion-curve envelopes based on the data in Table 6.2 plotted against a logarithmic timescale (Ballantyne 2002).

hanging fluvial domains and influence the morphology of the landscape. It seems probable that these first order landforms will persist until the end of the present interglacial period. Further, by plotting channel slope against corresponding area-based process domains, it has been shown that glacial macro-forms drive the active processes that dominate mass transport at the local level, for example, channel reach scale (Plate 6.1) (Brardinoni & Hassan 2006).

6.6 PROGLACIAL LANDSCAPES: MEDIUM-SCALE

As pointed out by Embleton-Hamann (2004), proglacial means literally "in front of the glacier" and so can scarcely be used to describe a macro-scale landscape. At medium scale one conceives of the proglacial zone moving with the ice edge. As the ice retreats, the proglacial zone follows the retreating margin and former subglacial sediment–landform assemblages are eroded, redeposited, or buried.

There are transitions from proglacial (glacifluvial), through fluviglacial to fluvial or proglacial (glacilacustrine) to lacustrine and proglacial (glacimarine) to marine. Proglacial processes and landforms are classified according to distance from the ice margin and surrounding environment (Table 6.3).

6.6.1 GLACIFLUVIAL LANDFORMS

The classic study by Church (1972) of Baffin Island sandurs identifies valley sandurs as characteristic of proglacial environments (Fig. 6.9). Conditions favorable for sandur development are the presence of an abundant supply of detrital material and the occurrence of relatively frequent floods competent to move the material. Nival floods, summer icemelt floods, and summer storm runoff generate significant sediment transport events.

Sediment is transported mainly as bedload though solution and suspended load are significant. Transport is much greater than the rate of primary detritus production as large volumes of glacially derived material are being redistributed. Sandur streams are wide and shallow; the sandur surface is an aggradational feature. Because changes in river stage are so rapid, deposition is

TABLE 6.3 Classification of proglacial landforms according to different land-forming environments (after Embleton-Hamann 2004).

Environment	Process	Landform	See also:
Terrestrial, ice-marginal	Meltwater erosion	Ice-marginal meltwater channels	Meltwater and meltwater channel Urstromtäler
	Mass movement/meltwater deposition	Ice-marginal ramps and fans Dump and push moraines Recessional moraines	Moraine
	Glacitectonics	Composite ridges and thrust block moraines Hill-hole pairs and cupola hills	Glacitectonics Kame Kettle and kettle-hole
	Meltwater deposition	Kame and kettle topography	Ice stagnation topography
Subaquatic, ice-marginal	Mass movement/meltwater deposition	Morainal banks De Geer moraines	Glacilacustrine Glacimarine Moraine
	Meltwater deposition	Grounding-line fans Ice-contact (kame-) deltas Grounding-line wedge	
	Debris flows		
Transitional from ice-marginal to fluvial	Meltwater erosion	Scabland topography Spillways	Meltwater and meltwater channel Outburst flood
	Meltwater deposition	Outwash plain (sandur) Outwash fan Valley train Pitted outwash Kettle hole/pond	Glacifluvial Kettle and kettle-hole
Transitional from ice-marginal to lacustrine and marine	Meltwater deposition/mass movement	Deltas	Glacideltaic Glacilacustrine Glacimarine
	Deposition from suspension settling and iceberg activity	Cyclopels, cyclopsams, varves Dropstone mud and diamicton Iceberg dump mounds Iceberg scour marks	

often chaotic and patterns are found mainly in the mean size and variance of coarse materials. Most proglacial rivers carry large amounts of suspended sediment and bedload and this is characteristically deposited in extensive, gently sloping outwash plains known by the Icelandic term sandur (Fig. 6.9). Narrow tracts of outwash hemmed in by valley sides in mountainous terrain are termed valley trains (Church 1972; Church & Gilbert 1975). Pitted sandar are sandar which are cratered by hollows left by the melt-out of isolated buried blocks of glacier ice (Maizels 1977).

Glacifluvial sediments occur in supraglacial, englacial, subglacial, ice marginal, and proglacial environments (Brennand 2004). Such sediments differ from nonglacial fluvial sediments in four main ways: (a) They are typically coarser grained (boulder through sand sizes) as the flow velocity is generally too high for settling of the finest particles; (b) coarse grained lithofacies may have relatively poor sorting and rudimentary bedding;

FIG. 6.9 The proglacial landscape, Ekalugad Valley (from Church 1972).

(c) they typically exhibit abrupt changes in lithofacies due to rapid changes in flow regime; and (d) ice-contact glacifluvial deposits frequently include diamictons and eroded ice deposits as well as shearing, faulting, slumping, and subsidence.

6.6.2 THE CHANNELED SCABLANDS

The Channeled Scablands resulted from catastrophic drainage of Glacial Lake Missoula. They occupy an area of approximately 40,000 km² in Washington State

(Fig. 6.10). They are "a great anastomosis of flood channels and recessional gorges replete with a bizarre assemblage of erosional and depositional landforms such as rock basins, giant cataract alcoves, large residual islands, great bars of gravel and giant current dunes" (Baker et al. 1987). They resulted from catastrophic drainage of Glacial Lake Missoula, a large ice-dammed lake ponded up by the margin of the late Pleistocene Cordilleran ice sheet (Bretz 1923, 1925). Various calculations of the flood discharges range from 2 to 20 times

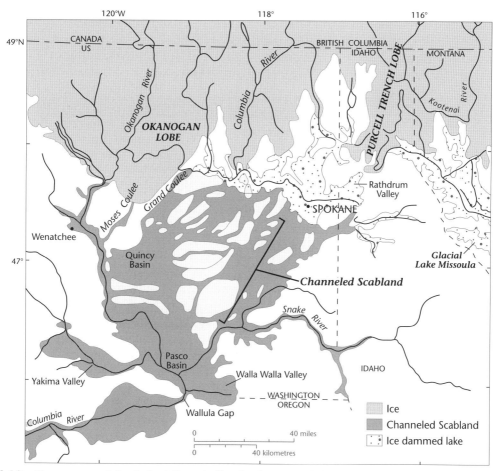

FIG. 6.10   The course of the Lake Missoula floods (horizontal dash pattern) and the location of the Channeled Scablands, Washington. The Cordilleran Ice Sheet (dot pattern) and ice-dammed lakes (coarse stipple pattern) are also shown (after Baker et al. 1987).

the mean flow of all the world's rivers (Baker 1973a). Up to 60 m of loess was stripped and in confined channels, water depths reached 100–200 m, culminating in nearly 300 m in the entrance to the Columbia gorge. Baker (1973a,b, 1981) concluded that most of this erosion was done in one or a few exceptionally large floods; Waitt (1980, 1985) has argued for up to 40 flood events. Baker et al. (1987) grouped the erosional forms into three categories: (a) scabland erosion complexes formed by the fluvial incision of bedrock; (b) streamlined erosional residuals which are lemniscate shaped loess hills; and (c) scour marks. Similar features have subsequently been identified from Siberia (Rudoy et al. 1989), the Canadian Prairies (Kehew & Lord 1989), Indiana (Fraser et al. 1983), Norway (Longva & Thoresen 1991), and Sweden (Elfstrom & Rossbacher 1985).

6.6.3 SUBGLACIAL CHANNELS

Of the various medium-scale erosional features, we will consider here only sub-glacial channels (see Benn & Evans 1998, pp. 323–42). Channel forms, cut by glacial meltwater, whether subglacial, ice-marginal, or proglacial, are widespread features of glaciated landscapes. The literature is divided into two broad categories: (a) that concern with the understanding glacial floods and jökulhlaups from a geophysical and hydro-logic perceptive (Fig. 6.11; Table 6.4); and (b) the characterization of the form and origin of subglacial channels. The literature on the contemporary jökulhlaups is well established but debate centers around the size and mechanisms involved in generating mega-floods in the past. Correspondence by Clarke et al. (2004) provides a stimulating discussion of various sources of misunder-standing. The main problem surrounding the subglacial mega-flood hypothesis by Shaw (1989) is the problem of water storage and

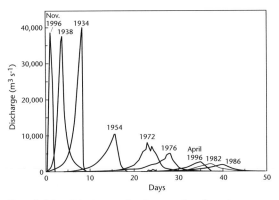

FIG. 6.11 Discharge hydrograph of various jökulhlaups from Grímsvötn, Iceland (Björnsson 2004).

meltwater supply (Table 6.5). Nevertheless, Shaw's work has drawn attention to the limits of our understanding of mega-scale, apparently subglacial, channel features. In the second literature on subglacial chan-nels, two types of subglacial channel have been distinguished solely on the basis of scale: Nye channels (up to a few thousand meters long and a few tens of meters wide) and tunnel valleys (up to 100 km long and 4 km wide).

Water filled subglacial channels are con-trolled by the hydraulic gradient (roughly parallel with the ice surface slope), subglacial topography, and glacier-sliding velocity. The long profiles of these channels are typically undulatory and the fact that they cross topographic barriers is one of the clearest criteria for their recognition (Mannerfelt 1945). The influence of both ice thickness and basal topography on subglacial drainage has been examined by Booth and Hallet (1993) who compared channel networks cut below the late glacial Puget Lobe of the Cordilleran Ice Sheet with theoretical patterns predicted by Shreve (1972) (Fig. 6.12a). Nye channels, typically up to 8 km long, 150 m wide, and 100 m deep are well developed in this area and the former ice surface can be reconstructed with considerable confidence.

TABLE 6.4 Sources of glacial floods and characteristics of drainage. P_w stands for water pressure and P_i for ice overburden pressure (after Björnsson 2004).

Source/type of lake	Flood initiation and drainage routes				
	Subglacial drainage	Englacial drainage	Subaerial drainage (overtopping, downcutting, dam failure)		
Marginal lake[*]	Through ice tunnels at $P_w < P_i$ or in a sheet flow at $P_w = P_i$, propagating a flood wave at $P_w > P_i$	Upwards from the base, creating supraglacial outlets by hydrofacturing the ice and retrofeeding moulins at $P_w > P_i$	Drainage over a subaerial breach	Fluvial erosion of sediment dums; piping	Mechanical failure caused by tectonic activity; rockfalls; landslides
Subglacial eruption often in the absence of significant storage[†]		Downward to the base through englacial tunnels			
Subglacial lakes in hydrothermal areas[‡]					
Supraglacial lakes in cauldrons and sink holes[§]					
Englacial storage[‖]					
Linked subglacial cavities[¶]	Distributed drainage System switches to a tunnel system				

[*] Bretz (1925, 1969); Thorarinsson (1939); Liestol (1956); Stone (1963); Post and Mayo (1971); Clague and Mathews (1973); Björnsson (1976); Lliboutry et al. (1977); Haeberli (1983); Sturm and Benson (1985); Yamada (1998).

[†] Thorarinsson (1957, 1958); Sturm et al. (1986); Björnsson (1988); Pierson et al. (1990); Thouret (1990); Trabant and Meyer (1992); Gudmundsson et al. (1997).

[‡] Björnsson (1974, 1988, 2002); Gudmundsson et al. (1995).

[§] Björnsson (1976); Russell (1993).

[‖] Haeberli (1983); Driedger and Fountain (1989); Walder and Driedger (1995).

[¶] Kamb et al. (1985); Kamb (1987); Björnsson (1998).

TABLE 6.5   Estimated volume of real and conjectured freshwater reservoirs (after Clarke et al. 2004).

Reservoir	Volume (10³ km³)	Notes
Subaerial reservoirs		
Altay flood reservoirs	1	Baker et al. (1993, Note 24)
Glacial Lake Missoula	2.184	Clarke et al. (1984)
Modern Lake Superior	12.1	www.greatlakes.net/lakes/ref/supfact.html
Glacial Lake Agassiz* final flood	151	Clarke et al. (2004)
Subglacial reservoirs		
Lake Vostok, East Antarctica	7	Maximum (Studinger et al. 2004)
All Antarctic subglacial lakes	12	Maximum (Dowdeswell & Siegert 2002)
Lake Livingstone A	84	Shaw (1989)
Lake Livingstone B	724	Blanchon and Shaw (1995)
Cumulative basal melt of LIS	1,860	Clarke et al. (2005)

* Refer to the superlake formed by the coalescence of glacial lakes Agassiz and Ojibway as "Lake Agassiz" (Clarke et al. 2004).

Although there are several sources of uncertainty in such a reconstruction, the fit is remarkably good, supporting the idea that drainage below the late glacial Puget Lobe was through a dendritic river network of water filled Nye channels directed by ice pressure and local topography.

Tunnel valleys are larger by an order of magnitude, usually have wide, relatively flat bottoms and steep sides and sometimes have lakes occupying troughs along their lengths. There is a tendency for tunnel valleys to terminate abruptly at major moraines, where they may grade into large subaerial ice-contact fans (e.g. Ussing 1903). The surfaces of these fans may lie up to 100 m above the tunnel valley bottom reflecting deposition from pressurized meltwater emerging from under the ice. In the case of the Gwaun–Jordanston subglacial network in Pembrokeshire, west Wales, the steep-sided and flat-floored channels are cut in bedrock, are up to 45 m deep, and have a combined maximum length of 20 km (Fig. 6.12b; Sugden & John 1976). Tunnel valleys may be completely infilled by glacigenic, glacifluvial, glacilacustrine, glacimarine, or nonglacial deposits and may not have any clear topographic expression on the surface (O Cofaigh 1996). In southern Alberta, preglacial river valleys are buried under later-glacial sediments. Chains of elongate depressions show the position of subglacial channels. The long profiles of the buried valley bottoms show that they are normal fluvial forms, but the superimposed chains of depressions indicate that subglacial meltwater excavated some of the glacial sediments (Stalker 1961).

6.6.4   SUBGLACIAL, ICE-MARGINAL, AND SUPRAGLACIAL SEDIMENT–LANDFORM ASSOCIATIONS

Subglacial landforms include flutings, drumlins, mega-flutings, mega-scale lineations, Rogen moraines, eskers, and tuyas. They are longitudinal or transverse accumulations of sediment formed below active ice (Rose 1987). The distinction between drumlins, flutings, and mega-flutings is based on the length and elongation ratio

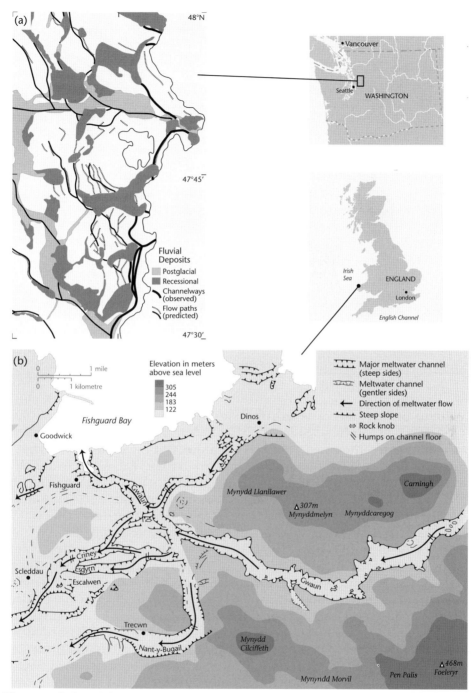

Fig. 6.12 (a) Subglacial drainage of the eastern part of the late Pleistocene Puget lobe of the Cordilleran Ice Sheet. Comparison of observed channelways and flow paths predicted from a numerical model and (b) subglacial drainage network in Pembrokeshire, west Wales. The steep-sided and flat-floored channels are cut in bedrock, are up to 45 m deep, and have a combined maximum length of 20 km (after Sugden & John 1976; Booth & Hallet 1993).

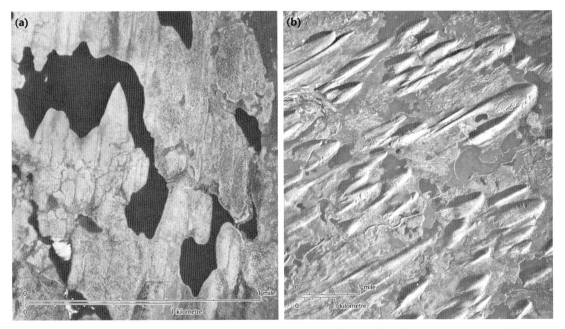

Fig. 6.13 Drumlins and macrolineations. (a) Near Snare Lake, Saskatchewan (NTS Map Sheet 74 J; Air photo designation A14509) and (b) south of Mashik Lake, Quebec (NTS Map Sheet 22 L; Air photo designation A295). Photos courtesy of the National Air Photo Library, Ottawa.

of the landform or the ratio of length to width (Fig. 6.13). Drumlins tend to have a higher, wider, and blunter form on the side of the drumlin from which the glacier advanced. Therefore, in the area of northern Saskatchewan illustrated in Fig. 6.13a, the last glacier appears to have advanced from the northeast and moved in a southwesterly direction (Mollard 1986). In general, drumlins (1–4 km long) tend to be more common in areas of hard, well-jointed sandstone and carbonate rocks, in this case Athabasca sandstone of Precambrian age. The large amount of outwash (flat areas) and ice-contact drift sediments (eskers) that can be seen between the drumlins indicates that sands are the common weathering products of this sandstone. In the example from Quebec (Fig. 6.13b), the drumlinoid ridges and drumlins are somewhat smaller (up to 1.5 km long), but there are also negative relief forms such as flutings and grooves aligned parallel to the drumlins.

Mega-scale lineations are very large elongate forms many tens of kilometer long, hundreds of meter wide, and >25 m high (Clark 1993). These forms are all aligned parallel to ice flow. Rogen moraines, by contrast, occur in fields of coalescent crescentic ridges up to 30 m high and 100 m wide, lying transverse to the direction of former ice flow. Eskers are elongate sinuous ridges of glacifluvial sand and gravel. They are the infillings of ice-walled river channels. A final subglacial landform is the tuya, a flat-topped volcano built during subglacial eruptions. The feature was first described by Mathews (1947) in north-central British Columbia, but has since been reported from Iceland and Antarctica.

6.7 Periglacial landscapes: medium-scale

As indicated in Chapter 3, there are three groups of landforms that are diagnostic of periglacial landscapes: patterned ground, palsas, and pingoes (Figs. 6.14–6.16). The Associate Committee on Geotechnical Research of the National Research Council of Canada (1988) defines the water–ice phase change as essential to the processes of cryoturbation. These processes include frost heave (Penner 1973), thaw settlement/consolidation, and all differential movements due to temperature changes and the growth and disappearance of segregated ice bodies (Gold 1985). The net effect of these processes is to produce patterned ground (French 1996). Patterned ground is classified according to (a) its geometric form and (b) the absence or presence of sorting. Circles, polygons, stripes, nets, and steps, all of which may be either sorted or unsorted, define ten principal categories of patterned ground. The illustration provided (Fig. 6.14) shows low-centered unsorted polygons on Bathurst Island in the Canadian Arctic. Low-center polygons are usually associated with poorly drained, peat covered, silty soil materials

occurring in depressions (Mollard 1986). Figure 6.14 should be compared with Fig. 3.9 showing sorted stone circles in Svalbard. For an extended discussion on patterned ground, Washburn (1979) remains the best source.

Central Bathurst Island, Nunavut shows a classic expression of low-center polygons in the northeast quarter of the photograph (Fig. 6.14). In most of the remaining part of the photograph, shallow melting of the permafrost has formed thermokarst lakes, usually from the coalescence of many individual low-center polygons. The site is known as Polar Bear Pass and is ecologically significant because it provides favored habitat for polar bears, muskoxen, and waterfowl.

Palsas are small mounds of peat rising out of mires in the subarctic region characteristic of the discontinuous permafrost zone (where the peat layer is thick enough) (Seppala 2004). In northern Ontario, palsas are close to their southernmost extent (Fig. 6.15). Palsas contain a permanently frozen core of

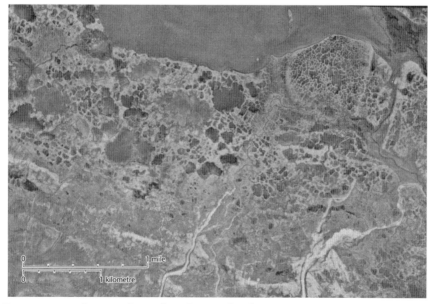

FIG. 6.14 Low center polygons. Central Bathurst Island, Nunavut (NTS Map Sheet 68 H; Air photo designation A16203). Photo courtesy of the National Air Photo Library, Ottawa.

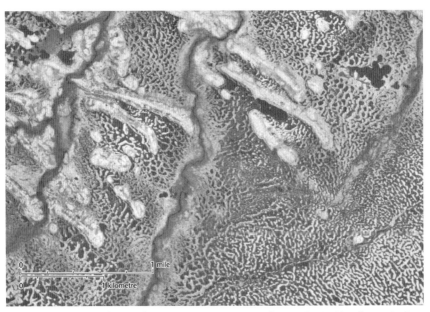

Fig. 6.15   Palsas. Southeast of Winisk, Ontario (NTS Map Sheet 43 N; Air photo designation A14136). Photo courtesy of the National Air Photo Library, Ottawa.

peat and/or silt, small ice crystals, and thin layers of segregated ice. An insulating peat layer is essential to preserve the frozen core during the summer. It is also important that the peat be dry in summer (low thermal conductivity) and wet in autumn (higher thermal conductivity). This allows the cold wave to penetrate deeply enough into the peat layer that it does not thaw in summer. Although the individual feature is small, they commonly are found in large numbers, such as to give the name "palsa mire" to the medium-scale expression of this feature.

The photograph from Ontario (Fig. 6.15) shows a web-like pattern of mature palsas, coalesced palsas, and peat plateaus in an area of extensive fen peatland (Mollard 1986). The linear ridges are raised marine sand and gravelly sand beaches. The area shown is near the boundary between continuous and widespread permafrost in the discontinuous zone. It is also located in the boreal forest–tundra transition (ecotone).

A pingo is a perennial permafrost mound or hill, formed through the growth of a body of ice in the subsurface (Gurney 2004). A pingo can reach heights of 50 m and basal diameters of over 600 m. There may be as much as 1 million cubic meter of ice in a large pingo.

Pingoes are classified as either hydraulic (or open system) or hydrostatic (or closed system). Hydraulic pingoes are initiated by water under pressure of a hydraulic head in a valley bottom and are features of a high-relief environment. Hydrostatic pingoes are initiated by the draining of a lake in a continuous permafrost environment. Following lake drainage, the unfrozen saturated sediments are aggraded by permafrost and the pore water is progressively squeezed out. The highest concentration of pingoes is found in the Tuktoyaktuk Peninsula in the Mackenzie delta region (Mackay 1972, 1998). The reason for their unusual concentration in this region is that the coastline

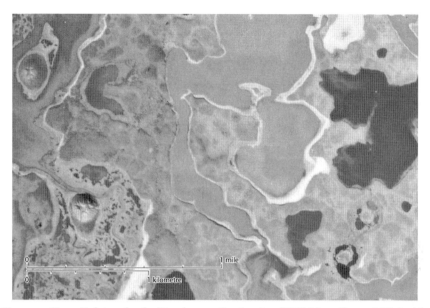

FIG. 6.16 Pingoes. Near Tuktoyaktuk, Northwest Territories (NTS Map Sheet 107 C; Air photo designation A19811). Photo courtesy of the National Air Photo Library, Ottawa.

is rapidly receding and lakes are regularly drained as the coastline intersects the lake shoreline. Closed system pingoes are associated with progressive freezing and contraction and extrusion of unfrozen ground in lakes whose bottoms are composed of silt, sand, and organic material. The beds of drained lakes are their preferred location (Fig. 6.16).

6.7.1 THE TRANSITION FROM PERIGLACIAL TO FLUVIAL DOMINANCE

French (1996) has raised a parallel issue in noting that many areas of current periglacial conditions have only recently emerged from beneath continental ice sheets and few periglacial regions can be regarded as being in geomorphic equilibrium. He identifies central and eastern Siberia, interior Alaska, northern Yukon, and some of the Canadian Arctic islands as regions where periglaciation has been continuous throughout

the Quaternary. French provides two case studies to illustrate periglacial terrain that has not been glaciated during the past 2 Ma: (a) the Beaufort Plain of northwestern Banks Island and (b) the Barn Mountains of interior northern Yukon. His conclusion is that a periglacial landscape of low relief (Beaufort Plain) is dominantly fluvial with braided rivers, highly variable discharge, thermal contraction cracks, asymmetrical valleys, and mass wasting beneath snow banks (cf. Church 1972). Periglacial landscapes of high relief (Barn Mountains) consist of structurally controlled upland massifs surrounded by extensive pediments, dissected by braided rivers draining to the Arctic coast. Angular rock rubble veneers many of the upland surfaces and large, nonsorted polygons are common. Tors and cryoplanation surfaces exist at higher elevations, but there is little sign that they, or the angular rock rubble surfaces, are forming under today's climate. French and Harry (1992) have speculated that higher moisture levels would have been necessary to intensify weathering and remove the sediments.

6.8   PARAGLACIAL LANDSCAPES: MEDIUM-SCALE

The paraglacial episode produced distinctive sediment–landform associations on slopes and on valley floors. When glaciers retreated from a drainage basin, unstable sediments were made available to rivers resulting in rapid aggradation of thick valley fills and alluvial fans at the mouths of tributary valleys or gullies (Fig. 6.17a). Ryder (1971a) was the first to attract attention to this feature from her detailed studies of early Holocene fans and terraced valley fills in south-central British Columbia.

To the extent that such processes are conditioned by glaciation, all of the following categories may be included in paraglacial effects (Benn & Evans 1998): (a) falls; (b) gelifluction; (c) slumps and slides; (d) debris flows and torrents; (e) turbulent flows; and (f) glacilacustrine and glacimarine processes. However they recognize that mass movements commonly evolve from one type to another during transport. Mass movement deposits may therefore show lateral and vertical changes in structure, grain size, and sorting. It is also true that debris may go through numerous cycles of movement and redeposition, especially in ice contact environments. Paraglacial slope deposits often form wedge- or cone-shaped slope foot accumulations (Fig. 6.17b). Talus deposits may be unmodified rockfall material or may be modified by snow avalanches, debris flows, or internal creep deformation.

Thick sequences of diamictons can result from debris flows, shallow landslides, and slower mass movements such as gelifluction and soil creep (Ballantyne & Harris 1994).

FIG. 6.17   (a) Landforms and sediments associated with alluvial fans in B.C. (Ryder 1971a) and (b) three stages in the paraglacial reworking of steep drift slopes in glaciated valleys (after Ballantyne & Benn 1996).

Such deposits are usually called "head" and are widely exposed in cliff sections around the coasts of lowland Britain (De La Beche 1839). Because the period of most active formation of head was the immediate postglacial period, their presence is readily reinterpreted as evidence of paraglacial conditions. Glacigenic sediments reworked by debris flows are common in glaciated mountain environments; they are distinguished from

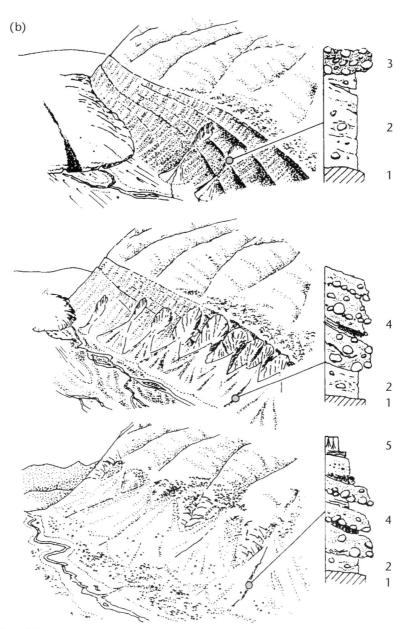

Fig. 6.17   (Cont'd)

nonglacial slope deposits by the presence of erratics and from *in situ* tills by slope parallel bedding (Benn & Evans 1998).

6.9 GLACIAL LANDSCAPES: LOCAL-SCALE

Much of the work of glacial geomorphologists is focused at the local scale, both in terms of small scale erosional forms and, even more intensively, on the interpretation of sediment facies and sediment–landform associations. Again, the reader is advised to turn to Benn and Evans (1998) for a comprehensive discussion. For the purpose of this book, in which transient cryospheric systems and transitional landscapes are being emphasized, it will suffice to illustrate a small number of these forms and facies. The essential point is that: "sediment sequences in glacial environments can be bewildering in their variety and complexity, and reflect the ever-changing character of their processes of deposition" (Benn & Evans 1998, p. 378). As a result, it is necessary to impose a framework for their interpretation, commonly called a debris cascade system (Chorley et al. 1984). A debris cascade system consists of debris sources, transport paths, and depositional processes. A glacial landscape will commonly have glacier bed, valley side, and atmospheric debris sources; subglacial, englacial, supraglacial, and subaerial transport paths and a whole range of depositional processes, of which only glacial processes are unique to glacial environments. In classifying sediments deposited in glacial environments, a useful distinction is that between primary deposits, laid down by uniquely glacial agencies, and secondary deposits, which have been reworked by nonglacial processes.

In this way it becomes apparent that at the local-scale also transitional forms and facies are more common than uniquely glacial erosional and depositional forms.

If we follow Lawson (1981) and insist that till is sediment deposited directly by glacier ice, and has not undergone subsequent disaggregation and resedimentation, then the transitional nature of much of the local-scale landscape becomes self-evident. Unfortunately, the literature is ambiguous on this point as many Quaternarists allow the incorporation of reworked glacial sediments in their definition of till (Dreimanis 1989).

6.9.1 PRIMARY GLACIGENIC DEPOSITS

Primary glacigenic deposits are defined as sediments laid down by uniquely glacial agencies which normally includes supraglacial deposition by melt out and sublimation of debris-rich ice and subglacial deposition by some combination of lodgment, deposition from a deforming layer, and melt out.

Lodgment till is "sediment deposited by plastering of glacial debris from a sliding glacier sole by pressure melting or other mechanical processes" (Dreimanis 1989). Lodgment tills are commonly overconsolidated because the pressure of the ice overburden induces dewatering. They usually have high bulk density and penetration resistance. Deformation till is sediment that has been disaggregated and largely homogenized by shearing in a subglacial deforming layer (Elson 1961). Melt out till, commonly called ablation till in the past, is sediment released by melting of slowly moving debris-rich glacier ice, and directly deposited without subsequent transport or deformation. Ice marginal moraines can be classified into four types: (a) glacitectonic landforms; (b) push and squeeze moraines; (c) dump moraines; and (d) latero-frontal ramps and fans. Glacitectonic landforms result from large-scale displacement of proglacial material due to stresses imposed by glacier ice. One of the larger examples is that of Herschel Island in Yukon Territory,

Canada (Mackay 1959). The island, with an area >100 km² is composed of preglacial Pleistocene sediments thrust up from a now submerged depression. Deformation occurred from the southeast consistent with a glacier lobe during the last glaciation documented by Rampton (1982). Push moraines are small moraine ridges produced by minor glacier advances. Dump moraines are commonly formed adjacent to retreating glaciers which dump supraglacial material on to the former subglacial surface. When the retreating ice margin remains stationary for a while, the dump moraine will grow. Latero-frontal ramps and fans are formed by coalescent debris fans which descend from the glacier snout (Owen & Derbyshire 1989). Supraglacial landforms include: (a) medial moraines; (b) hummocky moraine; (c) kame and kettle topography; (d) pitted sandar. Medial moraines are prominent features on the surface of many glaciers, but they tend to produce rather small landforms. The most spectacular medial moraines, referred to as interlobate moraines occur on the Canadian Shield and document the coalescence of the margins of different sectors of the former Laurentide Ice Sheet (Dyke & Dredge 1989). They may be as much as 1,000 km long (Wilson 1938). Hummocky moraine refers strictly to moraines deposited during the melt out of debris-mantled glaciers (Harker 1901). Terrain is moundy, irregular morainic topography ranging from entirely chaotic assemblages to nested transverse ridges. Kame and kettle topography forms tracts of mounds and ridges (kames) and intervening hollows (kettles).

6.9.2 SMALL-SCALE EROSIONAL FORMS

Of the four commonly recognized small-scale forms (striae, rat-tails, chatter marks, and P-forms) we will consider here only striae or striations. Esmark (1824) was one of the earliest observers of the near universal presence of scratches incised into bedrock in his native Norway and recognized this as evidence of scouring by particles embedded in glacier ice.

The orientations of striae on a flat upper surface of an outcrop are commonly parallel with each other and deviate only slightly from the average ice-flow direction. But on uneven surfaces they may deviate markedly because of the irregularity of the basal flow of the glacier. Often it is possible to see two or more sets of striae, indicating separate ice-flow events. Cross-cutting relations may result from shifts in ice dispersal centers during a single glaciation or from separate glacial episodes. Rampton et al. (1984) and Stea (1994) have documented many Wisconsinan phases of ice flow in New Brunswick and Nova Scotia using complex patterns of striae.

6.10 PROGLACIAL LANDSCAPES: LOCAL-SCALE

The proglacial landscape is defined as the landscape that lies immediately adjacent to glaciers or ice sheets. The environment can therefore be terrestrial, fluvial, lacustrine, or marine. Depositional landforms can be moraines, sandurs, deltas, and sediments deposited in lakes or the adjacent ocean. The proglacial zone migrates with the ice edge: as the ice advances, the proglacial environment moves forward; as the ice retreats formerly subglacially formed landforms are exposed, eroded, or buried (Embleton-Hamann 2004; Table 6.3).

Subaqueous debris-flow deposits are more commonly divided into cohesive and cohesionless flows. Cohesive debris-flow deposits form sheet like or lobate beds, with planar or slightly scoured bases. Such deposits, frequently kilometers long, are common in Antarctica (Wright & Anderson 1982) and in glacial lakes (Eyles 1987). Cohesionless debris-flow deposits typically form dipping sheet like or lobate masses. They are gravel, sand, or pebbly sand and may be well or poorly sorted, depending on the source material. Bases are commonly erosive and can form deep channelized scours aligned downslope (Lonne 1993).

Deposits from suspension settling and iceberg activity include the results of glacimarine and glacilacustrine sedimentation (e.g. Dowdeswell & Scourse 1990; Eyles & Eyles 1992; Menzies 1995; Fig. 6.18).

Glacimarine sediments in the Canadian Cordillera were deposited on the sea floor in front of temperate tidewater glaciers. They consist of material that was introduced into the sea by meltwater streams, flows of supraglacial debris directly from ice, and by the melting of icebergs. The most common type of glacimarine sediment is massive to stratified mud. By contrast, some muds are very gravelly and cannot be easily distinguished from till. Stratified gravel

and sand of glacimarine origin are restricted to deltas, subaqueous outwash fans, and beaches, contain less than 5% gravel-size material, and some are stone-free (Clague 1989). Subaqueous moraines are transverse moraines deposited at or close to the grounding lines of water terminating glaciers. They are also known as De Geer moraines (Hoppe 1959; Booth 1986).

Glacifluvial sediments associated with jökulhlaup events are illustrated in Fig. 6.19. This figure shows how facies associations deposited by jökulhlaups vary systematically with flow type and grain size distribution of the available sediment. A continuum of flow types is shown along the horizontal axis, ranging from high-concentration, cohesive debris flows to low-concentration, turbulent fluid flows. Sediment grain size is plotted on the vertical axis, which shows the relative abundance of a coarse gravel mode (c) and a finer grained pumice

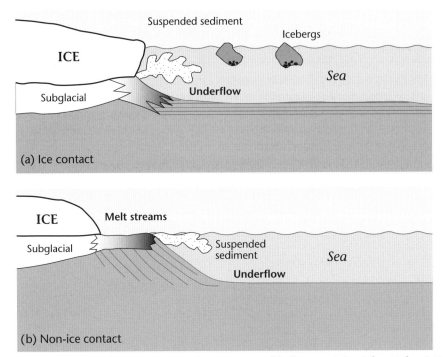

Fig. 6.18 Glacier contact and noncontact environments and iceberg activity (after Eyles & Eyles 1992).

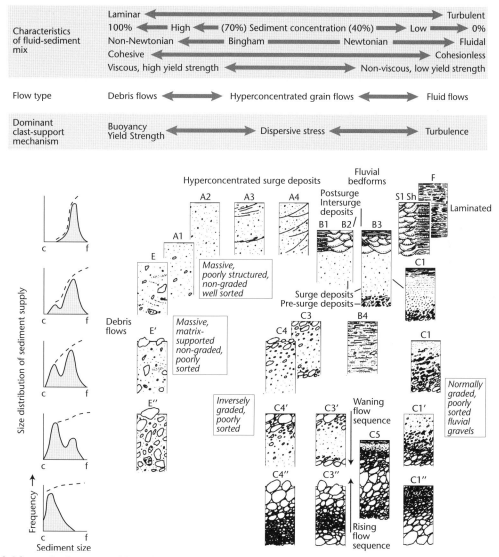

FIG. 6.19 Lithofacies profiles of Icelandic sandur deposits (after Maizels 1993).

fragment mode (f). This diagram demonstrates that deposits will increase in coarseness as the source materials get coarser, so that the granular sediments of volcano-glacial sandar plot toward the top of the diagram and the gravelly sediments of the limno-glacial and nonjökulhlaup sandar plot toward the base. Also the profiles show increasing organization toward the right of the diagram, with decreasing sediment concentration. At these lower sediment concentrations, particles are more able to move relative to one another, allowing the development of inverse or normal grading and other sedimentary structures. Glacier forefields such as those exposed in Plate 6.2 provide examples of small-scale proglacial deposition (Slaymaker 1974).

Glacier fed deltas receive sediment from glacial meltwater streams; ice-contact deltas are built out directly from glacier margins. Glacier fed deltas are commonly divided into Hjulström type (shallow water) and Gilbert type (intermediate to deep water) deltas. Because of the steepness of foreset beds on the Gilbert type deltas remobilization of debris is common and subaqueous debris cones are often produced. Ice-contact deltas commonly display slumping and normal faults. A particularly active glacier and Quaternary volcanics fed delta is illustrated in Fig. 6.20. Gilbert (1975) has described the usually active sedimentation processes at the head of Lillooet Lake. Jordan and Slaymaker (1991) established the sources of sediment and developed a sediment budget for the Lillooet River watershed (Fig. 6.21) and estimated the

average clastic sediment yield draining into upper Lillooet Lake from 1914 to 1969 to be $1.1 \times 10^6 \text{ m}^3 \text{ a}^{-1}$. Dissolved sediment yield is estimated to be an additional $0.23 \times 10^6 \text{ m}^3 \text{ a}^{-1}$.

6.11   PERIGLACIAL LANDSCAPES: LOCAL-SCALE

Perhaps the most important feature of the local-scale periglacial landscape is the nature and extent of ground ice (see Yershov 2004). Air temperature is important; ground temperature is even more important. But if the "ice content" of the soil is low or the amount of "excess ice" is small, the potential morphological change or volumetric ground loss under thawing conditions will be small. French (1996) cites three reasons for the importance of ground ice in

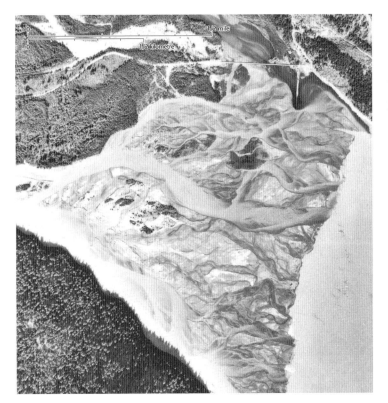

FIG. 6.20 Lillooet River delta (Gilbert 1975). September 13, 1967. Lillooet Lake, British Columbia (NTS Map Sheet 92 J; Air photo designation BC 5271). Photo courtesy of the National Air Photo Library, Ottawa.

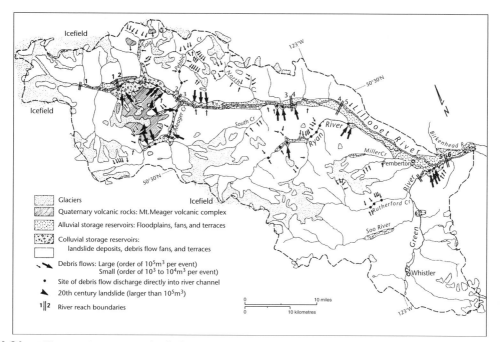

FIG. 6.21   Lillooet River watershed showing principal sediment sources and storage reservoirs (Jordan & Slaymaker 1991).

periglacial landscapes: (a) the thaw of ice rich permafrost terrain has major geotechnical implications; (b) the stratigraphic analysis of ground ice bodies and the cryostructures of their enclosing sediments are important tools in reconstructing Quaternary environments (Fig. 6.22); and (c) the recognition of ground ice pseudomorphs in regions that are presently beyond the limit of permafrost provides evidence of the former extent of permafrost (Figs. 6.23 and 6.24) (Mackay & Slaymaker 1989).

For the purpose of our present consideration of transitional landscapes, the phenomenon of thermokarst illustrates the point well. Thermokarst is the process of ground ice melt accompanied by local collapse or subsidence of the ground surface. The extent of morphological change associated with the thermokarst process depends on (a) the magnitude of the increase in depth of the active layer; (b) the ice content

of the soil; and (c) the tectonic regime of the region (French 1996).

Changes in the thickness of the active layer may result from either climatic, vegetational, geomorphic, or anthropogenic factors. The distribution of ground ice, by contrast with segregated or pore ice, determines the pattern of ground subsidence. Finally, thermokarst forms develop best in stable lowland areas, because tectonically subsiding regions tend to hide the thermokarst forms due to active sedimentation. A transition from periglacial to fluvial landscape is illustrated by the Horton River delta. Dendrochronological analysis and radiocarbon dating show that breakthrough to the sea occurred along the lower course of the Horton River after AD 1640. At the breakthrough site, the Horton River flowed 10 m above sea level so that river erosion and delta building would have been rapid. By 1826, when the coast was mapped by the

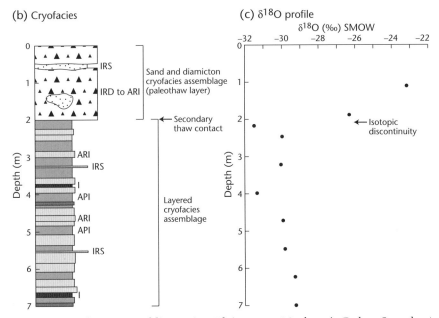

FIG. 6.22 Cryostratigraphy at Crumbling Point, Pleistocene Mackenzie Delta, Canada. (a) Sand wedge penetrating both a layered and a sand and diamicton cryofacies assemblage. Arrows mark the contact between the assemblages. Person is 1.8 m. (b) Generalized cryofacies log. (c) $\delta^{18}O$ profile of ground ice within both cryofacies assemblages. At their contact is an isotopic discontinuity. SMOW, standard mean ocean water (Murton & French 1994).

Fig. 6.23 Horton River delta (Mackay & Slaymaker 1989). Franklin Bay, Northwest Territories (NTS Map Sheet 97 C; Airphoto designation A119216). Photo courtesy of the National Air Photo Library, Ottawa.

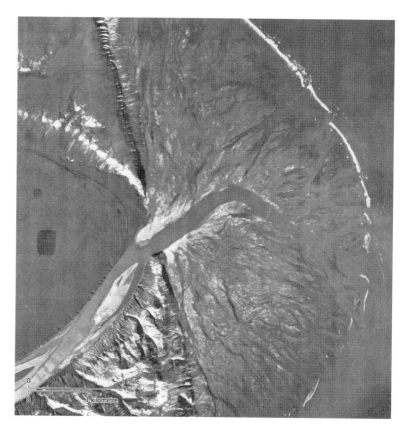

expedition of Dr. John Richardson, the Horton Delta had a surface area of about 10 to 15 km², approximately one third of the present area. Construction of the delta has protected the adjacent sea cliffs from wave erosion. Attention is paid to the nature of geomorphic change in a permafrost environment and the role of river and sea ice in affecting changes.

6.12 Paraglacial landscapes: local-scale

Although a number of studies have explicitly identified distinctive paraglacial facies in terrestrial Quaternary successions, the topic remains at a nascent stage (Ballantyne 2002). Interest has focused primarily on the

reinterpretation of diamict units. Identification of paraglacial facies has been effectively limited to sequences that postdate the LGM. As knowledge of the significance of paraglacial facies becomes more widely appreciated, it seems likely that they will be identified more frequently in the Quaternary record.

Of special interest because of their widespread distribution in glaciated mountain environments are debris flow deposits, both subaerially and subaqueously formed (Fig. 6.25). Subaerial debris flow deposits are usually diamictons with a fine-grained matrix and variable clast content. In section, individual flows form tabular or lens-shaped units and if flows are channelized, units can have concave-up erosional bases

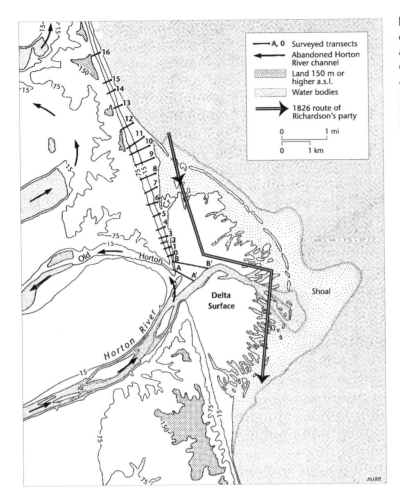

Fig. 6.24 The Horton River delta showing radiocarbon and dendrochronological dated sites, surveyed transects, and the 1826 route of Richardson's party plotted from an unpublished 1826 field survey (Mackay & Slaymaker 1989).

and flat tops. "Where successive flows are similar in texture, individual units are difficult to distinguish. Boundaries are usually marked by basal concentrations of clasts, upper washed horizons, interbeds of silt, sand or gravel or more subtle bedding structures" (Benn & Evans, p. 407). It should also be noted that debris torrent deposits, common in glaciated British Columbia and South Island, New Zealand, are characterized by very little fine-grained matrix and significant proportions of organic material (Slaymaker 1988).

6.13 LANDSCAPE RESISTANCE, COLLAPSE, AND RECOVERY

The data presented by Ballantyne bring into question the possibility of any equilibrium or balanced condition in landscapes that have undergone Quaternary glaciation. This discussion raises the question of landscape resistance, collapse, and recovery (Fig. 6.26). Brunsden (1990) has discussed these concepts in his propositions 4 through 10. Central to his discussion is his sense of geomorphically relevant time and the

FIG. 6.25 Examples of different structures from debris flows (from Benn & Evans 1998).

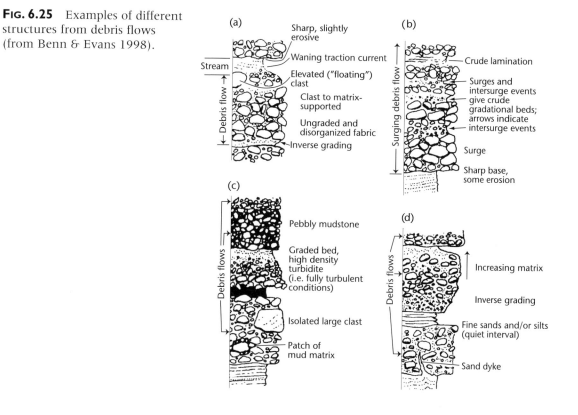

nature of landscape resistance as follows: (a) recurrence interval, that length of time which elapses between impulses of change; (b) reaction time, the time needed before the system notices that a perturbation has occurred (Fig. 6.26); (c) relaxation time, which is the time required for a system to respond and to reach a new characteristic state; (d) characteristic form time, that time during which a characteristic form persists; (e) transient form time, that time during which the landscape attempts to reach a characteristic form but is interrupted by new perturbations (in the preceding discussion of the landscape of British Columbia would be the time between glacial advances); (f) transient form ratio, which is the ratio of the relaxation time to the recurrence interval (in the British Columbia case, this ratio

would be >1 if the Church and Slaymaker analysis is correct; the implication is that the landscape is a transient landscape).

6.14 TRANSITIONAL LANDSCAPES AT QUATERNARY, HOLOCENE, AND ANTHROPOCENE TIMESCALES

Finally, we can discuss the relations between global environmental change at Quaternary, Holocene, and Anthropocene timescales. The Quaternary stratigraphic record of British Columbia, Canada is largely a product of brief depositional events separated by long periods of erosion or nondeposition (Clague 1986). The paraglacial episode, lasting from 12 ka to 6 ka BP (calendar years), is an example of a geomorphically active period. Long periods of comparative

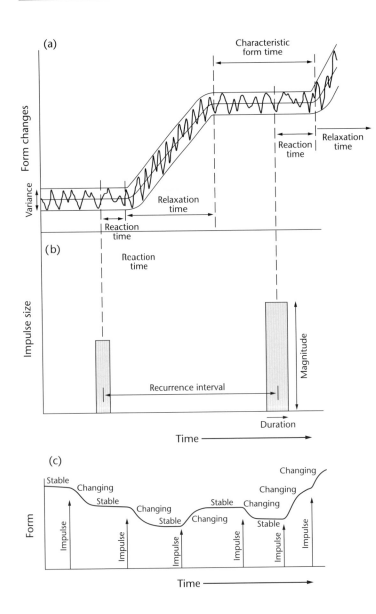

Fig. 6.26 The definitions of geomorphological time, impulse recurrence, reaction, relaxation, and characteristic form (after Brunsden 1990).

quiescence have alternated with shorter periods of intense geomorphic activity. This is not dissimilar to the geologic norm for all erosion and sedimentation (Gage 1970; Ager 1981). Schumm's complex response model emphasizes erosion following threshold exceedances at specific locations alongside slopes and whole basins that are inactive (Schumm 1973). Starkel (1987) showed a similar alternation of geomorphic and soils processes through one

glacial–interglacial cycle. The ice sheet decay process, incorporating Heinrich and Dansgaard–Oeschger events, seems to behave with a similar alternation (Marshall & Clarke 1999) and binge-purge models of cold-based and warm-based ice (MacAyeal 1993) are predicated on general ice sheet retreat conditions.

It seems probable that ice sheet covered landscapes, in which geomorphic work almost ceases during ice sheet cover, glacierized

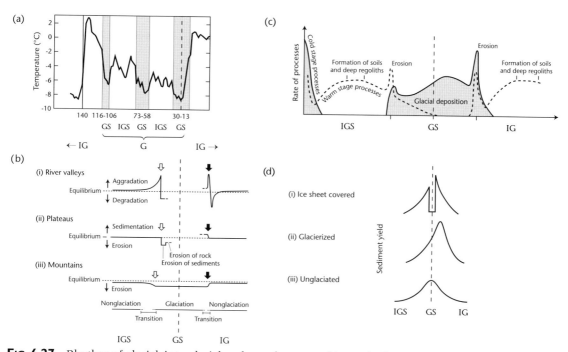

FIG. 6.27 Rhythm of glacial–interglacial cycles and geomorphic work (from Owens & Slaymaker 2004).

landscapes, which are geomorphically active throughout and landscapes that were unglaciated during the Quaternary all display elements of punctuated equilibrium, but with differing magnitudes and frequencies of operation (Fig. 6.27). These magnitudes and frequencies of operation depend very much on Brunsden's five possible barriers to geomorphic change: (a) strength resistance (e.g. rock type); (b) morphological resistance (e.g. slope gradient); (c) system state resistance (e.g. patterns of historical disturbance); (d) structural resistance (e.g. channel–hillslope cou-

pling); and (e) filter resistance (e.g. storage elements).

However, when considering cryospheric change, especially over the last 20 ka the ecological definition of vulnerability may be more useful. Ecologists define vulnerability as a function of both sensitivity and resilience, where sensitivity is the rate and magnitude of response to system disturbance, and resilience is the rate at which an ecosystem recovers from disturbance (Westman 1978). Application of these concepts to the cryosphere will be the emphasis of the final chapter.

7

CRYOSPHERIC CHANGE AND VULNERABILITY AT QUATERNARY, HOLOCENE, AND ANTHROPOCENE TIMESCALES

7.1 INTRODUCTION

The transience of components of the cryosphere and the idea of transitional landscapes constantly undergoing change are interesting concepts. But what is their relevance to contemporary global environmental change? The geomorphic concepts of landscape resistance, collapse, and recovery as developed by Brunsden (1990) and others have been applied primarily to geological timescales of landscape evolution and seem, on the face of it, to have little relevance to contemporary processes of change. The suggestion in this chapter is that these geomorphic concepts are not dissimilar to recent models in ecology and cultural geography, in which the concepts of environmental change and collapse are central to an understanding of contemporary change. We will first describe Holling's idea of panarchy (Holling 2001) and Diamond's idea of cultural collapse (Diamond 2005) preparatory to examining their relevance to cryospheric change at Quaternary, Holocene, and Anthropocene timescales.

7.2 PANARCHY

Holling is an ecologist who has developed a model of how ecosystems adapt or fail to adapt to disturbance (Holling 2001). His model, which he calls panarchy, proposes that complex systems are driven through an adaptive cycle of resource exploitation, conservation, collapse, and reorganization and are sensitive to the wealth of the ecosystem, its connectedness (Fig. 7.1a) and its resilience (Fig. 7.1b). Many such cycles operate within each region, and each level of this panarchy operates at its own speed and is influenced by cycles at other time and space scales.

The three key properties of an adaptive cycle are wealth, connectedness, and resilience. Wealth is defined as the inherent potential of a system that is available for change, for example, snow or ice mass in a cryospheric system or biomass in a forest ecosystem. Connectedness is defined as the degree to which a system can control external forces, for example, an ice sheet has greater connectedness than sea ice and a

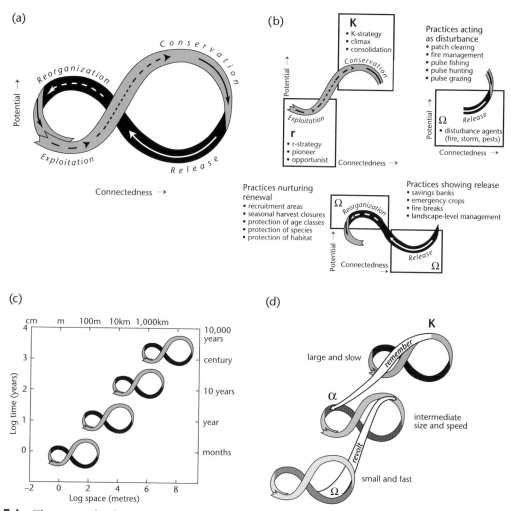

Fig. 7.1 The panarchy framework (after Gunderson & Holling 2002; composite modified by Slaymaker 2005). (a) The adaptive cycle; (b) Components of the adaptive cycle; (c) The panarchy in time and space; (d) The drivers and the constraints of the panarchy.

dense forest has greater connectedness than a loosely stocked plantation. Resilience is defined as the adaptive capacity of a system, for example, a diverse system can tolerate a wider range of environmental conditions than a simple system. During an "exploitation" stage, connectedness and resilience increase and wealth accumulates. At a "conservation" stage, resilience is reduced and the system becomes vulnerable to external events and collapses. Opportunity for

"reorganization" then exists, leading to a repeat of the adaptive cycle or, alternatively, the system collapses.

Adaptive cycles operate at a range of time and space scales (Fig. 7.1c). Larger cycles operate more slowly and smaller cycles operate more rapidly. Large-scale, slow cycles set the conditions within which the smaller cycles operate. For example, snowpack microclimates are controlled by the snow; the snow cover itself is controlled by the meso- and macro-regional climate. When one level of this panarchy collapses, this collapse can cascade to the next larger scale. In the case of an ice sheet such as Antarctica, a local perturbation can activate an ice stream which could destabilize the whole ice sheet.

7.2.1 PANARCHY, SUSTAINABILITY, AND TRANSFORMABILITY

The panarchy model introduces a new approach to the understanding of sustainability of ecosystems and cultures. The incorporation of collapse and creative reorganization as two of the normal stages of an adaptive cycle is at odds with the idea that sustainability is a condition of maintaining options for future generations. Whereas resilience is the capacity of a system to absorb disturbance and reorganize while undergoing change so as to still retain essentially the same function, structure, identity, and feedbacks, Holling introduced a new concept of "transformability." Transformability is the capacity to create a fundamentally new system when ecological, socioeconomic, and/or cultural conditions make the existing system untenable.

Holling does not claim that there are no differences between ecosystems and cultural systems but he does point to fruitful analogues between them. Perhaps the most seminal insight in this respect is the comment that "ecosystems are not only ecosystems and socio-economic systems are not only socio-economic systems" (Gunderson & Holling 2002), the implication being that we cannot reduce either ecosystems or socioeconomic systems to their theoretical frameworks, but we can learn from a host of examples of parallel behavior of self-organizing systems so as to improve our attempts to manage our environment.

The concepts and terminology of the panarchy model are consistent with concepts introduced into geomorphology by Schumm (1973), Brunsden and Thornes (1979), and Brunsden (1993) and which we have discussed in Chapter 6. Complex response, threshold exceedance, landscape sensitivity, vulnerability, and barriers to change are concepts that are readily applicable to cryospheric processes. Complex response can be analyzed in the framework of complex adaptive cycles; vulnerability is a combination of sensitivity and resilience, and resistance to change in the panarchy context plays a similar role to that of barriers to geomorphic change. One of the new features of the panarchy model is the identification of a system of adaptive cycles in which the influence of the states and dynamics of the subsystems at scales above and below the scale of interest can be sought.

7.2.2 COLLAPSE AND THE VULNERABILITY OF SOCIOECONOMIC SYSTEMS

Jared Diamond (2005) has raised the question of what causes the collapse of cultures and civilizations. Inter alia, he quotes the well known but poorly interpreted case of the Viking expansions during the ninth to eleventh centuries and the strong environmental factors that were, at least in part, responsible for their failures in Greenland and Vinland. He suggests that the process of colonization had many of the characteristics of a natural experiment designed to control for the effects of sailing time from the hearthland, resistance offered by local inhabitants, suitability for agriculture, and environmental fragility. The success of the first three settlements (Orkney, Shetland, and Faeroe islands) he ascribes to short sailing time, no local inhabitants, and relatively favorable environment. The colonization of Iceland was also a success story, in spite of significant environmental fragility. But the collapse and failure of the Greenland Norse after 450 years and the Vinland colony,

FIG. 7.2 The collapse framework illustrated from the Viking expansions between ninth and four-teenth centuries (after Diamond 2005).

which was abandoned in its first decade, can be interpreted in terms of the combined negative effects of all four major factors (Fig. 7.2). In cryospheric perspective, the growth and collapse of ice sheets with a 100 ka frequency during the Quaternary Period provides an interesting analogue (Fig. 6.27).

7.3 CHANGING ICE COVER AND BIOMES SINCE THE LAST GLACIAL MAXIMUM

A summary record of changing ice cover and biomes has been provided by international scientists working within the PAGES program and under the general direction of the Commission for the Geological Map of the World (Petit-Maire 1999; Plate 7.1). The authors stress that the maps are tentative, but contain the best information that was available in 1999. For the purpose of the present discussion, the maps are ideal in that they depict the state of the globe during the two most contrasted periods of the last 20 ka. The LGM is the coldest (*ca.* 18 ka ± 2 ka BP) and the Holocene Optimum is the warmest (*ca.* 8 ka ± 1 ka BP). It is a remarkable thing to note that these periods were only 10 ka apart and yet there was a dramatic reorganization of the shorelines, ice cover, permafrost, arid zones, surface hydrology, and vegetation at the Earth's surface. Within a 20 ka time span, the two vast ice sheets of Canada and Eurasia, which reached a height of 4 km and covered about 25 million km² have disappeared; 20 million km² of continental platform have been submerged by the sea; biomes of continental scale have been transformed and replaced by new ones; and *Homo sapiens sapiens* can no longer go by foot from Asia to America, nor from New Guinea to Australia or France to England.

7.3.1 THE LAST GLACIAL MAXIMUM

During the LGM, mean global temperature was about 4.5°C colder than present. Mean sea level was approximately at the position of the 125 m isobath. Large areas of continental shelves were above sea level, particularly off eastern Siberia and Alaska, Argentina, eastern and southern Asia, New Guinea was connected to Australia and the Persian Gulf had dried up. The Black Sea was cut off from the Mediterranean Sea and had become a lake. Maximum glacier and ice sheet extent is shown, but this hides a technical problem in that maximum glacial extent occurred at various times between 22 and 14 ka BP. Permafrost extended southwards to latitudes of 40–44°N in the Northern Hemisphere. In the Southern Hemisphere, only Patagonia and the South Island of New Zealand experienced permafrost.

There was a general decrease in rainfall near the tropics. Loess was widespread in periglacial areas and dunes in semiarid and arid regions. All desert areas were larger than today but in the Sahara there was the greatest southward extension of about 3–400 km. Surface hydrology reflected this global aridity except in areas that received meltwaters from major ice caps, such as the Caspian and the Aral seas. At lower latitudes, vegetation colonized the exposed continental shelves, especially in southeast Asia and the Caribbean. Grasslands, steppes, and savannas expanded at the expense of forests.

7.3.2 THE HOLOCENE OPTIMUM

During the Holocene Optimum, the mean global temperature was about 2°C warmer than today. Mean sea level was close to that of the present day except in two kinds of environments:

1 the Canadian Arctic and the Baltic Sea where isostatic and eustatic adjustments were at a maximum;
2 deltas of large rivers such as Mississippi, Amazon, Euphrates-Tigris, and Yangtze, which had not reached their present size.

The glacier and ice-sheet cover cannot be distinguished from that of today at this global scale. Permafrost, both continuous and discontinuous was within the present boundary of continuous permafrost in the Northern Hemisphere. Significantly wetter conditions were experienced in the Sahara, the Arabian Peninsula, Rajasthan, Natal, China, and Australia, where many lakes which have subsequently disappeared were formed. In Canada, the Great Lakes were formed following the melting of the ice sheet and the isostatic readjustment of the land. Rain forest had recolonized extensive areas and the taiga and boreal forest had replaced a large part of the tundra and areas previously covered by ice sheets.

7.4 THE FIRST EXPLORERS IN NORTH AMERICA

The early immigrants into the North American continent must have been far more sensitive to the presence of the cryosphere than we are today (McGee 1974). Whether they were conscious of the cryosphere's continuous changes is a moot point but they appear to have found their way along newly ice-freed corridors and along recently emerged coastlines from Alaska to British Columbia (Fig. 7.3). The first inhabitants emigrated from Eurasia by crossing the Bering Land Bridge connecting Siberia and Alaska some time between 40 and 12 ka BP (Ray 1996). During part of that

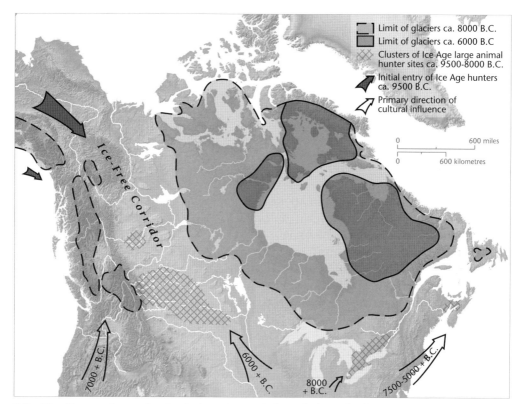

FIG. 7.3 The first explorers in North America and the relation of migratory routes to the receding ice sheets (from Ray 1996).

time much of the land was covered with massive continental ice sheets and vast meltwater lakes (Mathews 1979). The so-called Fluted Point people, who left the fluted stone points of their lances as archeological evidence of their existence, were hunters living close to the receding ice sheets (both Laurentide and Cordilleran) around 12–10 ka BP. Fluted Point sites are often found along the shores of former glacial lakes and appear to have been small hunting groups hunting mammoth, mastodon, bison, and caribou. By 10 ka BP, the Cordilleran Ice Sheet had broken into many sections, the Laurentide Ice Sheet had retreated from the Gulf of St. Lawrence, and the first indications of temperate rain forest on the British Columbia coast and the replacement of lichen woodland of

the interior of British Columbia by the Columbia–Montane forest have been documented (McAndrews et al. 1987). Several cultures moved into British Columbia during the early Holocene advancing both from north and south. By 4 ka BP, the Early Nesikep culture of the Interior of British Columbia, the Early Northwest Coast culture and the Early Paleo-Eskimo cultures of the Canadian North were all emerging and the environment had become quite similar to that of the present (Wright 2001).

The everyday lives of the early pioneers in Canada as described above were lived in the consciousness of the presence of a cryosphere which was in reality a determinant of their livelihoods and opportunities for development (12–10 ka BP). Rapidly thereafter the ice sheet became more distant, and life became more strongly influenced by the successive biomes that made eking out an existence so much more possible. But the evidence from archeology and from palynology seems to be that environmental change and cultural change went hand in hand. Commencing with the reality of circumventing ice sheets and glacial meltwater floods, early cultures succeeded in adapting to the changing environment. But there is also the unwritten account of those cultures that did not adjust and were eliminated. The account is unwritten because it is impossible to argue from lack of evidence. Ecologists, archeologists, and cultural geographers have wrestled with the problem of interpreting the records of changing ecosystems and changing cultures.

Pleistocene and Holocene landscapes, before substantial modification by human activity, were evidently not even then "sustainable" in the sense advocated by Gro Harlem Brundtland in "Our Common Future" (WCED 1987). The essence of sustainability in Brundtland's terms is the handing on to future generations an environment that offers the same or improved range of options (environmental, economic, social, cultural, and political) as those that are available at present. Is it even possible to reconcile the concept of global cryospheric change, a process that we have been describing as an inexorable force over space and through time, with the idea of sustainability? The concept of sustainability seems to ignore the phenomena of transience, of transitional states, and of collapse of the cryosphere.

Not only have ice sheets and glaciers collapsed on numerous occasions during the Quaternary Period but the cryosphere, in common with all ecosystems, has been exposed to changes in climate, nutrient loading, habitat fragmentation, and biotic exploitation throughout the Holocene. Studies on lakes, coral reefs, glaciers, and oceans have shown that smooth changes are often interrupted by sudden drastic switches to a contrasting state. The Vostok ice core record confirms this for temperature and carbon dioxide variations over 420,000 years; the sediment transfer and landscape evolution record for one glacial cycle confirms this pattern (Fig. 6.27) and the panarchy framework for ecosystems and the collapse framework for cultural systems suggest that a pattern of punctuated change is the norm.

7.5 IMPLICATIONS OF CRYOSPHERIC CHANGE/COLLAPSE

Human-driven changes to many features of the cryosphere have become so significant in magnitude that the increased vulnerability and possible collapse of cryospheric systems is greatly accentuated. Many of these changes are large in magnitude at the global scale, sometimes even exceeding the "natural" rates. In addition, anthropogenic changes invariably occur at rates that are higher than those of natural variability (Table 7.1). The magnitudes and rates of these changes, coupled with the fact that changes to each compartment of the cryosphere are occurring simultaneously has led to the suggestion that the Earth is now operating in a "no analogue" state (Steffen et al. 2004b). Hence the term Anthropocene has been proposed to describe an epoch in which human disturbance of the major biogeochemical cycles has become significant (Crutzen & Stoermer 2000). Some would say that the Anthropocene commenced several millennia ago (e.g. Ruddiman 2004) but everyone would agree that since 1990 (Turner et al. 1990), the scale of human perturbation of the global system has become quantitatively of the same order as natural flows (Fig. 7.4).

TABLE 7.1 Critical processes in the Quaternary system and likely changes in the future (Schellnhuber et al. 2004).

Process	Timescale of change (years)	Probability of major change	Uncertainty of our knowledge	Impact of these potential changes on society: an educated guess
Inland ice	100–10,000	Low	Low	High
$CO_2 + CH_4$ realse from permafrost	10–1,000	Moderate	High	Low
Gas hydrates	10–100	Low	High	High
CH_4 release from wetlands	10–100	High	Moderate	Low
CO_2 release from land biosphere	10–1,000	Moderate	High	High
Dust feedback	10–100	Moderate	High	Moderate
Sea ice	10–100	High	Moderate	Moderate
Thermohaline circulation thresholds	10–100	High	Moderate	Moderate
Atmospheric circulation (ENSO, NAO, monsoon)	1–100	High	High	Moderate
Biogeophysical feedback	10–100	High	High	Low
Denitrification and phosphorus recycling	100–1,000	Moderate	High	Low

FIG. 7.4 Building settlement on permafrost in Alaska (from Skinner & Porter 1995).

Characteristics of this "no analogue" state include the following (Steffen et al. 2004b):

1 connectivity between different cryospheric processes and regions of the planet is greater than previously thought, and the connectivity of human activities is increasing rapidly;

2 the impacts of global changes on human–environment systems can no longer be understood by simple cause–effect relations, but rather in terms of cascading effects that result from multiple, interacting stresses; and

3 new forcing functions arising from human activity (e.g. synthetic chemicals) and rapid rates of change imply that the natural resilience of cryospheric systems may be insufficient to cope with the change.

7.5.1 SNOW QUANTITY

Chapin et al. (2005) have shown that pronounced terrestrial summer warming in arctic Alaska correlates with the lengthening of the snow-free season. They calculated that the resultant reduced albedo has lead to an increase in atmospheric heating by 3 Wm^{-2} per decade (an amount comparable with the effect of doubling of atmospheric carbon dioxide over 40 years). A decrease in spring snow extent over Eurasia has occurred since 1915 (McCarthy et al. 2001). Most scenarios predict increasing precipitation in the Arctic and Antarctic as temperatures rise, but the proportion of this precipitation which falls as snow will vary regionally and locally. Snowfall amounts in extra-tropical mountains are also quite unpredictable. The duration of winter will presumably be reduced if warming continues but Woo (1996) envisages a scenario with increased snow accumulation at higher levels and reduced snowmelt at intermediate elevations because of increased summer storminess. At lower elevations however, rainfall and rain-on-snow events will likely

increase. Brown et al. (2000) demonstrated a consistent increase in snow accumulation in the winter and reduced snow accumulation in the spring in his discussion of Northern Hemisphere snow cover from 1915 to 1997. Implications for ski resorts include the necessity to link low-lying resorts to slopes higher up by using mass carrier systems.

7.5.2 SNOW QUALITY

Jones et al. (2001) provided one of the first systematic attempts to link the anticipated physical changes in snow cover to microbiological processes and associated invertebrate populations and vegetation. Combinations of thinner snow packs, increased freezing and thawing during winter leading to ice layer formation in the snow pack and wind action, leading to harder packed snow are the three physical changes that are most frequently referenced by the local Inuit population. Second, the concentration of pollutants accumulated in thinner snow packs releases a more polluted pulse of spring meltwater. Not only persistent organic pollutants (POPs), both airborne and waterborne, but other dissolved constituents which have accumulated over the winter are released over a comparatively short period. Changes in the subnivean runoff pathways may also change the downstream water quality.

7.5.3 RIVER AND LAKE ICE

Assel et al. (1995) have demonstrated the reduction in ice cover duration in the North American Great Lakes region from 1823 to 1994. The broad systemic signal of ice-free lakes in the 1990s (as much as 30 days longer compared with the 1830s) is confirmed. The duration of river ice cover can also be expected to be reduced. Many rivers within the more temperate regions would tend to become ice-free whereas in colder regions the present ice cover season could be shortened by as much as one month by 2050 (McCarthy et al. 2001). Prowse (2000) has drawn attention to the highly variable local effects of changing freeze-up and break-up dates of northern rivers on river ecology, especially on the conditions for salmon incubation in their gravel beds.

7.5.4 PERMAFROST

On the other hand, trends in the degradation of permafrost are relatively unambiguous and predictions of changing permafrost distributions under varying climate scenarios are widely available (Fig. 7.5). In Siberia, it is predicted that there will be a 10% reduction in permafrost extent over a 50 year period under a $+2°C$ warming scenario (Street & Melnikov 1990; Fig. 7.5a). Anisimov and Nelson (1995) have compiled global permafrost maps on the basis of which they predict a 16% reduction by 2050. This amounts to a loss of $4 \cdot 10^6$ km^2 of permafrost terrain. Although it is true that degradation will be gradual, it does mean that extensive areas of thermokarst activity will be exposed in which subsidence, erosion, increased runoff, and slope instability will ensue. As the active layer deepens, subsurface flow will increase in importance and with increased groundwater flow, more rainfall and less-intense snowmelt, the streamflow regimes will be altered. A similar sequence of events is predicted for Canada (Fig. 7.5b).

Osterkamp (2005) reports that permafrost is warming throughout Alaska north of the Brooks Range (3–4°C from 1977 to 2003); interior Alaska (1–2°C); and areas south of the Yukon River (0.3–1°C). Effects in the continuous permafrost zone can be expected to be localized thawing, slope instability, and ice-wedge thawing. But in the continuous permafrost zone,

most of it will begin thawing, from the top downward first and eventually from the bottom upward. Because thawing rates are slow (on the order of 10 cm a^{-1} near the surface and, theoretically, less than 2 cm a^{-1} at the base) (Lunardini 1996) decades to millennia will be required to complete the thawing, all other factors remaining constant. Harris et al. (2003) have provided similar data for the European mountains. In the discontinuous permafrost zone, thermokarst pits have appeared at sites where none existed prior to 1977.

7.5.5 GLACIERS

Oerlemans (2001) has shown the consistent reduction of glacier length since the Little Ice Age (1850–1990) from 48 glaciers located in 9 different regions. Mean rates of retreat of individual glaciers varied from 1.3 to 86 m a^{-1}. By introducing a scaling factor, adjusting for the greater sensitivity of maritime and low-gradient glaciers, the mean scaled rates varied from 6.3 to 14.9 m a^{-1}. Glaciers throughout the Arctic and in the Antarctic Peninsula are melting. Alaska's glaciers account for about half of the estimated loss of ice mass by glaciers worldwide. But the practical and direct implications of the complete disappearance of glaciers is most evident in the extra-polar regions, especially in East Africa, South America, and in the central Asian highmountain region. In these regions, a significant water supply problem is emerging because adjacent arid/semiarid regions have relied on glacier melt to supply late summer runoff and reservoir water recharge. Further depletion of ground water resources will result.

There is a discrepancy between the evidence that tropical glaciers are receding at an increasing rate in all tropical mountain areas and the lack of warming in the lower troposphere since 1979 indicated by satel-

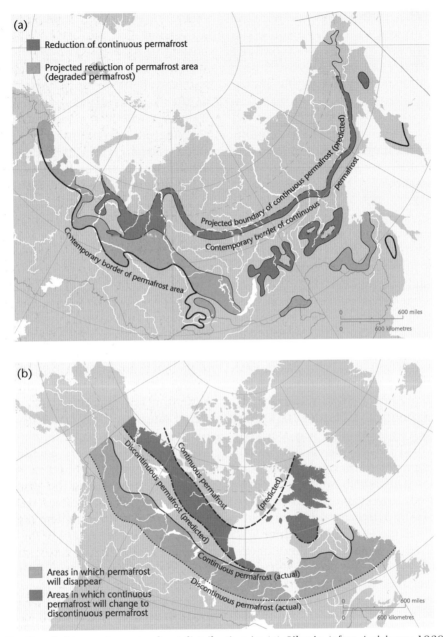

Fig. 7.5 Predicted changes in permafrost distribution in (a) Siberia (after Anisimov 1989) and (b) Canada (after French 1996).

lites and radiosondes in the tropics. This observation applies to the tropical Andes (Ames 1998); Mount Kenya and Kilimanjaro (Hastenrath & Greischar 1997); and to the glaciers in Irian Jaya (Peterson & Peterson 1994). The available data suggest that

glacier retreat started later at high latitudes but that in low and mid-latitudes, the retreat generally started in the mid-nineteenth century. This timing of the onset of retreat implies that a significant global warming is likely to have started not later than the mid-nineteenth century. This conflicts with the Jones et al. (2001) global land instrumental temperature data and the combined hemispheric and global land and marine data where clear warming is not seen until the beginning of the 20th century. These discrepancies are currently unexplained.

In a few regions such as western Norway and New Zealand, a considerable number of glaciers are advancing. In Norway, this is probably due to increases in precipitation associated with a positive phase of the North Atlantic Oscillation and in the Southern Alps of New Zealand due to wetter conditions with little warming since about 1980.

7.5.6 RIVER BASINS

Precipitation in the form of snow is expected to decrease. In North America, based on the Boer and Koster (1992) scenario, a 40% decrease in snow cover duration over the Canadian Prairies and a 70% decrease over the Great Plains (Brown et al. 1995) can be anticipated. For the whole Northern Hemisphere, the area of seasonal snow in February may decline by 6–20% or a mean decrease of 1.2 (10^7) km^2 (Henderson-Sellers & Hansen 1995).

By 2050, it is thought that up to a quarter of mountain glacier mass will have melted. An upward shift of the equilibrium line altitude of 2–300 m and annual ice thickness losses of 1–2 m are expected for temperate glaciers. Haeberli and Hoelzle (1995) estimate that glacier mass in the European Alps could be reduced to a few percent within decades if present trends continue. The implications of these changes for glacier melt runoff will be profound. Small glaciers will provide enhanced runoff for a short time and then cease to exist; large glaciers may produce extra runoff for a century or more.

Berezovskaya et al. (2004) have examined the compatibility of precipitation and

runoff trends in the three largest Siberian watersheds, the Lena, the Yenisei, and the Ob and concluded that there was great uncertainty in the quality of precipitation and runoff data sets (Fig. 7.6). They do, however, acknowledge that the Ob and the Yenisei are extensively influenced by land use changes and dam construction. The inconsistency that they identified (increased runoff with minimal increase in precipitation) could however possibly be related to a change in the proportion of runoff that is produced by surface runoff following change of land use.

On the other hand, more detailed studies of changes in the regime of the Lena River (Yang et al. 2002; Ye et al. 2003) provide the following results: the upper part of the basin shows increasing discharge in winter, spring, and summer, and a decrease in the fall. This is characteristic of earlier snowmelt, climate warming, and permafrost degradation. But as a result of a large dam on the west Lena river, the runoff in summer is underestimated and the winter and fall seasons are overestimated. Nevertheless there are strong linkages between precipitation, temperature, and runoff.

The Yenisei and Ob basins, which are more heavily impacted by human activity, fail to show a significant streamflow increase over the period 1935–99 (Nilsson et al. 2005). At the same time, studies of the Mackenzie basin (Woo & Thorne 2003) show that annual runoff has not changed in the past three decades despite a strong warming trend. Tributary basins, such as the Liard (Woo & Marsh 2005), show earlier snowmelt hydrograph rise but a decline in the summer months.

Two general points of interest emerge from these conclusions: (a) the one basin which has unambiguously increased runoff over the past 30 years is the Lena. Coincidentally, this is the basin with the greatest extent of permafrost and (b) it would repay effort to consider the implications of

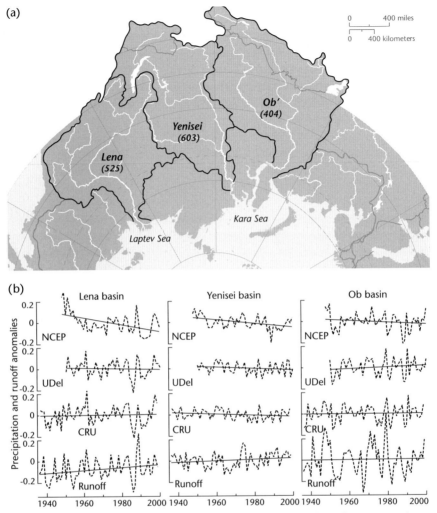

FIG. 7.6 Changing river basins tributary to Arctic Ocean. (a) The three largest Arctic watersheds. (b) Mean annual precipitation (upper three graphs) and runoff (lower graph) anomalies (normalized deviation from the mean) (after Berezovskaya et al. 2004).

any future increase in runoff that might result from the warming climate and increased precipitation. Predictions of future changes in precipitation over the large northern basins are not consistent. Many rivers in temperate regions will tend to become ice-free and a longer open water period together with warmer summer conditions will increase evaporative loss from lakes.

The major arctic rivers, like the Yenisei, the Ob, the Lena, and the Mackenzie, originate in more temperate latitudes, where population densities are markedly greater than those within the polar regions. Land use impacts associated with accelerating resource development influence the runoff and sediment discharge regimes (Yang et al. 2004a,b). Present runoff to the Arctic Ocean is approximately twice that produced by precipitation minus evaporation over the Arctic Ocean (Berezovskaya et al. 2004). The implication is that any change in runoff caused by climate change will influence the salinity of the Arctic Ocean; intensification of land use activities will influence sediment and solute loads delivered to the Arctic Ocean and a series of feedback effects from decreasing albedo and advancing treeline will tend to accentuate regional warming.

7.5.7 SEA ICE

According to Barber et al. (2005), approximately 74,000 km^2 of sea ice extent have been lost annually in the Canadian Arctic since 1979. The practical implication of this is that the reduction represents a change from multiyear ice to first year ice. Given that the barriers to navigation through the Northwest Passage have historically been caused by multiyear sea ice, the current trend is particularly significant. The complete disappearance of multiyear sea ice from the Arctic Ocean is now being confidently predicted by mid-twenty-first century (if GCM simulations are to be trusted). This will aid navigation and open up new shipping routes such that the sea route between Asia and Europe will be 7,000 km shorter than the route through the Panama Canal.

7.5.8 ICE SHEETS

Systematic changes in ice sheet extent are still impossible to confirm, and their direct implications at the scale of a human life time are, at all events, small. But consider indirect implications for sea level change below. The issue of ice sheet collapse remains in the realm of the improbable on a century time scale. Although it is improbable, there is a small, nonzero possibility that the West Antarctic Ice Sheet and the Greenland Ice Sheet could collapse under

circumstances of continuing global warming. Such an eventuality would spell disaster for major port cities of the world, such as New York, New Orleans, London, and Amsterdam, to name only a few.

7.5.9 SEA LEVEL CHANGE

Thermal expansion of the ocean and melting of glaciers, ice caps, and ice sheets are thought to be jointly responsible for the 10–25 cm rise of sea level that has been documented over the past century. Most of the projected 50 cm sea level rise by 2100 is accounted for by thermal expansion, followed by increased melting of glaciers and ice caps. The likelihood of a major sea level rise by the year 2100 due to the collapse of the West Antarctic Ice Sheet is considered low.

7.5.10 CARBON SEQUESTRATION

Most wetlands in permafrost regions are peatlands, which may absorb or emit carbon (as carbon dioxide or as methane) depending on the depth of the water table. There are huge uncertainties as to whether the peatlands would switch from emitting carbon dioxide to absorbing it or, if the rate of decomposition of organic matter were to increase faster than the rate of photosynthesis, there would be an increase in the rate of emissions. A combination of temperature increase and elevated ground water levels could result in increased methane emissions.

7.5.11 VEGETATION

We know that treeline is advancing toward higher altitudes in Europe, New Zealand, and North America (Kullman 2001); that alpine plants in the European Alps are experiencing an elevation shift of 1–4 m per decade (Grabherr et al. 1994); and that arctic shrub vegetation in Alaska is expanding into previously shrub-free areas (Sturm et al. 2001). In Antarctica there have been distribution changes of plants and invertebrates (Kennedy 1995), especially the colonization by mosses of previously bare ground and the commensurate colonization by soil invertebrates. There remains considerable uncertainty about community and ecosystem trajectories (Walther et al. 2002).

7.5.12 POLAR BEARS

Polar bears (*Ursus maritimus*) range over large areas in search for food. They move south with the ice in the autumn and winter and then north as the pack ice melts in the spring and summer. These seasonal movements of the sea ice also influence the distribution of their primary prey, ringed and bearded seals. Polar bears often eat only the blubber from the seal where the highest concentrations of organochlorine contaminants (OCs) are found. As an apex predator, the polar bear has among the highest levels of POPs. Some of the highest levels of POPs such as polychlorinated biphenyls (PCBs) are found around Svalbard and Franz Josef Land (Fig. 7.7). There are indications that polar bears with high PCB levels have impaired defense against infection and that cub survival may be affected.

7.5.13 HUMAN HEALTH

Impact of cryospheric change on the health of arctic residents will vary depending on such factors as age, gender, socioeconomic status, lifestyle, culture, location, and the capacity of the local health infrastructure to adapt. It is likely that populations living in close association with the land, in remote communities and those that already face a variety of health-related challenges will be most vulnerable (Berner & Furgal 2005). The World Health Organization (WHO 1967)

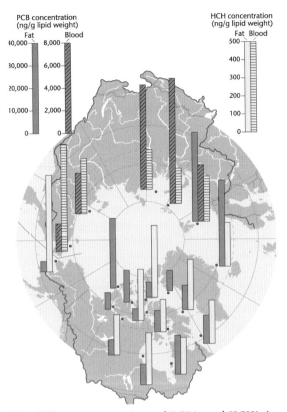

FIG. 7.7 Concentrations of PCB's and HCH's in blood from adult female bears from Svalbard and the Russian Arctic (Andersen et al. 2001) and in fat from polar bears from the Bering Sea, Canada, eastern Greenland, and Svalbard (de March et al. 1998).

defines health to include aspects of physical, mental, and social well-being and human health status is a result of the complex interaction of genetic, nutritional, and environmental factors. Berner and Furgal (2005) point out that cryospheric change is likely to have a large impact on arctic residents for three reasons: (a) many arctic residents live in small, isolated communities with a fragile system of economic support, dependence on subsistence hunting and fishing, and little or no economic infrastructure; (b) rural arctic public health and acute care systems are often marginal and sometimes nonexistent; (c) culture is often critical to community and individual health and the loss of, for example, a traditional subsistence food source can result in severe stress.

Blood and fatty tissue samples taken from Inuit in southern Baffin Island and southern Quebec in the late 1980s showed surprisingly high levels of certain POPs, including PCBs and dichlorodiphenyl-trichloroethane (DDT) (Dewailly et al. 1989; Fig. 7.8). The findings were highlighted in Canada's 1991 national state of the environment report and in June, 1991, the 8 arctic states (Canada, USA, Russia, Sweden, Norway, Finland, Denmark, and Iceland) met in Rovaniemi to mandate the Arctic Monitoring and Assessment Program (AMAP), with a secretariat in Norway, to examine six pollution issues, including POPs. The AMAP process has produced the most intensive data on human health in the Arctic as influenced by global change and imported pollutants from the south. Moving from the raw data to resource management policy is

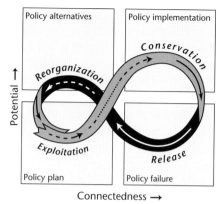

FIG. 7.9 General adaptive cycle applied to resource management policy (after Gunderson & Holling 2002).

proving to be a more complex process and can be envisaged in a way as suggested by Gunderson and Holling (2002) (Fig. 7.9). Human health concerns also include increased accident rates due to environmental changes such as sea ice thinning and health problems caused by adverse impacts on sanitation infrastructure due to thawing permafrost.

7.5.14 PERSISTENT ORGANIC POLLUTANTS

The existence of high levels of organic contaminants in the Canadian Arctic food chain has been studied since 1987 (Downie & Fenge 2003). The primary group of environmental contaminants that poses a risk to human health is the organochlorines, including substances such as PCBs and chlorinated pesticides. Many of these contaminants are transported via atmosphere, oceans, and rivers, as well as migratory birds (Blais et al. 2005). They find their way into terrestrial, marine, and freshwater food chains. Because the Inuit traditionally feed off the land, they are more exposed to these contaminants than southern Canadians. Potential health effects include attacks on the immune system leading to infections,

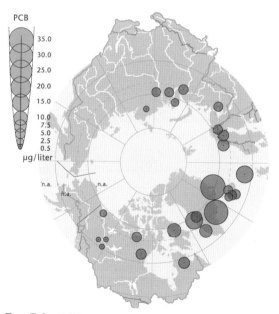

FIG. 7.8 PCB concentrations (as Alochlor 1260) in blood of mothers and women of child bearing age (AMAP 2004).

complications of neurobehavioral development, reproductive loss, chronic exposure to POPs, and possible links to cancer and osteoporosis. The problem is complicated by the fact that these foods are a valuable source of key nutrients which act to reduce risk of contaminant exposure. A further complication is that by comparison with most other areas of the world, the Arctic remains a clean environment. "For some pollutants, combinations of different factors give rise to concern in certain ecosystems and for some human populations" (AMAP 1998).

Figure 7.10(a) shows the average annual flux of DDTs and HCHs in Russian tributaries to the Arctic Ocean and Figure 7.10(b) fluxes of DDTs and PCBs across dated slices of lake sediment cores in the Canadian Arctic and Svalbard (AMAP 2002). Two patterns are immediately obvious: the comparatively undeveloped Lena River basin shows low fluxes and, from the lake sediments, POPs concentrations and fluxes drop off significantly toward the pole. There are four major pathways for the transport of POPs (Fig. 7.11). The atmosphere, ocean currents, transpolar ice pack, and large Arctic rivers are the major pathways but it is also worth noting that POPs can be transported into the Arctic via pelagic organisms (crustaceans, fish, marine mammals) and migratory birds and animals.

Data are available for organo-chlorine contaminants in Arctic air, snow, sea water, zooplankton, amphipods, Arctic cod, beluga, ringed seal, and polar bears. Hexachlorocyclohexanes (HCHs) are a less persistent chlorinated pesticide though they do resist degradation in cold water; polychlorinated biphenyl PCBs are extremely persistent in the environment with half lives ranging from 3 weeks to 2 years in air and essentially nonbiodegradable in aerobic soils or sediments; Hexachlorobenzenes (HCBs) have an estimated half life of 2.7–5.7 years and are especially persistent in abiotic environments; toxaphene, chlordane,

and DDT are persistent pesticides and are of particular concern because of the high levels of their toxic metabolites in top predators. Figure 7.12 illustrates the way in which each of these six major classes of OCs has differing bioaccumulation rates and differing degrees of ease of transfer from one ecosystem compartment to another.

7.5.15 SOCIOCULTURAL CONDITIONS AND HEALTH STATUS

Social conditions and lifestyles vary widely throughout the Arctic. Many indigenous peoples rely on the food that they hunt and harvest from the land and sea and this is a critical component of their cultural identity. Other, more urban residents have lives that are not unlike those of other parts of Europe and North America and cryospheric change affects primarily conditions for infrastructure development and transport. As indicated in AMAP (2003) indigenous as well as nonindigenous residents are involved in the process of change toward a modern wage earning economy.

Nevertheless, traditional foods collected from the land, sea, lakes, and rivers continue to be important sources of health and well-being to many indigenous communities. Such foods continue to contribute significant amounts of protein to the total diet. By encouraging the eating of all the animal parts, northern indigenous diets provided the nutrients and essential elements required to sustain life. Two examples would be the provision of nearly 50% of the weekly protein, iron, and vitamin A intake in Nunavik women under 45 years of age (Jette 1992) and nearly 50% in Labrador Inuit (Lawn & Langer 1994). In addition to the nutritional benefits, traditional foods provide many cultural, social, and economic benefits to individuals and communities (Berner & Furgal 2005).

Cryospheric changes may affect the consumption of traditional foods by changes in

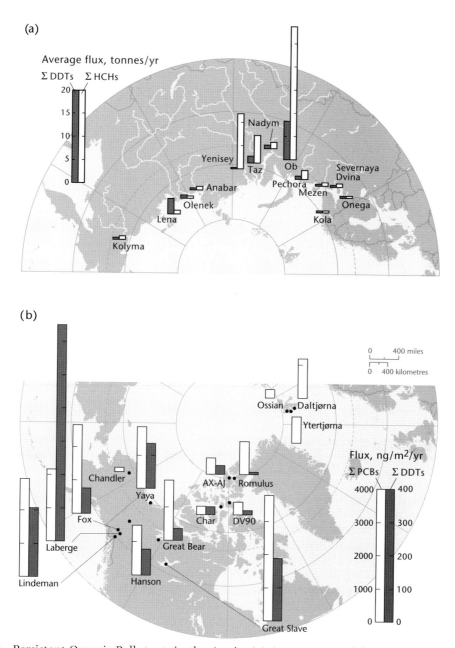

Fig. 7.10 Persistent Organic Pollutants in the Arctic. (a) Average annual fluxes of DDT's and HCH's from Russian rivers to the Arctic Ocean (Alexeeva et al. 2001). (b) Fluxes of DDTs and PCBs in lake sediment cores from the Canadian Arctic and Svalbard (AMAP 2004).

access to food sources, such as: (a) changes in the distribution of important food species; (b) unpredictable weather; (c) low water levels in lakes and rivers, the timing of snow, and ice extent, and stability; (d) a shorter winter season and increased snowfall; (e) vegetation change may influence caribou health; (f) deeper snow and incidence of freezing rain can affect the ability of caribou and reindeer to forage in winter;

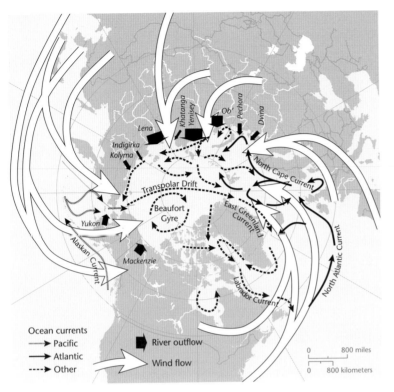

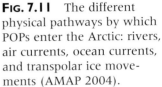

FIG. 7.11 The different physical pathways by which POPs enter the Arctic: rivers, air currents, ocean currents, and transpolar ice movements (AMAP 2004).

(g) increase in insects, pests, and parasites; and (h) changed migration and breeding patterns.

At the same time, a decline in the proportion of traditional foods consumed by northern peoples is thought to be associated with cardiovascular disease, diabetes, vitamin deficiency disorders, dental cavities, anemia, obesity, and lower resistance to infection (Berner & Furgal 2005).

7.5.16 LIVELIHOODS AND SOCIOECONOMIC CONDITIONS

The Arctic includes part or all of the territories of eight nations: Norway, Sweden, Finland, Denmark, Iceland, Canada, Russia, and the United States. It includes the home lands of dozens of indigenous groups (Fig. 7.13) that encompass distinct subgroups and communities. Indigenous people currently make up roughly 10% of the total Arctic population of 4 million, though in Canada they represent half the nation's Arctic population, and in Greenland they are the majority. Nonindigenous residents also include many different people with distinct ways of life.

A whole range of impacts on society result from cryosphere change. The loss of the indigenous hunting culture follows from reductions in sea ice and loss of the animals on which they depend. Declining food security may result from reduced quality of food sources, such as diseased fish and dried-up berries. Shifting to a more Western diet carries risk of increased diabetes, obesity, and cardiovascular disease. Caribou and reindeer herds will face a variety of changes in their migration routes, calving grounds, and forage availability as snow and river ice conditions change.

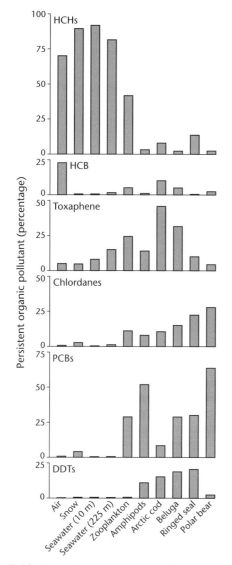

FIG. 7.12 Distribution of organochlorine contaminants (OCs) in Arctic air, snow, sea water, and the marine mammals food chain (Norstrom & Muir 1994). Data are plotted as the percentage of OCs in each category.

Transportation routes (Brigham 2001a, b) and pipelines on land are already being disturbed in some places by thawing ground. Oil and gas extraction and forestry will be increasingly disrupted by the shorter period when ice roads and tundra are sufficiently

frozen to allow industrial operations. Northern communities that rely on frozen roadways to truck in supplies are also affected. Decreased abundance and local and global extinctions of arctic-adapted fish species are predicted for this century. Arctic char, broad whitefish, and cisco, which are major contributors to local people's diets, are threatened by a warming climate. In spite of the above litany, a number of benefits can also be identified. Marine access to offshore oil and gas and some minerals is likely to be enhanced by the reduction in sea ice, bringing new opportunities as well as environmental concerns.

Expansion of tourism and marine transport of goods are likely outcomes as the summer navigation season lengthens both in the Northern Sea Route and in the Northwest Passage (Fig. 7.14). But one should also consider the probability of enhanced marine fisheries and agriculture and forestry.

7.5.17 GOVERNANCE

Unfortunately, international policy making is complex, has few robust institutions and where institutions do exist, they tend to respond to the larger interests and the more powerful nations. The low economic productivity (Fig. 7.13), the small numbers of residents in the polar regions and their varied interests all combine to make it difficult for the international and global community to hear their voice.

Environmentalists view the "atlantification" of the Arctic Ocean as a tragic loss to the global ecosystem. Neoliberals already see it as a potential Klondike (Barber et al. 2005). There is little doubt that the unregulated exploitation of the new resources of the Arctic will end in catastrophe. For this reason, the assertion of political control (e.g. DFAIT 2002) over the arctic territories of the 8 arctic states either unilaterally or, preferably, in concert is an urgent priority (Fig. 7.15).

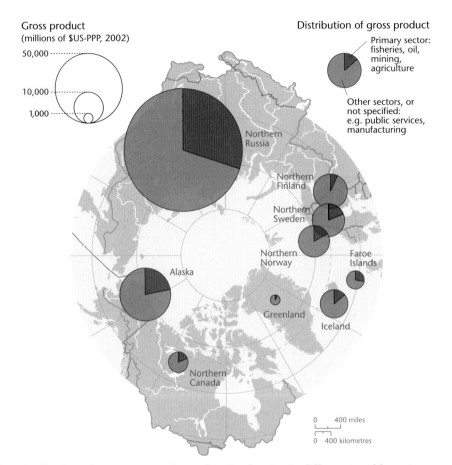

FIG. 7.13 Distribution of gross economic product in the Arctic differentiated by primary sector (fisheries, oil, mining, agriculture) and other sectors (such as public services and manufacturing) (Arctic Human Development Report 2004).

The Antarctic solution is attractive, but completely impractical given the range of legal and political constraints governing international and territorial waters vis-à-vis uninhabited terrestrial territory (Fig. 7.16).

7.6 CONCLUDING THOUGHTS

The results of the second AMAP reports (AMAP 2004) have shown that some indigenous peoples of the Arctic, especially those living mainly from the marine food web and consuming large quantities of marine mammals are the most POPs exposed people on Earth. It was also the case that the Sami people of Lappland whose traditional diet includes mainly reindeer meat have been among the most exposed to radioactivity from global nuclear weapons tests and accumulation of radionucleides from Chernobyl in their lichen–reindeer–human food chain. These examples pose profound ethical questions. People harvesting traditional food from some of the cleanest environments on Earth receive some of the highest contaminant exposures. Chemicals of little benefit to them are brought in by

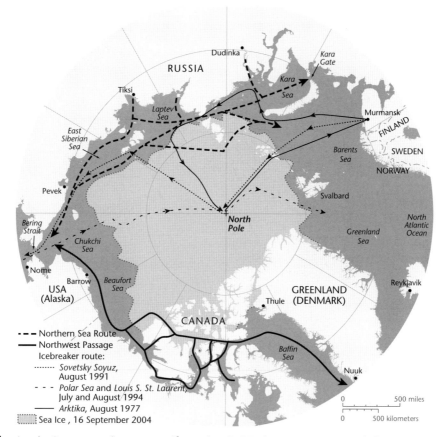

FIG. 7.14 Arctic Ocean marine routes (from Arctic Marine Transport Workshop 2004). The Northern Sea Route encompasses all routes across the Russian Arctic coastal seas from Kara Gate at the southern tip of Novaya Zemlya) to Bering Strait. The Northwest Passage includes all marine routes between Atlantic and Pacific oceans along the northern coast of North America, spanning the straits and sounds of the Canadian Arctic Archipelago. Three historic polar voyages and the ice edge for September 16, 2004 are also shown.

atmospheric transport, ocean currents, and rivers, accumulate in their foods and potentially damage their health. In many areas of the Arctic, alternative foods are either not available or not affordable. This is a dilemma that calls for international policy making.

The conceptual models of landscape resistance (Brunsden 1990), transience and transition (this volume), panarchy (Holling 2001), and collapse (Diamond 2005), drawn from geomorphology, ecology, and cultural geography are rich in many ways. They link biological and physical factors with socioeconomic and cultural changes in such a way as to emphasize the two-way causality and interdependence of humankind with its environment. They also face squarely the historical and contemporary evidence that socioecological systems collapse; some reorganize themselves creatively while others disappear.

This evidence has been used widely to argue in favor of a sustainable development policy which carefully avoids collapse, both

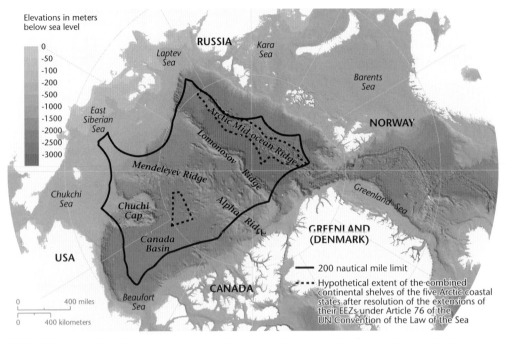

FIG. 7.15   Map showing the five coastal Arctic states, their joint Exclusive Economic Zones, and the "natural" prolongation of their land territories.

environmental and socioeconomic through a process of adaptive management. Adaptive management encourages planners to approach their work with the expectation that they may be incorrect. "Expect the unexpected and learn from it" is the underlying principle. But there are major institutional, participatory, and communication problems associated with this strategy and there are as many examples of ecosystem collapse under adaptive management as of creative reorganization. It therefore seems that the emphasis on sustainability may be theoretically valid but unachievable in practice. Perhaps a policy of sustainability should be replaced by a strategy of detecting unsustainability (Head 1991). This is not a matter of semantics. Unsustainable practices can, in principle, be legislated against; sustainable practices in the sense of environmentally, economically, socially, culturally, and politically sustainable practices are

impossible to define adequately for legislative purposes. A constantly changing social order, interacting with a continuously changing cryospheric environment makes for even greater difficulties in defining sustainability (Fig. 7.17). The dynamism of the cryosphere which we have been describing in this book makes the objective of defining a sustainable cryosphere almost self-contradictory. Viewed in the short-term, what is the meaning of a sustainable snow cover? in the medium-term, what is sustainable permafrost or a sustainable glacier? and in the long-term, is there such a thing as a sustainable Antarctic Ice Sheet?

These phenomena are inherently unsustainable and planners need to take into account the fact that change and uncertainty about the rates of change have to be incorporated into their plans. When we factor into the cryospheric system the interaction with a society that has multiple value

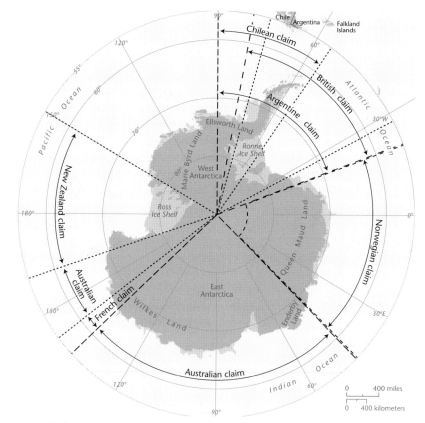

Fig. 7.16 Territorial claims in Antarctica.

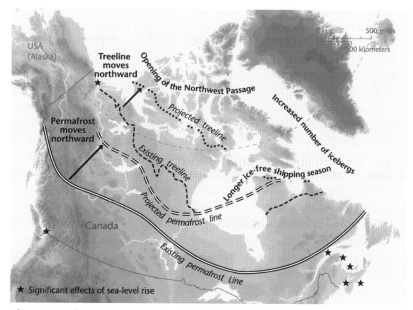

Fig. 7.17 Implications of cryospheric change for Canadian ecosystems and socioeconomic systems (French & Slaymaker 1993).

TABLE 7.2 Regional environmental changes observed by Inuit and Cree in the Hudson Bay bioregion (Fenge 2001).

	Eastern James Bay	Eastern Hudson Bay	Hudson Strait	Northwestern Hudson Bay	Western Hudson Bay	Western James Bay
Weather	• Shorter spring and fall seasons • Greater variability in fall • Colder winters in reservoir areas • Increased snowfall	• Persistence of cold weather into spring • Snow melts later • Spring and summer cooling trend • Less rain: fewer thunder-storms	• Greater variability: less predictable • Cooling trend • New snowfall cycle • Longer winters: snow melts later • Less rainfall	• Greater variability • Warmer and shorter winters • Snow falls and melts earlier • Cool summers in early 1990s	• Longer winters • Colder springs • Snow melts faster	• Shorter and warmer winters • Spring wind shifts several times a day
Atmosphere	• Change in sky color	• Change in sky color • Sun's heat blocked by haze	• Change in sky color	• Change in sky color	• Change in sky color	• Change in sky color
Sea level	• Salinity changing along northeast coast • More freshwater ice forming in the bay • Less solid in La Grande River area: freezes later, breaks earlier	• Freezes faster • Solid ice cover is larger and thicker • Fewer Polynyas • Floe edge melts before breaking up	• Freezes faster • Poorer quality • Landfast ice extends farther offshore • Polynyas freeze • Floe edge melts before breaking up	—	—	—
Currents	• Weaker in Eastmain area Swifter and less predictable north of La Grande River	• Weakening currents	• Weakening currents	• Weaker currents in Roes Welcome Sound	—	—

Rivers	• Seasonal reversal in levels and flow • Decline in water quality • Unstable ice conditions on La Grande River: freezes later, breaks earlier • Vegetation dying along diverted rivers	• Decreased water levels and river flow	• Decreased water levels and river flow	• Decreased water levels and river flow	• Seasonal reversal in water levels and flow • Increased salinity, erosion, and sediment in Nelson River • Decline in water quality	• Decreased water levels and river flow in southern James Bay Rivers • Increased erosion and mud slides
Canada and Snow Geese	• Coastal and inland habit changes • Coastal flyways have shifted eastward • Fewer being harvested in spring and fall • Large flocks of nonnesting/molting geese along coastal flyway	• Small flocks of Canada geese arrive in Belcher islands since 1984 • Increase in nonnesting/molting geese in Belcher and Long islands	• New snow goose migration routes • Increase in number of molting snow geese • Canada geese no longer nest in Soper River area	• More Canada geese in Repulse Bay area during summers of 1992 and 1993	• More snow geese migrating to and from the west • Habitat changes and Mash Point staging area • Earlier and shorter fall migrations	• Habitat changes in Moose Factory area • More snow geese flying in from the west • Canada geese arrive from the north first part of June • Change in fall migration patterns
Beluga whale	• Decrease in numbers	• Decrease in numbers along coast • Moved to and traveling in current currents farther offshore	• Decrease in numbers in Salluit area	• Decrease in numbers in Repulse Bay and Arviat area	• Decrease in numbers in Fort Severn and Winisk estuaries • Decrease in numbers in Nelson River estuary	—

(Continued)

TABLE 7.2 (Cont'd)

	Eastern James Bay	Eastern Hudson Bay	Hudson Strait	Northwestern Hudson Bay	Western Hudson Bay	Western James Bay
Fish	• Mercury contamination • Loss of adequate habitat for several species, for example, whitefish • Morphological changes in sturgeon	• Decrease in Arctic char and Arctic cod in Inukjuak area	—	• Decrease in Arctic cod in near-shore areas • Arctic cod no longer found in rear-shore areas off Cape Smith and Repulse Bay	• Mercury contamination • Loss of habitat including spawning grounds • Change in taste of fish: some are inedible	• Morphological changes in sturgeon • Dried river channels
Polar bear	—	• Increase in numbers since 1960s	• Decrease in numbers in Ivujivik area	• Increase in numbers • Appear leaner and more aggressive	• Thin-looking bears in York Factory area • Drink motor oil • Change in behavior	• Recent increase in reproduction rates • Fearless of humans
Walrus	• No longer present in Wemindji area	• Shift away from Belcher Islands	• Increase in numbers around Nottingham Island	• Decrease in numbers near Arviat and Whale Cove • Increase in numbers in Coral Harbor and Chesterfield	—	• Decrease in numbers in Attawapiskat area

Moose				
• Loss of habitat • Decrease in numbers • Change in body condition • Change in taste of meat	• In-migration from southeastern James Bay	—	—	• Change in taste of meat • Greater number drowning • No moose at March Point
Caribou				
• Change in body condition and behavior • Increase in number of diseased livers and intestines • Change in taste of meat • More caribou along the coast	• Caribou from different areas mingle together • Very large herds • Traveling closer to coast • Change in diet • Change in taste of meat	• Increase in number • Increase in abnormal livers, for example, spots and lumps • Change in diet	• Increase in number • Not intimidated by exploration activity • Feed close to exploration caps • Change in diet	• Increase in number • Pin Island herds is mixing with Wood and herd

systems, but at one basic level is an overconsuming society, then it becomes important to identify the elements of unsustainability in that relation. When it is further recognized that the cryosphere is an open system which is impacted by globalization as well as global environmental change it must become apparent that sustainability in the Brundtland sense is unachievable. The precautionary principle, applied specifically to unsustainable human interventions in the cryosphere, together with an acceptance of the inevitability of change would seem to be a more realistic approach.

There were three key recommendations for improving future assessments provided by ACIA (2004):

1 subregional impacts, possibly at the local level will have the greatest relevance and usefulness for residents in the Arctic;
2 socioeconomic impacts including oil and gas production, mining, transportation, fisheries, forestry, and tourism need more careful analysis;

3 assessing vulnerabilities or degrees of susceptibility to adverse effects of multiple interacting stresses is needed. Involves knowledge of capacity of the system to adapt as well as consequences of stresses and their interactions.

Research in these directions will go a long way to recognizing unsustainability and will make more realistic policy making possible. What kind of a world do northern residents want in the face of certain inevitable cryospheric and socioeconomic changes that are beyond their control? Global environmental change and globalization in its broadest sense have set in motion multiple interacting stresses, the medium- and long-term outcomes of which are inadequately understood. Residents of the Hudson Bay bioregion, for example, are conscious of many environmental changes (Table 7.2); does the global community care? And if so, how much "collapse" can be absorbed and how much of the "transience" can be accommodated to allow an orderly and ethical transition?

REFERENCES

Abdalati, W. & Steffen, K. (1995) Passive microwave-derived snow melt regions on the Greenland Ice Sheet. *Geophysical Research Letters* **22**, 787–90.

Abdalati, W., Krabill, W., Frederick, E., Manizade, S., Martin, C., Sonntag, J., Swift, R., Thomas, R., Wright, W., & Yungel, J. (2001) Outlet glacier and margin elevation changes: near coastal thinning of the Greenland Ice Sheet. *Journal of Geophysical Research* **106**, 33729–41.

Abshire, J.B., Sun, X., Riris, H., Sirota, J.M., McGarry, J.F., Palm, S., Yi, D., & Liiva, P. (2005) Geoscience Laser Altimeter System (GLAS) on the ICESat Mission: on-orbit measurement performance. *Geophysical Research Letters* **32**, L21S02, doi:10.1029/2005GL024028.

ACIA (2004) *Impacts of a Warming Arctic: Arctic Climate Impact Assessment*. Cambridge University Press, Cambridge.

ACIA (2005) *Arctic Climate Impact Assessment*. Cambridge University Press, Cambridge.

Adger, W.N. & Brown, K. (1994) *Land Use and the Causes of Global Warming*. John Wiley & Sons, Chichester.

Ager, D.G. (1981) *The Nature of the Stratigraphic Record*. Halsted Press, New York.

Ahlmann, W.H. (1948) *Glaciological Research on the North Atlantic Coasts*, Research Series 1. Royal Geographical Society, London.

Ahlmann, H.W. & Thorarinsson, S. (1938) Vatnajokull: scientific results of the Swedish-Icelandic investigations 1936–38, Chapter V – The ablation. *Geografiska Annaler* **20**, 171–233.

Alley, R.B. & Clark, P.U. (1999) The deglaciation of the Northern Hemisphere: a global perspective. *Annual Reviews of Earth and Planetary Sciences* **27**, 149–82.

Alley, R.B. & MacAyeal, D.R. (1993) West Antarctic Ice Sheet collapse: chimera or clear danger? *Antarctic Journal Review* **1993**, 59–60.

Alley, R.B. & Whillans, I.M. (1991) Changes in the West Antarctic Ice Sheet. *Science* **254**, 959–63.

Alley, R.B., Clarke, P.U., Huybrechts, P., & Joughin, I. (2005) Ice sheet and sea level changes. *Science* **310**, 456–60.

Alexeeva, L.B., Strachan, W.M.J., Shlychkova, V.V., Nazarova, A.A., Nikanorov, A.M., Korotova, L.G., & Koreneva, V.I. (2001) Organochlorine pesticide and trace metal monitoring of Russian rivers flowing to the Arctic Ocean: 1990–1996. *Marine Pollution Bulletin* **43**, 71–85.

Alsdorf, D., Birkett, C., Dunne, T., Melack, J., & Hess, L. (2001) Water level changes in a large Amazon lake measured with spaceborne radar interferometry and altimetry. *Geophysical Research Letters* **28**, 2671–4.

Alverson, K., Bradley, R., & Pedersen, T. (2001) *Environmental Variability and Climate Change*. IGBP Science 3, Stockholm.

AMAP (1998) *Assessment Report: Arctic Pollution Issues*. Arctic Monitoring and Assessment Program, Oslo, Norway.

AMAP (2002) *Arctic Pollution 2002: Persistent Organic Pollutants, Heavy Metals, Radioactivity, Human Health, Changing Pathways*. Arctic Monitoring and Assessment Program, Oslo, Norway.

AMAP (2003) *AMAP Assessment 2002: Human Health in the Arctic*. Arctic Monitoring and Assessment Program, Oslo, Norway.

AMAP (2004) *AMAP Assessment 2002: Persistent Organic Pollutants in the Arctic*. Arctic Monitoring Assessment Program. Oslo, Norway.

Ambach, W. (1979) Zum Warmehaushalt des Gronlandischen Inlandeises Vergleichende Studie im Akkumulations- und Ablationsgebiet. *Polarforschung* **49**, 44–54.

American Geophysical Union (1992) FIFE Special Issue. Reprinted from the *Journal of*

Geophysical Research **97**, DOI: 10.1029/92JD02189. American Geophysical Union, Washington, DC.

American Geophysical Union (1996) The Second FIFE Special Issue. Reprinted from the *Journal of Geophysical Research* **101**, DOI:10.1029/96WR00269. American Geophysical Union, Washington, DC.

American Geophysical Union (1997) BOREAS Special Issue. *Journal of Geophysical Research* **102(D24)**, DOI:10.1029/97JD02231.

American Geophysical Union (1999) BOREAS Special Issue II. *Journal of Geophysical Research* **104(D22)**, DOI:10.1029/1999JD901026.

American Geophysical Union (2001a) BOREAS Special Issue III. *Journal of Geophysical Research* **106(D24)**, DOI:10.1029/2001JD900056.

American Geophysical Union (2001b) PARCA: Mass balance of the Greenland Ice Sheet. *Journal of Geophysical Research* **106(D24)**, DOI:10.1029/2001JD900183.

Ames, A. (1998) A documentation of glacier tongue variations and lake development in the Cordillera Blanca, Peru. *Zeitschrift fur Gletscherkunde und Glazialgeologie* **34**, 1–36.

Andersen, B.G. (1979) The deglaciation of Norway 15,000 to 10,000 years BP. *Boreas* **8**, 79–87.

Andersen, M., Lie, E., Derocher, A.E., Belikov, S.E., Bernhoft, A., Boltunov, A.N., Garner, G.W., Skaare, J.U., & Wiig, O. (2001) Geographic variations of PCB congeners in polar bears (*Ursus maritimus*) from Svalbard east to the Chukchi Sea. *Polar Biology* **24**, 231–8.

Andrews, J.T. (1970) *A Geomorphological Study of Post-Glacial Uplift with Particular Reference to Arctic Canada*. Institute of British Geographers, Special Publication 2.

Anisimov, O.A. (1989) Changing climate and permafrost distribution in the Soviet Arctic. *Physical Geography* **10**, 282–93.

Anisimov, O.A. & Fitzharris, B.B. (2001) Polar regions (arctic and antarctic). In: McCarthy et al. (eds.) *Climate Change 2001*. Cambridge University Press, Cambridge, pp. 801–41.

Anisimov, O.A. & Nelson, F.E. (1997) Permafrost zonation and climate change in the northern hemisphere: results from transient general circulation models. *Climate Change* **35**, 241–58.

Anisimov, O.A., Shiklomanov, N.I., & Nelson, F.E. (1997) Effects of global warming on permafrost and active layer thickness: results from transient general circulation models. *Global and Planetary Change* **61**, 61–7.

Archer, D.A. (2003) Contrasting hydrological regimes in the Upper Indus Basin. *Journal of Hydrology* **274**, 198–210.

Arctic Human Development Report (2004) Akureyri, Stefansson Arctic Institute.

Arctic Marine Transport Workshop (2004) *Key Issues in the Future of Arctic Marine Transport*. Scott Polar Research Institute, Cambridge, 28–30 September, 2004. Sponsored by Institute of the North, U.S. Arctic Research Commission and International Arctic Science Committee, 18 pp. + 10 Appendices.

Armstrong, R.L. & Armstrong, B.R. (1987) Snow and avalanche climates of the western United States: a comparison of maritime, intermountain and continental conditions. *International Association of Hydrological Sciences Publication* **162**, 281–94.

Armstrong, R.L. & Brodzik, M.J. (2001) Recent Northern Hemisphere snow extent: a comparison of data derived from visible and microwave satellite sensor. *Geophysical Research Letters* **28(19)**, 3673–6.

Armstrong, R.L. & Brodzik, M.J. (2002) Hemispheric-scale comparison and evaluation of passive-microwave snow algorithms. *Annals of Glaciology* **34**, 38–44.

Armstrong, R.L., Brodzik, M.J., Knowles, K., & Savoie, M. (2005) *Global monthly EASE-Grid snow water equivalent climatology*. National Snow and Ice Data Center, Boulder, CO. Digital media.

Assel, R.A., Robertson, D.M., Hoff, H., & Segelby, G.H. (1995) Climatic change implications from long-term (1823–1994) ice records for the Laurentian lakes. *Annals of Glaciology* **21**, 383–6.

Associate Committee on Geotechnical Research (1988) *Glossary of Permafrost and Related Ground Ice Terms*. Technical Memorandum 142, National Research Council of Canada, Ottawa.

Baker, V.R. (1973a) Paleohydrology and sedimentology of Lake Missoula flooding in eastern Washington. *Gelogical Society of America*, Special Paper 144.

Baker, V.R. (1973b) Erosional forms and processes for the catastrophic Pleistocene Missoula floods in eastern Washington. In: Morisawa, M. (ed.) *Fluvial Geomorphology*. Allen and Unwin, London, pp. 123–48.

Baker, V.R. (1981) *Catastrophic Flooding: The Origin of the Channeled Scablands*. Dowden, Hutchinson and Ross, Stroudsburg.

Baker, V.R., Benito, G., & Rudoy, A.N. (1993) Paleohydrology of late Pleistocene superflooding, Altay Mountains, Siberia. *Science* **259**, 348–50.

Baker, V.R., Greeley, R., Komar, P.D., Swanson, D.A., & Waitt, R.B. (1987) Columbia and Snake River plains. In: Graf, W.L. (ed.) *Geomorphic Systems of North America*. Geological Society of America, Centennial Special Volume 2, pp. 403–68.

Bales, R.C., McConnell, J.R., Mosley-Thompson, E., & Csatho, B. (2001) Accumulation over the Greenland Ice Sheet from historical and recent records. *Journal of Geophysical Research* **106(D24)**, 33813–26.

Ballantyne, C.K. (1990) The late Quaternary glacial history of the Trotternish escarpment, Isle of Skye: and implications for icesheet reconstruction. *Proceedings of the Geologists' Association* **101**, 171–86.

Ballantyne, C.K. (1995) Paraglacial debris cone formation on recently deglaciated terrain. *The Holocene* **5**, 25–33.

Ballantyne, C.K. (2002) Paraglacial geomorphology. *Quaternary Science Reviews* **21**, 1935–2017.

Ballantyne, C.K. & Benn, D.I. (1994) Paraglacial slope adjustment and resedimentation following recent glacier retreat, Fabergstolsdalen, Norway. *Arctic and Alpine Research* **26**, 255–69.

Ballantyne, C.K. & Benn, D.I. (1996) Paraglacial slope adjustment during recent deglaciation: implications for slope evolution in formerly glaciated terrain. In: Brooks, S. & Anderson, M.G. (eds.) *Advances in Hillslope Processes*. Wiley, Chichester, pp. 1173–95.

Ballantyne, C.K. & Harris, C. (1994) *The Periglaciation of Great Britain*. Cambridge University Press, Cambridge.

Ballantyne, C.K. & Mathews, J.A. (1982) The development of sorted circles on recently deglaciated terrain, Jotunheimen, Norway. *Arctic and Alpine Research* **14**, 341–54.

Ballantyne, C.K. & Mathews, J.A. (1983) Desiccation cracking and sorted polygon development, Jotunheimen, Norway. *Arctic and Alpine Research* **15**, 339–49.

Bamber, J.L. & Gomez-Dans, J.L. (2005) The accuracy of digital elevation models of the Antarctic continent. *Earth & Planetary Science Letters* **237**, 516–23.

Bamber, J.L. & Kwok, R. (2004) Remote sensing techniques. In: Bamber, J.L. & Payne, A.J. (eds.) *Mass Balance of the Cryosphere: Observations and Modelling of Contemporary and Future Changes*. Cambridge University Press, Cambridge, pp. 59–113.

Bamber, J.L. & Payne, A.J. (eds.) (2004) *Mass Balance of the Cryosphere: Observations and Modelling of Contemporary and Future Changes*. Cambridge University Press, Cambridge.

Barber, D.C., Dyke, A., Hillaire Marcel, C., Jennings, A.E., Andrews, J.T., Kerwin, M.W., Bilodeau, G., McNeely, R., Southon, J., Morehead, D., & Gagnon, J.M. (1999) Forcing of the cold event of 8,200 years ago by catastrophic drainage of Laurentide lakes. *Nature* **400**, 344–48.

Barber, D., Fortier, L., & Byers, M. (2005) The incredible shrinking sea ice. *Policy Options*. December 2005–January 2006, pp. 66–71.

Bard, E., Hamelin, B., & Fairbanks, R.G. (1990) U–Th ages obtained by mass spectrometry in corals from Barbados sea level during the past 130,000 years. *Nature* **346**, 456–8.

Barrett, P.J. (1991) Antarctica and global climatic change: a geological perspective. In: Harris, C.M. & Stonehouse, B. (eds.) *Antarctica and Global Climatic Change*. Belhaven Press, London, pp. 35–50.

Basist, A., Garrett, D., Ferraro, R., Grody, N., & Mitchell, K. (1996) A comparison between snow cover products derived from visible and microwave satellite observations. *Journal of Applied Meteorology* **35**, 163–77.

Baulin, V.V. & Danilova, N.S. (1984) Dynamics of late Quaternary permafrost in Siberia. In: Velitchko, A.A. (ed.) *Late Quaternary Environments of the Soviet Union.* Longman, London, pp. 69–78.

Beaty, C. (1978) The causes of glaciation. *American Journal of Science* **66**, 452–9.

Beaudoin, A.B. & King, R.H. (1994) Holocene palaeoenvironmental record preserved in a paraglacial alluvial fan, Sunwapta Pass, Jasper National Park, Alberta, Canada. *Catena* **22**, 227–48.

Beget, J.E. (1985) Tephrochronology of antiscarp slopes on an alpine ridge near Gamma Peak, Washington, USA. *Arctic and Alpine Research* **17**, 143–52.

Benn, D.I. & Evans, D.J.A. (1998) *Glaciers and Glaciation.* Arnold, London, 734 pp.

Benson, C.S. (1961) Stratigraphic studies in the snow and firn of the Greenland Ice Sheet. *Folia Geographica Danica* **9**, 13–37.

Benson, B. & Magnusson, J. (2000) *Global Lake and River Ice Phenology Database.* Boulder, CO, National Snow and Ice Data Center/World Data Center for Glaciology. Digital media.

Bentley, C.R. (2004) Mass balance of the Antarctic Ice Sheet: observational aspects. In: Bamber, J.L. & Payne, A.J. (eds.) *Mass Balance of the Cryosphere: Observations and Modelling of Contemporary and Future Changes.* Cambridge University Press, Cambridge, pp. 459–90.

Berezovskaya, S., Yang, D., & Kane, D. (2004) Compatibility analysis of precipitation and runoff trends over the large Siberian watersheds. *Geophysical Research Letters* **31**, 1–4.

Berger, A.R. & Iams, W.J. (1996) *Geoindicators: Assessing Rapid Environmental Change in Earth Systems.* Balkema, Rotterdam.

Berner, J. & Furgal, C. (2005) *Human Health in ACIA.* Cambridge University Press, Cambridge, pp. 863–906.

Bindschadler, R., Vornberger, P., Blankenship, D., Scambos, T., & Jacobel, R. (1996) Surface velocity and mass balance of Ice Streams D and E, West Antarctica. *Journal of Glaciology* **42**, 461–75.

Birkett, C.M. (1998) Contribution of the TOPEX NASA radar altimeter to the global monitoring of large rivers and wetlands. *Water Resources Research* **34**, 1223–39.

Bishop, M.P., Barry, R.G., Bush, A.B.G., Copland, L., Dwyer, J.L., Fountain, A.G., Haeberli, W., Hall, D.K., Kaab, A., Kargel, J.S.,

Molnia, B.F., Olsenholler, J.A., Paul, F., Raup, B.H., Shroder, J.F., Trabant, D.C., & Wessels, R. (2004) Global land ice measurements from space (GLIMS): remote sensing and GIS investigations of the Earth's cryosphere. *Geocarto International* **9**, 57–84.

Björnsson, H. (1974) Explanation of jökulhlaups from Grímsvötn, Vatnjökull, Iceland. *Jökull* **24**, 1–26.

Björnsson, H. (1976) Marginal and supraglacial lakes in Iceland. *Jökull* **26**, 40–51.

Björnsson, H. (1988) *Hydology of Ice Caps in Volcanic Regions.* Reykjavik, Societas Scientarium Islandica 45.

Björnsson, H. (1998) Hydrological characteristics of the drainage system beneath a surging glacier. *Nature* **395**, 771–4.

Björnsson, H. (2002) Subglacial lakes and jökulhlaups in Iceland. *Global and Planetary Change* **35**, 255–71.

Björnsson, H. (2004) Glacial lake outburst floods in mountain environments. In: Owens, P.N. & Slaymaker, O. (eds.) *Mountain Geomorphology.* Arnold, London, pp. 165–84.

Blair, R.W. (1994) Mountain and valley wall collapse due to rapid deglaciation in Mount Cook National Park, New Zealand. *Mountain Research and Development* **14**, 347–58.

Blais, J.M., Kimpe, L.E., McMahon, D., Keatley, B.E., Mallory, M.L., Douglas, M.S.V., & Smol, J.P. (2005) Arctic seabirds transport marine-derived contaminants. *Science* **309**, 449.

Blanchon, P. & Shaw, J. (1995) Reef drowning during the last deglaciation: evidence for catastrophic sea-level rise and ice-sheet collapse. *Geology* **23**, 4–8.

Blunier, T., Chappellaz, J., Schwander, J., Dallenbach, A., Stauffer, B., Stocker, T.F., Raynaud, D., Clausen, H.B., Hammer, C.U., & Johnsen, S.J. (1998) Asynchrony of Antarctic and Greenland climate change during the last glacial period. *Nature* **394**, 739–43.

Boer, M.M. & Koster, E.A. (eds.) (1992) *Greenhouse Impact on Cold Climate Ecosystems and Landscape.* Catena Supplement 22.

Bond, G., Broecker, W., Johnsen, S., McManus, J., Labeyrie, L., Jouzel, J., & Bonani, G. (1993) Correlations between climate records from North Atlantic sediments and Greenland ice. *Nature* **365**, 143–7.

Bond, G., Showers, W., Cheseby, M., Lotti, R., Almasi, P., deMenocal, P., Priore, P., Cullen, H., Hajdas, I., & Bonani, G. (1997) A pervasive millennial scale cycle in North Atlantic Holocene and glacial climates. *Science* **278**, 1257–66.

Booth, D.B. (1986) Mass balance and sliding velocity of the Puget Lobe of the Cordilleran ice sheet during the last glaciation. *Quaternary Research* **25**, 269–80.

Booth, D.B. & Hallet, B. (1993) Channel networks carved by subglacial water: observations and reconstruction in the eastern Puget Lowland of Washington. *Geological Society of America, Bulletin* **105**, 671–82.

Boulton, G.S. & Dent, D.L. (1974) The nature and rate of post-depositional changes in recently deposited till from south-east Iceland. *Geografiska Annaler* **56A**, 121–34.

Boulton, G.S., Smith, G.D., Jones, A.S., & Newsome, J. (1985) Glacial geology and glaciology of the last mid-latitude ice sheets. *Journal of the Geological Society of London* **142**, 447–74.

Bovis, M.J. (1990) Rock-slope deformation at Affliction Creek, southern Coast Mountains, British Columbia. *Canadian Journal of Earth Sciences* **27**, 243–54.

Bowen, D.Q. (1978) *Quaternary Geology*. Pergamon, Oxford, 221.

Bowen, D.Q. (2004) Ice ages (interglacials, interstadials and stadials). In: Goudie, A.S. (ed.) *Encyclopedia of Geomorphology (vol. 1)*. Routledge, London, pp. 549–54.

Bradley, R.S. (1999) *Paleoclimatology: Reconstructing Climates of the Quaternary*. Academic Press, San Diego and London.

Braithwaite, R.J. (2002) Glacier mass balance: the first 50 years of international monitoring. *Progress in Physical Geography* **26(1)**, 76–95.

Braithwaite, R.J. (1981) On glacier energy balance, ablation and air temperature. *Journal of Glaciology* **27**, 381–91.

Brardinoni, F. & Hassan, M.A. (2006) Glacial erosion, evolution of river long-profiles and the organization of process domains in mountain drainage basins of coastal British Columbia. *Journal of Geophysical Research* **11**, DOI:10.1029/2005JF000358.

Braun, L.N. & Slaymaker, O. (1981) Effect of scale on the complexity of snowmelt systems. *Nordic Hydrology* **12**, 225–34.

Braun, L.N., Weber, M., & Schulz, M. (2000) Consequences of climate change for runoff from alpine regions. *Annals of Glaciology* **31**, 19–25.

Brennand, T.A. (1994) Macroforms, large bedforms and rhythmic sedimentary sequences in subglacial eskers, south-central Ontario: implications for esker genesis and meltwater regime. *Sedimentary Geology* **91**, 9–55.

Brennand, T.A. (2004) Glacifluvial (glaciofluvial). In: Goudie, A.S. (ed.) *Encyclopedia of Geomorphology*. Routledge, London and New York, vol. 1, pp. 459–65.

Bretz, J.H. (1923) The Channeled Scabland of the Columbia Plateau. *Journal of Geology* **31**, 617–49.

Bretz, J.H. (1925) The Spokane flood beyond the Channeled Scablands. *Journal of Glaciology* **12**, 501–4.

Bretz, J.H. (1969) The Lake Missoula floods and the Channeled Scabland. *Journal of Geology* **77**, 505–43.

Brigham, L.W. (2001a) The northern sea route, 1999–2000. *Polar Record* **37**, 329–36.

Brigham, L.W. (2001b) Developing scenarios of the future: the Arctic in 2002–32. In: Erickson, K. (ed.) *Proceedings of the Arctic Global Change Workshop*, December, 2001. University of Alaska, Fairbanks.

Brink, V.C., Mackay, J.R., Freymand, S., & Pearce, D.C. (1967) Needle ice and seedling establishment in southwestern British Columbia. *Canadian Journal of Plant Science* **47**, 135–9.

Broecker, W.S. (1986) Oxygen isotope constraints on surface ocean temperatures. *Quaternary Research* **26**, 121–34.

Broecker, W.S. (1994) Massive iceberg discharges as triggers for global climatic change. *Nature* **372**, 421–24.

Broecker, W.S. (1996) Glacial climate in the Tropics. *Science* **272**, 1902–4.

Broecker, W.S. & Denton, G.H. (1990) The role of ocean–atmosphere reorganizations in glacial cycles. *Quaternary Science Reviews* **9**, 305–41.

Broecker, W.S. & Van Donk, J. (1970) Insolation changes, ice volumes and the oxygen 18 record in deep sea cores. *Reviews of Geophysics and Space Physics* **8**, 169–98.

Broecker, W.S., Thurber, D.I., Goddard, J., Ku, T.-L., Mathews, R.K., & Mesolella, K.J. (1968) Milankovitch hypothesis supported by precise dating of coral reefs and deep sea sediments. *Science* **159**, 297–300.

Brown, R.D. (1997) Historical variability in Northern Hemisphere spring snow covered area. *Annals of Glaciology* **25**, 340–6.

Brown, R.D. & Braaten, R.O. (1998) Spatial and temporal variability of Canadian monthly snow depths, 1946–95. *Atmosphere-Ocean* **36**, 37–45.

Brown, J., Ferrians, O.J., Jr., Heginbottom, J.A., & Melnikov, E.S. (1997) *International Permafrost Association Circum-Arctic Map of Permafrost and Ground Ice Conditions*. U.S. Geological Survey Circum Pacific Map Series, Map CP 45, Washington, DC.

Brown, J., Ferrians, O.J., Jr., Heginbottom, J.A., & Melnikov, E.S. (1998, revised February 2001) *Circum-Arctic Map of Permafrost and Ground Ice Conditions*. Boulder, CO, National Snow and Ice Data Center/World Data Center for Glaciology. Digital media.

Brown, J., Nelson, F.E., & Hinkel, K.M. (2000) The circumpolar active layer monitoring (CALM) program research designs and initial results. *Polar Geography* **3**, 165–258.

Brun, E., Martin, E., Simon, V., Gendre, C., & Col, C. (1989) An energy and mass model of snow cover suitable for operational avalanche forecasting. *Journal of Glaciology* **35**, 333–42.

Brunsden, D. (1990) Tablets of stone: toward the ten commandments of geomorphology. *Zeitschrift fur Geomorphologie Supplementband* **79**, 1–37.

Brunsden, D. (1993) The persistence of landforms. *Zeitschrift fur Geomorphologie Supplementband* **93**, 13–28.

Brunsden, D. & Thornes, J.B. (1979) Landscape sensitivity and change. *Transactions of the Institute of British Geography* **4**, 463–84.

Burn, C.R. (1997) Cryostratigraphy, paleogeography and climate change during the early Holocene warm interval, western Arctic coast. *Canadian Journal of Earth Sciences* **34**, 912–25.

Callaghan, T.V. & Jonasson, S. (1995) Implications for changes in arctic plant biodiversity from environmental manipulation experiments. In: Chapin III, F.S. & Korner, C.H. (eds.) *Arctic and Alpine Biodiversity: Patterns, Causes and Ecosystem Consequences*. Springer-Verlag, Heidelberg, pp. 151–64.

Campbell, J.B. (2002) *Introduction to Remote Sensing*, 3rd edn. The Guilford Press, New York.

Carter, R.M. & Gammon, P. (2004) New Zealand maritime glaciation: millennial scale southern climate change since 3.9 Ma. *Science* **304**, 1659–62.

Cavalieri, D.J. & Comiso, J. (2004) *AMSR-E/Aqua daily L3 12.5 km Tb, sea ice conc., & snow depth polar grids V001*, March to June 2004. National Snow and Ice Data Center, Boulder, CO, USA. Digital media – updated daily.

Cavalieri, D.J., Gloersen, P., & Campbell, W.J. (1984) Determination of sea ice parameters with the NIMBUS-7 SMMR. *Journal of Geophysical Research* **89(D4)**, 5355–69.

Cavalieri, D.J., Gloersen, P., Parkinson, C.L., Comiso, J.C., & Zwally, H.J. (1997a) Observed hemispheric asymmetry in global sea ice changes. *Science* **278**, 1104–6.

Cavalieri, D.J., Parkinson, C., Gloersen, P., & Zwally, H.J. (1997b, updated 2004) *Sea ice concentrations from Nimbus-7 SMMR and DMSP SSM/I passive microwave data*, June to September 2001. Boulder, CO, USA, National Snow and Ice Data Center. Digital media.

Cavalieri, D.J., Parkinson, C.L., Gloersen, P., Comiso, J.C., & Zwally, H.J. (1999) Deriving long-term time series of sea ice cover from satellite passive-microwave multisensor data sets. *Journal of Geophysical Research* **104**, 15803–14.

Cavalieri, D.J., Parkinson, C.L., & Vinnikov, K.Y. (2003) 30-year satellite record reveals contrasting Arctic and Antarctic decadal sea ice variability. *Geophysical Research Letters* **30**, 4-1–4-4.

Cayan, D.R. (1996) Interannual climate variability and snowpack in the western United States. *Journal of Climate* **9**, 928–48.

Chagnon, R. (2002) *Sea Ice Climatic Atlas: Northern Canadian Waters (1971–2000)*. Canadian Ice Service, Environment Canada, Ottawa.

Chang, A.T.C., Foster, J.L., & Hall, D.K. (1987) Nimbus-7 SMMR derived global snow cover parameters. *Annals of Glaciology* **9**, 39–44.

Chapin, F.S. et al. (20 other authors) (2005) Role of land surface changes in arctic summer warming. *Science* **310**, 657–60.

Chapman, W.L. & Walsh, J.E. (1993) Recent variations of sea ice and air temperature in high latitudes. *Bulletin of the American Meteorological Society* **74**, 34–47.

Chorley, R.J., Schumm, S.A., & Sugden, D.E. (1984) *Geomorphology*. Methuen, London.

Christian, C.S. (1957) The concept of land units and land systems. *Proceedings of the 9th Pacific Science Congress* **20**, 74–81.

Church, M. (1972) *Baffin Island Sandurs: A Study of Arctic Fluvial Processes*. Geological Survey of Canada Bulletin 216.

Church, M. (2002) Fluvial sediment transfer in cold regions. In: Hewitt, K., Byrne, M.-L., English, M., & Young, G. (eds.) *Landscapes of Transition*. Kluwer Academic, Dordrecht, pp. 93–117.

Church, M. & Gilbert, R.E. (1975) Proglacial fluvial and lacustrine sediments. In: Jopling, A.V. & McDonald, B.C. (eds.) *Glaciofluvial and Glaciolacustrine Sedimentation*. SEPM, Special Publication 23, pp. 22–100.

Church, M. & Ryder, J.M. (1972) Paraglacial sedimentation, a consideration of fluvial processes conditioned by glaciation. *Geological Society of America Bulletin* **83**, 3059–72.

Church, M. & Slaymaker, O. (eds.) (1985) *Field and Theory: Lectures in Geocryology*. University of British Columbia Press, Vancouver.

Church, M. & Slaymaker, O. (1989) Disequilibrium of Holocene sediment yield in glaciated British Columbia. *Nature* **337**, 452–4.

Church, J.A., Godfrey, J.S., Jackett, D.R., & McDougall, T.J. (1999) A model of sea level rise caused by ocean thermal expansion. *Journal of Climate* **4**, 438–56.

Clague, J.J. (1986) The Quaternary stratigraphic record of British Columbia: evidence for periodic sedimentation and erosion controlled by glaciation. *Canadian Journal of Earth Sciences* **23**, 885–94.

Clague, J.J. (1989) The Cordilleran Ice Sheet. In: Fulton, R.J. (ed.) *Quaternary Geology of Canada and Greenland*. Geological Survey of Canada, Ottawa, pp. 40–2.

Clague, J.J. & Mathews, W.H. (1973) The magnitude of jökulhlaups. *Journal of Glaciology* **12**, 501–4.

Clare, G.R., Fitzharris, B.B., Chinn, T.J.H., & Salinger, M.J. (2002) Interannual variation in end-of-summer snowlines of the Southern Alps of New Zealand, and relationships with Southern Hemisphere atmospheric circulation and sea surface temperature patterns. *International Journal of Climatology* **22**, 107–20.

Clark, C.D. (1993) Mega scale lineations and cross-cutting ice flow landforms. *Earth Surface Processes and Landforms* **18**, 1–29.

Clark, P.U., Alley, R.B., & Pollard, D. (1999) Northern Hemisphere ice sheet influences on global climate change. *Science* **286**, 1104–11.

Clarke, G.K.C., Mathews, W.H., & Pack, R.T. (1984) Outburst floods from Glacial Lake Missoula. *Quaternary Research* **22**, 289–99.

Clarke, G.K.C., Leverington, D.W., Teller, J.T., & Dyke, A.S. (2004) Paleohydraulics of the last outburst flood from glacial Lake Agassiz and the 8,200 BP cold event. *Quaternary Science Reviews* **23**, 389–407.

Clarke, G.K.C., Leverington, D.W., Teller, J.T., Dyke, A.S., & Marshall, S.J. (2005) Fresh arguments against the Shaw megaflood hypothesis – A reply to comments by David Sharpe, Correspondence. *Quaternary Science Reviews* **24**, 1533–41.

CLIMAP Project Members (1981) Seasonal reconstruction of the Earth's surface at the Last Glacial Maximum. *Geological Society of America*, Map Chart Series, MC-36.

Cogley, J.G. & Adams, W.P. (1998) Mass balance of glaciers other than ice sheets. *Journal of Glaciology* **44**, 315–25.

Colbeck, S.C., Akitiya, E., Armstrong, R., Gubler, H., Lafeuille, J., Lied, K., McClung, D., & Morris, E. (1990) *The International Classification for Seasonal Snow on the Ground*. International Association of Hydrological Sciences, International Commission on Snow and Ice, Wallingford, UK.

Collins, D.N. (1990) Glacial processes. In: Goudie, A.S. (ed.) *Geomorphological Techniques*, 2nd edn. Routledge, London, pp. 302–20.

Collins, D.N. & MacDonald, O.G. (2004) Year-to-year variability of solute flux in meltwaters draining from a highly glacierized basin. *Nordic Hydrology* **35**, 359–67.

Comiso, J.C. (2003) Warming trends in the Arctic from clear-sky satellite observations. *Journal of Climate* **16**, 3498–510.

Comiso, J.C. & Kwok, R. (1996) Surface and radiative characteristics of the summer arctic sea ice cover from multi-sensor satellite observations. *Journal of Geophysical Research* **101(C12)**, 28397–416.

Cook, A.J., Fox, A.J., Vaughan, D.G., & Ferrigno, J.G. (2005) Retreating glacier fronts on the Antarctic Peninsula over the past half century. *Science* **308**, 541–4.

Coordinated Eastern Arctic Experiment (CEAREX) (1995) *Coordinated Eastern Arctic Experiment (CEAREX), 1988–9*. National Snow and Ice Data Center, Boulder, CO, USA. CD-ROM.

Croll, J. (1867) On the eccentricity of the Earth's orbit and its physical relations to the glacial epoch. *Philosophical Magazine* **33**, 119–31.

Cruden, A.M. & Hu, X.Q. (1993) Exhaustion and steady-state models for predicting landslide hazards in the Canadian Rocky Mountains. *Geomorphology* **8**, 270–85.

Crutzen, P.J. & Stoermer, E.F. (2000) The "Anthropocene." *Global Change Newsletter* **41**, 17–18.

Cubasch, U., Hasselman, K., Hock, H., Maierraimer, E., Mikolajewicz, U., Santer, B.D., & Sausen, R. (1992) Time dependent greenhouse warming computations with a coupled ocean–atmosphere model. *Climate Dynamics* **8**, 55–69.

Dahl, S.O. & Nesje, A. (1996) A new approach to calculating Holocene winter precipitation by combining glacier equilibrium line altitudes and pine tree limits: a case study from Hardangerjokulen, central south Norway. *The Holocene* **6**, 381–98.

Dahl-Jensen, D., Johnsen, S.J., Hammer, C.U., Clausen, H.B., & Jouzel, J. (1993) Past accumulation rates derived from observed annual layers in the GRIP ice core from Summit, central Greenland. In: Peltier, W.R. (ed.) *Ice in the Climate System*. Springer-Verlag, Berlin, Heidelberg, Germany, pp. 517–32.

Dansgaard, W., Clausen, H.B., Gundestrup, N., Hammer, C.U., Johnsen, S.F., Kristinsdottir, P.M., & Reeh, N. (1982) A new Greenland deep ice core. *Science* **218**, 1273–77.

Dansgaard, W., Johnsen, S.J., Clausen, H.B. et al. (1993) Evidence of general instability of past climate from a 250-kyr ice-core record. *Nature* **364**, 218–20.

Davis, C.H. (1995) Growth of the Greenland Ice Sheet – a performance assessment of altimeter retracking algorithms. *IEEE Transactions on Geoscience and Remote Sensing* **33**, 1108–16.

Davis, C.H., Li, Y., McConnell, J.R., Frey, M.M., & Hanna, E. (2005) Snowfall-driven growth in East Antarctic Ice Sheet mitigates recent sea-level rise. *Science* **308**, 1898–1901.

Dawson, A.G. (1992) *Ice Age Earth: Late Quaternary Geology and Climate*. Routledge, London.

De La Beche, H.T. (1839) *Report on the Geology of Cornwall, Devon and West Somerset*. Memoirs of the Geological Survey, London.

Denton, G.H. (2000) Does an asymmetric thermo-haline ice sheet oscillator drive 100,000 year glacial cycles? *Journal of Quaternary Science* **15**, 301–18.

Denton, G.H. & Hughes, T.J. (eds.) (1981) *The Last Great Ice Sheets*. Wiley-Interscience Publications, New York.

Department of Foreign Affairs and International Trade, Canada (2002) *The Northern Dimension of Canada's Foreign Policy*. Ottawa, pp. 1–19.

Derksen, C., Walker, A., LeDrew, E., & Goodison, B. (2003) Combining SMMR and SSM/I data for time series analysis of central North American snow water equivalent. *Journal of Hydrometeorology* **4**, 304–6.

Dewailly, E. et al. (1989) High levels of PCBs in breast milk of Inuit women from arctic Quebec. *Bulletin of Environmental Contamination and Toxicology* **43**, 641–6.

Diamond, J. (2005) *Collapse: How Societies Choose to Fail or Succeed*. Penguin Books, London.

Dietrich, W.E., Wilson, C.J., Montgomery, D.R., McKean, J., & Bauer, R. (1992) Erosion thresholds and land surface morphology. *Geology* **20**, 675–9.

Dingman, S.L. (2002) *Physical Hydrology*, 2nd edn. Prentice Hall, New Jersey.

Doeksen, N.J. & Judson, A. (1996) *The Snow Booklet: A Guide to the Science, Climatology, and Measurement of Snow in the United States*. Colorado Climate Center, Fort Collins, CO.

Donlon, C.J., Minnett, P., Gentemann, C., Nightingale, T.J., Barton, I.J., Ward, B., & Murray, J. (2002) Towards improved validation of satellite sea surface skin temperature measurements for climate research. *Journal of Climate* **15**, 353–69.

Dowdeswell, J.A. & Drewry, D. (1989) The dynamics of Austfonna, Nordaustlandet,

Svalbard: surface velocities, mass balance and subglacial meltwater. *Annals of Glaciology* **12**, 37–45.

Dowdeswell, J.A. & Hagen, J.O. (2004) Arctic ice caps and glaciers. In: Bamber, J.L. & Payne, A.J. (eds.) *Mass Balance of the Cryosphere*. Cambridge University Press, Cambridge, pp. 527–58.

Dowdeswell, J.A. & Scourse, J.D. (eds.) (1990) *Glacimarine Environments, Processes and Products*. Geological Society Special Publication 53, London.

Dowdeswell, J.A. & Siegert, M.J. (2002) The physiography of modern Antarctic subglacial lakes. *Global and Planetary Change* **35**, 221–36.

Dowdeswell, J.A., Hagen, J.O., Bjornsson, H., Glazovsky, A., Holmlund, P., Jania, J., Josberger, E., Koerner, R., Ommanney, S., & Thomas, B. (1997) The mass balance of circum-Arctic glaciers and recent climate change. *Quaternary Research* **48**, 1–14.

Dowlatabadi, H. (2002) Global change: much more than a matter of degrees. *Meridian*, Spring/Summer, Canadian Polar Commission, Ottawa, pp. 8–12.

Downie, D.L. & Fenge, T. (eds.) (2003) *Northern Lights Against POPs: Combatting Toxic Threats in the Arctic*. McGill-Queen's University, Montreal and Kingston.

Dreimanis, A. (1989) Tills, their genetic terminology and classification. In: Goldthwait, R.P. & Matsch, C.L. (eds.) *Genetic Classification of Glacigenic Deposits*. Balkema, Rotterdam, pp. 17–84.

Driedger, C.L. & Fountain, A.G. (1989) Recent climatic change and catastrophic geomorphologic processes in mountain environments. *Geomorphology* **10**, 107–28.

Duchkov, A.D. & Balobayev, V.T. (2001) Geothermal studies of permafrost response to global natural changes. In: Paepe, R. et al. (eds.) *Permafrost Response on Economic Development, Environmental Security and Natural Resources*. Kluwer, Dordrecht, pp. 317–32.

Duguay, C.R. & Lafleur, P.M. (2003) Determining depth and ice thickness of shallow sub-arctic lakes using space-borne optical and SAR data. *International Journal of Remote Sensing* **24**, 475–89.

Duguay, C.R. & Pietroniro, A. (eds.) (2005) *Remote Sensing in Northern Hydrology: Measuring*

Environmental Change. Geophysical Monograph 163. American Geophysical Union, Washington, DC.

Duguay, C.R., Pultz, T., Lafleur, P.M., & Drai, D. (2002) RADARSAT backscatter characteristics of ice growing on shallow sub-Arctic lakes, Churchill, Manitoba, Canada. *Hydrological Processes* **16**, 1631–44.

Duk-Rodkin, A. & Hughes, O.L. (1994) Tertiary-Quaternary drainage of the pre-glacial Mackenzie basin. *Quaternary International* **22/23**, 221–41.

Duplessy, J.-C., Arnold, M., Maurice, P., Bard, E., Duprat, J., & Moyes, J. (1986) Direct dating of the oxygen-isotope record of the last deglaciation by C-14 accelerator mass spectrometry. *Nature* **320**, 350–2.

Dyke, A.S. & Dredge, L.A. (1989) Quaternary geology of the northwestern Canadian Shield. In: Fulton, R.J. (ed.) *Quaternary Geology of Canada and Greenland*. Geological Survey of Canada, Geology of Canada No.1, pp. 178–214.

Dyke, A.S. & Prest, V.K. (1987) The late Wisconsin and Holocene history of the Laurentide Ice Sheet. *Géographie physique et Quaternaire* **36**, 5–14.

Dyke, A.S., Andrews, J.T., Clark, P.U., England, J.H., Miller, G.H., Shaw, J., & Veillette, J.J. (2002) The Laurentide and Innuitian ice sheets during the Last Glacial Maximum. *Quaternary Science Reviews* **21**, 9–31.

Dyunin, A.K., Kvon, Ya.D., Zhilin, A.M., & Komarov, A.A. (1991) Effect of snow drifting on large scale aridization. In: Kotlyakov, V.M., Ushakov, A., & Glazovsky, A. (eds.) *Glaciers–Ocean–Atmosphere Interactions*. International Association of Hydrological Sciences Publication 208, pp. 489–94.

Eldhuset, K., Andersen, P.H., Hauge, S., Isaksson, E., & Weydahl, D. (2003) ERS tandem InSAR processing for DEM generation, glacier motion estimation and coherence analysis on Svalbard. *International Journal of Remote Sensing* **24**, 1415–30.

Elfstrom, A. & Rossbacher, L. (1985) Erosional remnants in the Baldakatj area, Lappland, northern Sweden. *Geografiska Annaler* **67A**, 167–76.

Elson, J.A. (1961) The geology of tills. In: Penner, E. & Butler, J. (eds.) *Proceedings of the*

14th Canadian Soil Mechanics Conference: 5–36. Commission for Soil and Snow Mechanics Technical Memoir 69.

Embleton-Hamann, C. (2004) Proglacial landforms. In: Goudie, A.S. (ed.) *Encyclopedia of Geomorphology (vol. 2)*. Routledge, London, pp. 810–13.

Environment Canada (1999) *The Canadian National Report on Systematic Observations for Climate: The Canadian Global Climate Observing System Program*. Environment Canada/ Meteorological Services Submission to the UN Framework Convention on Climate Change, Ottawa.

Environment Canada (2002) *Sea Ice Climatic Atlas, Northern Canadian Waters, 1971–2000*. Canadian Ice Service, Ottawa.

EPICA (2004) Eight glacial cycles from an Antarctic ice core. *Nature* **429**, 623–8.

Esmark, J. (1824) Remarks tending to explain the geological history of the Earth. *Edinburgh New Philosophical Journal* **2**, 107–21.

Evans, D.J.A. (ed.) (1994) *Cold Climate Landforms*. Wiley, Chichester.

Evans, D.J.A. (2003) *Glacial Landsystems*. Arnold, London.

Evans, D.J.A. & Rogerson, R.J. (1986) Glacial geomorphology and chronology in the Selamiut Range/Nachvak Fiord area, Torngat Mountains, Labrador. *Canadian Journal of Earth Sciences* **23**, 66–76.

Evans, D.J.A., Clark, C.D., & Mitchell, W.A. (2005) The last British ice sheet: a review of the evidence utilised in the compilation of the Glacial Map of Britain. *Earth Science Reviews* **70**, 253–312.

Evans, I.S. (1977) World-wide variations in the direction and concentration of cirque and glacier aspects. *Geografiska Annaler* **59A**, 151–75.

Evans, I.S. (2004) Cirque, glacial. In: Goudie, A.S. (ed.) *Encyclopedia of Geomorphology*. Routledge, London & New York, vol. 1, pp. 154–8.

Evans, I.S. (2006) Local aspect asymmetry of mountain glaciation: a global survey of consistency of favoured directions for glacier numbers and altitudes. *Geomorphology* **73**, 166–84.

Evans, M. & Slaymaker, O. (2004) Spatial and temporal variability of sediment delivery from

alpine lake basins, Cathedral Provincial Park. *Geomorphology* **61**, 209–24.

Eyles, N. (1987) Late Pleistocene debris flow deposits in large glacial lakes in British Columbia and Alaska. *Sedimentary Geology* **53**, 33–71.

Eyles, N. & Eyles, C.H. (1992) Glacial depositional systems. In: Walker, R.G. & James, N.P. (eds.) *Facies Models: Response to Sea Level Change*. Geological Association of Canada, Toronto, pp. 73–100.

Eyles, N., Dearman, W.R., & Douglas, T.D. (1983) The distribution of glacial land systems in Britain and North America. In: Eyles, N. (ed.) *Glacial Geology*. Pergamon, Oxford, pp. 213–28.

Fairbanks, R.G. (1989) A 17,000 year glacioeustatic sea level record: influence of glacial melting rates on the Younger Dryas event and deep ocean circulation. *Nature* **342**, 637–43.

Fenge, T. (2001) The Inuit and climate change. *Isuma* **2**, 79–85.

Fischer, A.G. (1981) Climatic oscillations in the biosphere. In: Nitecki, M.H. (ed.) *Biotic Crises in Ecological and Evolutionary Time*. Academic Press, New York, pp. 103–31.

Fitzharris, B.B. (1996) The cryosphere: changes and their impacts. In: Watson, R.T., Zinyowera, M.C., Moss, R.H., & Dokken, D.J. (eds.) *Chapter 7, Climate Change 1995: Impacts, Adaptations and Mitigation of Climate Change: Scientific-Technical Analyses*. Cambridge University Press, Cambridge, pp. 243–65.

Flato, G.M. (2004) Sea-ice modeling. In: Bamber, J.L. & Payne, A.J. (eds.) *Mass Balance of the Cryosphere*. Cambridge University Press, Cambridge, pp. 337–92.

Flint, R.F. (1957) *Glacial and Pleistocene Geology*. Wiley, New York.

Forbes, D.L. & Syvitski, J.P.M. (1994) Paraglacial coasts. In: Carter, R.W.G. & Woodroffe, C.D. (eds.) *Coastal Evolution: Late Quaternary Shoreline Morphodynamics*. Cambridge University Press, Cambridge, pp. 373–424.

Ford, A.L.J., Forster, R.R., & Bruhn, R.L. (2003) Ice surface velocity patterns on Seward Glacier, Alaska/Yukon, and their implications for regional tectonics in the Saint Elias Mountains. *Annals of Glaciology* **36**, 21–8.

Foster, J.L., Chang, A.T.C., & Hall, D.K. (1997) Comparison of snow mass estimates from a prototype passive microwave snow algorithm, a revised algorithm and snow depth climatology. *Remote Sensing of Environment* **62**, 132–42.

Fountain, A.G., Krimmel, R.M., & Trabant, D.C. (1997) *A Strategy for Monitoring Glaciers.* U.S. Geological Survey Circular 1132, Washington, DC.

Frakes, L.A. & Francis, J.E. (1988) A guide to Phanerozoic cold polar climates from high latitude ice-rafting in the Cretaceous. *Nature* **333**, 547–9.

Fraser, G.S., Bleuer, N.K., & Smith, N.D. (1983) History of Pleistocene alluviation of the middle and upper Wabash valley: Field Trip 13. In: Shaver, R.H. & Sunderman, J.A. (eds.) *Field Trips in Midwestern Geology.* Indiana Geological Survey, Bloomington, pp. 197–224.

Frei, A. & Robinson, D.A. (1999) Northern Hemisphere snow extent: regional variability 1972–94. *International Journal of Climatology* **19**, 1535–60.

French, H.M. (1996) *The Periglacial Environment.* Addison Wesley Longman Limited, Harlow, 341 pp.

French, H.M. & Harry, D.G. (1992) Pediments and cold climate conditions, Barn Mountains, unglaciated northern Yukon, Canada. *Geografiska Annaler* **74A**, 145–57.

French, H.M. & Slaymaker, O. (eds.) (1993) *Canada's Cold Environments.* McGill Queen's University, Montreal and Kingston.

Friele, P.A., Ekes, C., & Hicken, E.J. (1999) Evolution of Cheekye fan, Squamish, British Columbia: Holocene sedimentation and implications for hazard assessment. *Canadian Journal of Earth Sciences* **36**, 2023–31.

Fulton, R.J. (ed.) (1989) Foreword to the Quaternary Geology of Canada and Greenland. In: *Geology of Canada and Greenland.* Geological Survey of Canada, Geology of Canada No. 1.

Gage, M. (1970) The tempo of geomorphic change. *Journal of Geology* **78**, 619–25.

Geiger, C.A. & Drinkwater, M.R. (2005) Coincident buoy- and SAR-derived surface fluxes in the western Weddell Sea during Ice Station Weddell 1992. *Journal of Geophysical Research* **100**, C04002, doi:10.1029/2003JC002112.

Gilbert, R.E. (1975) Sedimentation in Lillooet Lake B.C. *Canadian Journal of Earth Sciences* **12**, 1696–711.

Ginzburg, B.M., Polyakova, K.N., & Soldatova, I.I. (1992) Secular changes in dates of ice formation on rivers and their relation with climate change. *Soviet Meteorology and Hydrology* **12**, 57–64.

Godard, A. & André, M.-F. (1999) *Les Milieux Polaires.* Armand Colin, Paris.

Gogineni, S., Tammana, D., Braaten, D., Leuschen, C., Akins, T., Legarsky, J., Kanagaratnam, P., Stiles, J., Allen, C., & Jezek, K. (2001) Coherent radar ice thickness measurements over the Greenland Ice Sheet. *Journal of Geophysical Research* **106**, doi: 10.1029/2001JD900183.

Goita, K., Walker, A.E., & Goodison, B.E. (2003) Algorithm development for the estimation of snow water equivalent in the boreal forest using passive microwave data. *International Journal of Remote Sensing* **24**, 1097–102.

Gold, L.W. (1985) The ice factor in frozen ground. In: Church, M. & Slaymaker, O. (eds.) *Field and Theory.* University of British Columbia Press, Vancouver, pp. 74–95.

Goodison, B.E. (1978) Accuracy of Canadian snow gage measurements. *Journal of Applied Meteorology* **17**, 1542–8.

Goodison, B. & Walker, A. (1994) Canadian development and use of snow cover information from passive microwave satellite data. In: Choudhury, B.J., Kerr, Y.H., Njoku, E.G., & Pampaloni, P. (eds.), *Proceedings of the ESA/NASA International Workshop, St. Lary, France, 11–15 January 1993.* pp. 245–62.

Gordon, A.L. & Ice Station Weddell Group of Principal Investigators & Chief Scientists (1993) Weddell Sea Exploration from Ice Station. *EOS Transactions American Geophysical Union* **74**, 121 and 124–6.

Goudie, A.S. (ed.) (1990) *Geomorphological Techniques.* Routledge, London.

Goudie, A.S. (ed.) (2004) *Encyclopedia of Geomorphology* (2 vols.). Routledge, London.

Grabherr, G., Gottfried, M., & Pauli, H. (1994) Climate effects on mountain plants. *Nature* **369**, 448.

Gradinger, R. & Nurnberg, D. (1996) Snow algal communities on Arctic pack ice floes dominated by Chlamydomonas nivalis (Bauer) Wille. *Proceedings of a NIPR Symposium on Polar Biology* **9**, 35–43.

Gradstein, F.M., Ogg, J.G., & Smith, A.G. (eds.) (2005) *A Geologic Time Scale 2004*. Cambridge University Press, Cambridge.

Gray, D.M. & Male, D.H. (eds.) (1981) *Handbook of Snow: Principles, Processes, Management and Use*. Pergamon Press, Toronto.

Grenfell, T.C. & Perovich, D.K. (1984) Spectral albedos of sea ice and incident solar irradiance in the southern Beaufort Sea. *Journal of Geophysical Research* **89**, 3573–80.

Greuell, W. & Smeets, C.J.J. (2001) Variations with elevation in the surface energy balance on the Pasterze (Austria). *Journal of Geophysical Research* **106**, 31717–27.

Groisman, P.Ya & Davies, T.D. (2001) Snow cover and the climate system. In: Jones, H.G. et al. (eds.) *Snow Ecology*. Cambridge University Press, Cambridge, pp. 1–44.

Groisman, P.Ya, Karl, T.R., Knight, R.W., & Stenchikov, G.L. (1994) Changes of snow cover. Temperature and the radiative heat balance over the Northern Hemisphere. *Journal of Climate* **7**, 1633–56.

Gudmundsson, M.T., Björnsson, H., & Pálsson, F. (1995) Changes in jökulhlaup size in Grímsvötn, Vatnjökull, Iceland, 1934–91, deduced from in situ measurements of subglacial lake volume. *Journal of Glaciology* **41**, 263–72.

Gudmundsson, M.T., Sigmundsson, F., & Björnsson, H. (1997) Ice–volcano interaction of the 1996 Gjálp subglacial eruption. Vatnajökull, Iceland. *Nature* **389**, 954–7.

Guilcher, A., Bodere, J.-C., Coude, A., Hansom, J.D., Moign, A., & Peulvast, J.-P. (1986) Le problème des strandflats en cinq pays de hautes latitudes. *Revue de Géologie Dynamique et de Géographie Physique* **27**, 47–79.

Guilderson, T.P., Fairbanks, R.G., & Rubenstone, J.L. (1994) Tropical temperature variations since 20,000 years ago: modulating inter-hemispheric climate change. *Science* **263**, 663–5.

Gunderson, L.H. & Holling, C.S. (2002) *Panarchy: Understanding Transformation in Human and Natural Systems*. Island Press, Washington, DC.

Gurney, S.D. (2004) Pingo. In: Goudie, A.S. (ed.) *Encyclopedia of Geomorphology*. Routledge, London and New York, vol. 2, pp. 782–4.

Gurney, R., Foster, J.L., & Parkinson, C.L. (eds.) (1993) *Atlas of Satellite Observations Related to Global Change*. Cambridge University Press, Cambridge.

Guymon, G.L. (1974) Regional sediment yield analysis of Alaska streams. American Society of Civil Engineers Proceedings. *Journal of the Hydraulics Division* **100**, 41–51.

Haeberli, W. (1983) Frequency and characteristics of glacier floods in the Swiss Alps. *Annals of Glaciology* **4**, 85–90.

Haeberli, W. & Hoelzle, M. (1995) Applications of inventory data for estimating characteristics of and regional climate change effects on mountain glaciers: a pilot study in the European Alps. *Annals of Glaciology* **21**, 206–12.

Haeberli, W., Barry, R. & Cihlar, J. (2000) Glacier monitoring within the global climate observing system. *Annals of Glaciology* **31**, 241–6.

Hagen, J.O. & Reeh, N. (2004) In situ measurements: land ice. In: Bamber, J.L. & Payne, A.J. (eds.) *Mass Balance of the Cryosphere: Observations and Modelling of Contemporary and Future Changes*. Cambridge University Press, Cambridge, pp. 11–43.

Haggerty, C.D. & Armstrong, R.L. (1996) Snow trends within the former Soviet Union. EOS Supplement, *Transactions of the American Geophysical Union*, 77 (abstracts of the Fall Meeting of the AGU, San Fransico).

Hall, D.K., Kelly, R.E.J., Foster, J.L., & Chang, A.T.C. (2005a) Estimation of snow extent and snow properties. In: Anderson, M.G. (ed. in chief) *Encyclopedia of Hydrological Sciences*. John Wiley & Sons, Chichester, DOI: 10.1002/0470848944.hsa062.

Hall, D.K., Kelly, R.E.J., Foster, J.L., & Chang, A.T.C. (2005b) Hydrological application of remote sensing: surface states – snow. In: Anderson, M.G. (ed.) *Encyclopedia of Hydrological Sciences*. John Wiley & Sons, Ltd., Chichester, p. 3456.

Hall, D.K., Kelly, R.E.J., Riggs, G.A., Chang, A.T.C., & Foster, J.L. (2002a) Assessment of the relative accuracy of hemispheric-scale snow-cover maps. *Annals of Glaciology* **34**, 24–30.

Hall, D.K., Riggs, G.A., & Salomonson, V.V. (2003) *MODIS/Aqua Snow Cover Daily L3 Global 0.05Deg CMG V004*, January to March 2003. National Snow and Ice Data Center, Boulder, CO, USA. Digital media – updated daily.

Hall, D.K., Riggs, G.A., Salomonson, V.V., DiGirolamo, N.E., & Bayr, K.J. (2002b) MODIS snow-cover products. *Remote Sensing of Environment* **83**, 181–94.

Hall, D.K., Riggs, G.A., & Salomonson, V.V. (2006) MODIS snow and sea ice products. In: Qu, J. (ed.) *Earth Science Satellite Remote Sensing – Volume I: Science and Instruments*. Springer-Verlag, Berlin.

Hall, D.K., Williams Jr, R.S., & Sigurðsson, O. (1995) Glaciological observations of Brúarjökull, Iceland, using synthetic aperture Radar and thematic mapper satellite data. *Annals of Glaciology* **21**, 271–6.

Hallet, B., Hunter, L., & Bogen, J. (1996) Rates of erosion and sediment evacuation by glaciers: a review of field data and their implications. *Global and Planetary Change* **12**, 213–35.

Hand, D.J. (2004) *Measurement Theory and Practice: The World Through Quantification*. Hodder Arnold, London.

Hansen, J. & Lebedeff, S. (1988) Global surface temperatures: update through 1987. *Geophysical Research Letters* **15**, 323–6.

Harker, A. (1901) Ice erosion in the Cuillin Hills, Skye. *Transactions of the Royal Society of Edinburgh* **40**, 221–52.

Harms, S., Fahrbach, E., & Strass, V.H. (2001) *AWI moored ULS data, Weddell Sea (1990–98)*. National Snow and Ice Data Center/World Data Center for Glaciology, Boulder, CO, digital media.

Harris, C., Vonder-Muhl, D., Isaksen, K., Haeberli, W., Sollid, J.L., King, L., Holmlund, P., Dramis, F., Guglielmin, M., & Palacios, D. (2003) Warming permafrost in European mountains. *Global and Planetary Change* **39**, 215–25.

Harrison, S. & Winchester, V. (1997) Age and nature of paraglacial debris cones along the margins of the San Rafael glacier, Chilean Patagonia. *The Holocene* **7**, 481–7.

Hastenrath, S. & Greischar, L. (1997) Glacier recession on Kilimanjaro, East Africa, 1912–89. *Journal of Glaciology* **43**, 455–9.

Hastenrath, S. & Kruss, P.D. (1992) The dramatic retreat of Mt. Kenya's glaciers between 1963 and 1987. *Annals of Glaciology* **16**, 127–33.

Hay, J.E. & Fitzharris, B.B. (1988). A comparison of the energy balance and bulk-aerodynamic approaches for estimating glacier melt. *Journal of Glaciology* **34**, 145–53.

Hays, J.D., Lozano, J.A., Shackleton, N., & Irving, G. (1976) Reconstruction of the Atlantic and western Indian sectors of the 18,000 BP Antarctic Ocean. In: Cline, R.M. & Hays, J.D. (eds.) *Investigation of Late Quaternary Paleoceanography and Paleoclimatology*. Geological Society of America Memoir 145, Boulder, CO, pp. 53–76.

Head, I. (1991) *On A Hinge of History: The Mutual Vulnerability of South and North*. University of Toronto Press, Toronto.

Heginbottom, J.A. (2002) Permafrost mapping: a review. *Progress in Physical Geography* **26**, 623–42.

Heginbottom, J.A., Brown, J., Melnikov, E.S., & Ferrians, O.J.Jr. (1993) Circum-arctic map of permafrost and ground ice conditions. In: *Proceedings of the Sixth International Permafrost Conference*, Wushan, Guangzhou. South China University Press, vol. 2, p. 1132.

Heinrich, H. (1988) Origin and consequences of cyclic ice rafting in the northeast Atlantic Ocean during the past 130,000 years. *Quaternary Research* **29**, 143–52.

Henderson, F.M. & Lewis, A.J. (eds.) (1998) *Manual of Remote Sensing. Volume 2, Principles and Applications of Imaging Radar*, 3rd edn. John Wiley & Sons Ltd, Chichester.

Henderson, G.M. & Slowey, N.C. (2000) Evidence from U–Th dating against Northern Hemisphere forcing of the penultimate glaciation. *Nature* **404**, 61–6.

Henderson-Sellers, A. & Hansen, A.-M. (1995) *Climate Change Atlas*. Kluwer, Dordrecht.

Hewitt, G. (2000) The genetic legacy of the Quaternary ice ages. *Nature* **405**, 907–13.

Hewitt, K. (2002) Introduction. In: Hewitt, K. et al. (eds.) *Landscapes of Transition*. Kluwer Academic, Dordrecht, pp. 1–8.

Hewitt, K., Byrne, M.-L., English, M., & Young, G. (eds.) (2002) *Landscapes of Transition: Landform Assemblages and Transformations in Cold Regions*. Kluwer Academic, Dordrecht.

Hilmer, M., Harder, M., & Lemke, P. (1998) Sea ice transport: a highly variable link between Arctic and North Atlantic. *Geophysical Research Letters* **25**, 3359–62.

Hinchliffe, S. & Ballantyne, C.K. (1999) Talus accumulation and rockfall retreat, Trotternish,

Isle of Skye, Scotland. *Scottish Geographical Journal* **115**, 53–70.

Hoham, R.W. & Ling, H.U. (2000) Snow algae: the effects of chemical and physical factors on their life cycles and populations. In: Seekbach, J. (ed.) *Journey to Diverse Microbial Worlds: Adaptation to Exotic Environments*. Kluwer, Dordrecht, pp. 131–45.

Hollin, J.T. & Schilling, D.H. (1981) Late Wisconsin–Weichselian mountain glaciers and small ice caps. In: Denton, G.H. & Hughes, T.J. (eds.) *The Last Great Ice Sheets*. Wiley, New York, pp. 179–206.

Holling, C.S. (1986) The resilience of terrestrial ecosystems: local surprise and global change. In: Clark, W.C. & Munn, R.E. (eds.) *Sustainable Development of the Biosphere*. Cambridge University Press, Cambridge, pp. 292–320.

Holling, C.S. (2001) Understanding the complexity of economic, ecological and social systems. *Ecosystems* **4**, 390–405.

Holm, K., Bovis, M., & Jakob, M. (2004) The landslide response of alpine basins to post-Little Ice Age glacial thinning and retreat. *Geomorphology* **57**, 201–16.

Hopkins, D.M. & Karlstrom, T.D. (1955) *Permafrost and Groundwater in Alaska*. United States Geological Survey Professional Paper 264-F, pp. 113–46.

Hoppe, G.E. (1959) Glacial morphology and inland ice recession in northern Sweden. *Geografiska Annaler* **41A**, 193–212.

Houghton, J.T., Ding, Y., Griggs, D.J., Noguer, M., van der Linden, P.J., & Xiaosu, D. (2001) *Climate Change 2001: The Scientific Basis*. Cambridge University Press, Cambridge.

Houghton, J.T., Meira Filho, L.G., Callander, B.A., Harris, N., Kattenberg, A., & Maskell, K. (eds.) (1996) *Climate Change 1995: The Science of Climate Change*. Cambridge University Press, Cambridge.

Howat, I.M., Joughin, I., Tulaczyk, S., & Gogineni, S. (2005) Rapid retreat and acceleration of Helheim Glacier, east Greenland. *Geophysical Research Letters* **32**, L22502, doi:10.1029/2005GL024737.

Hulbe, C.L. & Whillans, I.M. (1994) A new method for determining ice thickness changes at remote locations. *Annals of Glaciology* **20**, 263–8.

Hulton, N., Sugden, D.E., Payne, A.J., & Clapperton, C.M. (1994) Glacier modelling and the climate of Patagonia during the Last Glacial Maximum. *Quaternary Research* **4**, 1–19.

Humboldt, A. von (1849) Cosmos: A sketch of a Physical Description of the Universe, vol. 1, translated from the German by Otte, E.C. Henry G. Bohn, London.

Huntington, H. & Fox, S. (2005) The changing Arctic: indigenous perspectives. In: *ACIA*. Cambridge University Press, Cambridge, pp. 61–98.

Huybrechts, P. (1992) The Antarctic Ice Sheet and environmental change: a three dimensional modelling study. Reports on Polar Research, 99, Alfred Wegener Institut fur Polar und Meeresforschung.

Huybrechts, P. (1999) The Antarctic Ice Sheet during the last glacial–interglacial cycle: a three dimensional experiment. *Annals of Glaciology* **14**, 115–19.

Iken, A., Echelmeyer, K., Harrison, W., & Funk, M. (1993) Mechanisms of fast flow in Jakobshavn Glacier, West Greenland: Part I. Measurements of temperature and water level in deep boreholes. *Journal of Glaciology* **39**, 15–25.

Jette, V. (ed.) (1992) Sante Quebec, 1992. *A Health Profile of the Inuit*. Report of the Sante Quebec health survey among the Inuit of Nunavut, vol. 1 & 2.

Jezek, K. & RAMP Product Team (2002) RAMP AMM-1 SAR Image Mosaic of Antarctica. Fairbanks, AK, Alaska Satellite Facility, in association with the National Snow and Ice Data Center, Boulder, CO. Digital media.

Jin, M. (2004) Analysis of land skin temperature using AVHRR observations. *Bulletin of the American Meteorological Society* **85**, 587–600.

John, B.S. & Sugden, D.E. (1971) Raised marine features and phases of glaciation in the South Shetland Islands. *British Antarctic Survey Bulletin* **24**, 45–111.

Jones, H.G. (1991) Snow chemistry and biological activity: a particular perspective on nutrient cycling. In: Davies, T.D., Tranter, M., & Jones, H.G. (eds.) *Seasonal Snowpacks: Processes of Compositional Change*. Springer-Verlag, Berlin, pp. 21–66.

Jones, H.G., Pomeroy, J.W., Walker, D.A., & Hoham, R.W. (eds.) (2001) *Snow Ecology: An Interdisciplinary Examination of Snow-Covered Ecosystems*. Cambridge University Press, Cambridge.

Jones, J.A.A. (1997) *Global Hydrology: Processes, Resources and Environmental Management.* Addison Wesley, Longman Limited, Harlow.

Jones, P.D. (1994) Hemispheric surface air temperature variations: a reanalysis and an update to 1993. *Journal of Climate* **7**, 1794–802.

Jones, P.D., Osborn, T.J., Briffa, K.R., Folland, C.K., Horton, E.B., Alexander, L.V., Parker, D.E., & Rayner, N.A. (2001) Adjusting for sampling density in grid box land and ocean surface temperature time series. *Journal of Geophysical Research* **106**, 3371–80.

Jordan, P. & Slaymaker, O. (1991) Holocene sediment production in Lillooet River basin: a sediment budget approach. *Géographie physique et Quaternaire* **45**, 45–57.

Jordan, R. (1991) *A One-Dimensional Temperature Model for a Snowcover: Technical Documentation for SNTHERM.89,* Spec. Rep. 657, User's guide. U.S. Army Cold Reg. Res. and Eng. Lab., Hanover, NH.

Joughin, I. (2002) Ice-sheet velocity mapping: a combined interferometric and speckle-tracking approach. *Annals of Glaciology* **34**, 195–201.

Judge, A. (1973) The prediction of permafrost thickness. *Canadian Geotechnical Journal* **10**, 1–11.

Kamb, B. (1987) Glacier surge mechanism based on linked cavity configuration of the basal water conduit system. *Journal of Geophysical Research* **92**, 9083–100.

Kamb, B., Raymond, C.F., Harrison, W.D., Engelhardt, H., Echelmeyer, K.A., Humphrey, N., Brugman, M.M., & Pfeffer, T. (1985) Glacier surge mechanism: 1982–3 surge of Veriegated Glacier, Alaska. *Science* **227**, 469–70.

Karl, T.R., Knight, R.W., & Christy, J.R. (1994) Global and hemispheric trends: uncertainties related to inadequate spatial sampling. *Journal of Climate* **7**, 1144–63.

Kehew, A.E. & Lord, M.L. (1989) Canadian landform examples 12: glacial lake spillways of the central interior plains, Canada-USA. *The Canadian Geographer* **33**, 274–7.

Kelly, R.E.J. (2002) Determination of the ELA on Hardangerjokulen, Norway during the 1995–6 winter season using repeat pass SAR coherence. *Annals of Glaciology* **34**, 349–54.

Kelly, R.E.J., Chang, A.T.C., Tsang, L., & Foster, J.L. (2003) A prototype AMSR-E global snow area and snow depth algorithm. *IEEE Transactions on Geoscience and Remote Sensing* **41**, 230–42.

Kennedy, A.D. (1995) Antarctic terrestrial ecosystem response to global environmental change. *Annual Review of Ecological Systems* **26**, 683–704.

Kennett, J.P. (1978) Cainozoic evolution of circum-antarctic paleo-oceanography. In: Zinderen Bakker, E.M. van (ed.) *Antarctic Glacial History and World Paleoenvironments.* Balkema, Rotterdam, pp. 41–56.

Kennett, J.P., Cannariato, K.G., Hendy, I.L., & Behl, R.J. (2000) Carbon isotopic evidence for methane hydrate instability during Quaternary interstadials. *Science* **288**, 128–33.

Key, J. & Haefliger, M. (1992) Arctic ice surface temperature retrieval from AVHRR thermal channels. *Journal of Geophysical Research* **97**, 5885–93.

Khalsa, S.J.S., Dyurgerov, M.B., Khromova, T., Raup, B.H., & Barry, R.G. (2004) Space-based mapping of glacier changes using ASTER and GIS tools. *IEEE Transactions on Geoscience and Remote Sensing* **42**, 2177–82.

Khromova, T.E., Dyurgerov, M.B., & Barry, R.G. (2003) Late-twentieth century changes in glacier extent in the Ak-shirak Range, Central Asia, determined from historical data and ASTER imagery. *Geophysical Research Letters* **30**, 1863, doi: 10.1029/2003GL017233.

Kiefer, H. et al. (41 others) (2000) New eyes in the sky measure glaciers and ice sheets. *EOS Transactions of American Geophysical Union* **81**, 265 and 270–1.

Kimball, J.S., McDonald, K.C., Frolking, S., & Running, S.W. (2004) Radar remote sensing of the spring thaw transition across a boreal landscape. *Remote Sensing of Environment* **89**, 163–75.

Klein, A.G., Hall, D.K., & Riggs, G.A. (1998) Improving snow-cover mapping in forests through the use of a canopy reflectance model. *Hydrological Processes* **12**, 1723–44.

Knox, J.C. (1984) Responses of river systems to Holocene climates. In: Wright, H.E. (ed.) *Late Quaternary Environments of the United States. Volume 2: The Holocene.* Longman, London, pp. 26–41.

Koç Karpuz, N. & Jansen, E. (1992) A high-resolution diatom record of the last deglaciation from the SE Norwegian Sea: documentation of rapid climatic changes. *Paleoceanography* **7**, 499–520.

König, M., Winther, J.-G., & Isaksson, E. (2001) Measuring snow and glacier ice properties from satellite. *Reviews of Geophysics* **39**, 1–27.

Konstantinov, Yu B. & Grachev, K.I. (2000) *High-Latitude Airborne Expeditions Sever (1937, 1941–93)*. Gidrometeoizdat Publishing House, St. Petersburg, Russia.

Kopanev, I.D. (1982) Climatological aspects of snow cover evolution studies. Gidrometeo. Izdat, Leningrad (in Russian; quoted in Groisman & Davies, 2001). In: Jones, H.G. et al. (eds.) *Snow Ecology*. Cambridge University Press, Cambridge, pp. 1–44.

Koren', V.N. (1991) Mathematical models for the forecast of runoff. Gidrometeo. Izdat, Leningrad (in Russian, quoted in Groisman & Davies, 2001). In: Jones, H.G. et al. (eds.) *Snow Ecology*. Cambridge University Press, Cambridge, pp. 1–44.

Koskinen, J.T., Pulliainen, J.T., & Hallikainen, M.T. (1997) The use of ERS-1 SAR data in snow melt monitoring. *IEEE Transactions on Geoscience and Remote Sensing* **35**, 601–10.

Kovanen, D.J. & Slaymaker, O. (2004a) Glacial imprints of the Okanogan Lobe, southern margin of the Cordilleran Ice Sheet. *Journal of Quaternary Science* **19**, 547–65.

Kovanen, D.J. & Slaymaker, O. (2004b) Relict shorelines and ice-flow patterns of the northern Puget Lowland from lidar data and digital terrain modeling. *Geografiska Annaler* **86A**, 385–400.

Kovanen, D.J. & Slaymaker, O. (2005) Fluctuation of the Deming Glacier and theoretical equilibrium line altitudes during the Late Pleistocene and Early Holocene on Mt. Baker, Washington. *Boreas* **34**, 157–75.

Krabill, W.B., Thomas, R.H., Martin, C.F., Swift, R.N., & Frederick, E.B. (1995) Accuracy of airborne laser altimetry over the Greenland Ice Sheet. *International Journal of Remote Sensing* **16**, 1211–22.

Krenke, A. (1998, updated 2004) *Former Soviet Union Hydrological Snow Surveys, 1966–96*. Edited by NSIDC. National Snow and Ice Data Center/World Data Center for Glaciology, Boulder, CO. Digital media.

Kudryavtsev, V.A. (1965) Temperature, thickness, and discontinuity of permafrost. In: USSR Academy of Sciences, 1959, pp. 219–73, *Principles of Geocryology, Chapter VIII* (in Russian, 1959). National Research Council Canada Technical Translation TT-1187, Ottawa.

Kukla, G.J. (1968) Comment. *Current Anthropology* **9**, 37–9.

Kullman, L. (2001) 20th century climate warming and tree limit rise in the southern Scandes of Sweden. *Ambio* **30**, 72–80.

Kurvonen, L. & Hallikainen, M. (1997) Influence of land-cover category on brightness temperature of snow. *IEEE Transactions on Geoscience and Remote Sensing* **35**, 367–77.

Kuzmin, P.P. (1961) *Melting of Snow Cover*. Gidrometeo. Izdat, Leningrad (Translated by E.Vilim, 1972. Israel Program for Scientific Translation, Jerusalem).

Kwok, R. & Baltzer, T. (1995) The geophysical processor system at the Alaska SAR Facility. *Photogramm Engd Remote Sensing* **61**, 1445–53.

Kwok, R., Maslowski, W., & Laxon, S.W. (2005) On large outflows of Arctic sea ice into the Barents Sea. *Geophysical Research Letters* **32**, L22503, doi:10.1029/2005GL024485.

Lambeck, K. (1995) Late Devensian and Holocene shorelines of the British Isles and North Sea from models of glacio-hydro-isostatic rebound. *Journal of the Geological Society* **152**, 437–48.

Lammers, R.B., Shiklomanov, A.I., Vörösmarty, C.J., Fekete, B.M., & Peterson, B.J. (2001) Assessment of contemporary Arctic runoff based on observational discharge records. *Journal of Geophysical Research* **106**, 3321–34.

Laumann, T. & Reeh, N. (1993) Sensitivity to climate change of the mass balance of glaciers in southern Norway. *Journal of Glaciology* **39**, 656–65.

Lawn, J. & Langer, N. (1994) Air stage subsidiary monitoring program. Final Report, v.2: Food Consumption Survey. Department of Indian Affairs and Northern Development, Ottawa.

Lawson, D.E. (1981) *Sedimentological Characteristics and Classification of Depositional Processes and Deposits in the Glacial Environment*. Cold Regions

Research and Engineering Laboratory, Report 81–27, Hanover, New Hampshire.

Laxon, S., Peacock, N., & Smith, D. (2003) High interannual variability of sea-ice thickness in the Arctic region. *Nature* **425**, 947–50.

Laxon, S.W., Walsh, J.E., Wadhams, P., Johannessen, O.M., & Miles, M. (2004) Sea ice observations. In: Bamber, J.L. & Payne, A.J. (eds.) *Mass Balance of the Cryosphere: Observations and Modelling of Contemporary and Future Changes.* Cambridge University Press, Cambridge, pp. 337–66.

Leonard, E.M. (1985) Glaciological and climatic controls on lake sedimentation, Canadian Rocky Mountains. *Zeitschrift fur Gletscherkunde und Glazialgeologie* **21**, 35–42.

Letreguilly, A., Reeh, N., & Huybrechts, P. (1991) The Greenland Ice Sheet through the last glacial–interglacial cycle. *Global and Planetary Change* **90**, 385–94.

Lian, O.B. & Hickin, E.J. (1996) Early postglacial sedimentation of lower Seymour Valley, southwestern British Columbia. *Géographie Physique et Quaternaire* **50**, 95–102.

Liestøl, O. (1956) Glacier dammed lakes in Norway. *Norsk Geografisk Tidsskrift* **15**, 122–49.

Lillesand, T.M., Kiefer, R.W., & Chipman, J.W. (2003) *Remote Sensing and Image Interpretation,* 5th edn. Wiley, New York.

Lliboutry, L., Arnao, B.M., Pautre, A., & Schneider, B. (1977) Glaciological problems set by the control of dangerous lakes in the Cordillera Blanca, Peru. I. Historical failures of morainic dams, their causes and prevention. *Journal of Glaciology* **18**, 239–54.

Lock, G.S.H. (1990) *The Growth and Decay of Ice.* Cambridge University Press, Cambridge.

Longva, O. & Thoresen, M.K. (1991) Iceberg scours, iceberg gravity craters and current erosion marks from a gigantic Preboreal flood in southeastern Norway. *Boreas* **20**, 47–62.

Lonne, I. (1993) Physical signatures of ice advance in a Younger Dryas ice-contact delta, Troms, northern Norway: implications for glacier terminus history. *Boreas* **22**, 59–70.

Lowe, J.J. & Walker, M.J.C. (1984) *Reconstructing Quaternary Environments.* Longman, London.

Luckman, A., Murray, T., & Strozzi, T. (2002) Surface flow evolution throughout a glacier surge measured by satellite radar interferometry. *Geophysical Research Letters* **29**, doi:10.1029/2001GL014570.

Luckman, B.H. & Fiske, C.J. (1995) Estimating long-term rockfall accretion rates by lichenometry. In: Slaymaker, O. (ed.) *Steepland Geomorphology.* Wiley, New York, pp. 233–55.

Lunardini, V.J. (1996) Climate warming and the degradation of warm permafrost. *Permafrost and Periglacial Processes* **7**, 311–20.

Mabbutt, J.A. (1968) Review of concepts of land classification. In: Stewart, G.A. (ed.) *Land Evaluation.* Macmillan, Melbourne, pp. 11–28.

MacAyeal, D.R. (1993) Binge/purge oscillations of the Laurentide Ice Sheet as a cause of North Atlantic's Heinrich Events. *Paleooceanography* **8**, 775–84.

Mackay, J.R. (1959) Glacier ice thrust features of the Yukon coast. *Geographical Bulletin* **13**, 5–21.

Mackay, J.R. (1970) Disturbances to the tundra and forest tundra environment of the western Arctic. *Canadian Geotechnical Journal* **7**, 111–24.

Mackay, J.R. (1972) The world of underground ice. *Annals of the Association of American Geographers* **62**, 1–22.

Mackay, J.R. (1975) Relict ice wedges, Pelly Island, N.W.T. (107 C/12) In: Report of Activities, part A. Geological Survey of Canada, Paper 75–1A, pp. 469–70.

Mackay, J.R. (1983) Downward water movement into frozen ground, western arctic coast, Canada. *Canadian Journal of Earth Sciences* **20**, 120–34.

Mackay, J.R. (1998) Pingo growth and collapse, Tuktoyaktuk Peninsula area, western Arctic coast, Canada: a long-term field study. *Géographie physique et Quaternaire* **52**, 271–323.

Mackay, J.R. & Black, R.F. (1973) Origin, composition and structure of perennially frozen ground and ground ice. *Second International Conference on Permafrost,* Yakutsk, 13–28 July, 1973. North American Contribution, National Academy of Science, Washington, DC, pp. 223–8.

Mackay, J.R. & Mathews, W.H. (1974) Needle ice striped ground. *Arctic and Alpine Research* **6**, 79–84.

Mackay, J.R. & Slaymaker, O. (1989) The Horton River breakthrough and resulting geomorphic changes in a permafrost environment, western Arctic coast. *Geografiska Annaler* **71A**, 171–84.

Magnusson, J.J., Robertson, D.M., Benson, B.J., Wynne, R.H., Livingstone, D.M., Arai, T., Assel, R.A., Barry, R.G., Card, V., Kuusisto, E., Granin, N., Prowse, T., Stewart, K.M., & Vuglinski, V.S. (2000) Historical trends in lake and river ice cover in the Northern Hemisphere. *Science* **289**, 1743–6.

Mahaffy, M.W. (1976) A three-dimensional numerical model of ice sheets: test on the Barnes Ice Cap, Northwest Territories. *Journal of Geophysical Research* **81**, 1059–66.

Maizels, J.K. (1977) Experiments on the origin of kettle holes. *Journal of Glaciology* **18**, 291–303.

Maizels, J.K. (1993) Lithofacies variations within sandur deposits: the role of runoff regime, flow dynamics and sediment supply characteristics. *Sedimentary Geology* **85**, 299–325.

Mann, M.E. & Jones, P.D. (2003) Global surface temperatures over the past two millennia. *Geophysical Research Letters* **30**, 1820, doi:10.1029/2003GL017814.

Mannerfelt, C.M. (1945) Nagra glasialmorfologiska formelement. *Geografiska Annaler* **27A**, 1–239.

de March, B.G.E., de Wit, C.A., Muir, D.C.G., Braune, B.M., Gregor, D.J., Norstrom, R.J., Olsson, M., Skaare, J.U., & Strange, K. (1998) Persistent organic pollutants. In: *AMAP Assessment Report, Arctic Pollution Issues, Chapter 6.* AMAP, Oslo, pp. 183–372.

Mark, B.G., Harrison, S.P., Spessa, A., New, M., Evans, D.J.A., & Helmens, K.F. (2005) Tropical snowline changes at the Last Glacial Maximum: a global assessment. *Quaternary International* **138–139**, 168–201.

Markus, T. & Cavalieri, D.J. (1998) Snow depth distribution over sea ice in the Southern Ocean from satellite passive microwave data. In: Jeffries, M.O. (ed.) *Antarctic Sea Ice Physical: Processes, Interactions and Variability.* Antarctic Research Series 74, American Geophysical Union, Washington, DC, pp. 19–39.

Markus, T. & Cavalieri, D.J. (2000) An enhancement of the NASA Team sea ice algorithm. *IEEE Transactions on Geoscience and Remote Sensing* **38**, 1387–98.

Marsh, P. & Woo, M.-K. (1981) Snowmelt, glacier melt and High Arctic streamflow regimes.

Canadian Journal of Earth Sciences **18**, 1380–4.

Marshall, S.J. & Clarke, G.K.C. (1999) Modeling North American freshwater runoff through the last glacial cycle. *Quaternary Research* **52**, 300–15.

Matear, R.J. & Hirst, A.C. (1999) Climate change feed-back on the future oceanic CO_2 uptake. *Tellus* **51B**, 722–33.

Mathews, W.H. (1947) Tuyas: flat-topped volcanoes in northern British Columbia. *American Journal of Science* **245**, 560–70.

Mathews, W.H. (1979) Late Quaternary environmental history affecting human habitation of the Pacific Northwest. *Canadian Journal of Archeology* **3**, 145–56.

Matthews, J.A. (1992) *The Ecology of Recently Deglaciated Terrain: A Geoecological Approach to Glacier Forelands and Primary Succession.* Cambridge University Press, Cambridge.

Matthews, J.A., Shakesby, R.A., Berrisford, M.S., & McEwen, L.J. (1983) Periglacial patterned ground in the Styggedalsbreen glacier foreland, Jotunheimen, southern Norway: microtopographical, paraglacial and geochronological controls. *Permafrost and Periglacial Processes* **9**, 147–66.

Mayewski, P.A. & Goodwin, I.D. (1997) *International Trans Antarctic Scientific Expedition (ITASE) "200 Years of Past Antarctic Climate and Environmental Change," Science and Implementation Plan, Proceedings of a workshop held at University of New Hampshire, 19–23 April 1997.* PAGES Workshop Report Series 97–1, 48 pp.

Mazaud, A., Vimeux, F., & Jouzel, J. (2000) Short fluctuations in Antarctic isotope records: a link with cold events in the North Atlantic? *Earth and Planetary Science Letters* **177**, 219–25.

McAndrews, J.H., Liu, K.-B., Manville, G.C., Prest, V.K., &Vincent, J.-S. (1987) Environmental change after 9,000 BC. In: Harris, R.C. (ed.) *Historical Atlas of Canada: v.1. From the Beginning to 1800.* University of Toronto Press, Toronto, plate 4.

McCarthy, J.J., Canziani, O.F., Leary, N.A., Dokken, D.J., & White, K.S. (eds.) (2001) Climate *Change 2001: Impacts, Adaptations and Vulnerability.* Cambridge University Press, Cambridge.

McClain, E.P., Pichel, W.G., & Walton, C.C. (1985) Comparative performance of AVHRR based multichannel sea surface temperatures. *Journal of Geophysical Research* **90**, 11587–601.

McClung, D. & Schaerer, P. (1993) *The Avalanche Handbook*. The Mountaineers, Seattle.

McConnell, J.R., Arthern, R.J., Mosley-Thompson, E., Davis, C.H., Bales, R.C., Thomas, R., & Kyne, J.D. (2000) Changes in Greenland Ice Sheet elevation attributed primarily to snow accumulation variability. *Nature* **406**, 877–9.

McCulloch, D. & Hopkins, D. (1966) Evidence for an early recent warm interval in north western Alaska. *Geological Society of America Bulletin* **77**, 1089–108.

McGee, R. (1974) The peopling of Arctic North America. In: Ives, J. & Barry, R.G. (eds.) *Arctic and Alpine Environments*. Methuen, London, pp. 831–5.

McGuffie, K. & Henderson-Sellers, A. (1997) *A Climate Modelling Primer*. Wiley, Chichester.

McKane, R.B., Rasetter, E.B., Shaver, G.R., Nadelhoffer, K.J., Giblin, A.E., Laundre, J.A., & Chapin III, J.S. (1997a) Climate effects of tundra carbon inferred from experimental data and a model. *Ecology* **78**, 1170–87.

McKane, R.B., Rasetter, E.B., Shaver, G.R., Nadelhoffer, K.J., Giblin, A.E., Laundre, J.A., & Chapin III, J.S. (1997b) Reconstruction and analysis of historical changes in carbon storage in Arctic tundra. *Ecology* **78**, 1188–98.

Meier, M.F. (1998) Monitoring ice sheets, ice caps and glaciers. In: Haeberli, W., Hoelze, M., & Suter, S. (eds.) *Into the second century of Worldwide Glacier Monitoring – prospects Strategies*. UNESCO, Studies and reports in hydrology no. 56, pp. 209–14.

Menzies, J. (ed.) (1995) *Modern Glacial Environments*. Butterworth-Heinemann, Oxford.

Mercer, J.H. (1968) Antarctic ice and Sangamon sea level. *International Association of Scientific Hydrology Symposium* **79**, 217–25.

Miller, M.F. & Mabin, M.C.G. (1998) Antarctic Neogene landscapes: in the refrigerator or in the deep freeze. *Geological Society of America Today* **8**, 1–8.

Mognard, N.M. & Josberger, E.G. (2002) Northern Great Plains 1996/97 seasonal evolution of snowpack parameters from satellite passive-microwave measurements. *Annals of Glaciology* **34**, 15–23.

Mollard, J.D. (1986) *Landforms and Surface Materials of Canada*, 8th edn. J.D. Mollard and Associates Limited, Regina.

Montgomery, D.R. (2001) Slope distributions, threshold hillslopes and steady state topography. *American Journal of Science* **301**, 432–54.

Morner, N.A. (ed.) (1980) The Fenno-scandian uplift: geological data and their geodynamical implications. In: *Earth Rheology, Isostasy and Eustasy*. Wiley, Chichester, pp. 251–84.

Muller, F. (1962) Zonation in the accumulation area of the glaciers of Axel Heiberg Island, N.W.T. *Journal of Glaciology* **4**, 302–13.

Mullins, H.T. & Hinchey, E.J. (1989) Origin of New York Finger Lakes: a historical perspective on the ice erosion debate. *Northeastern Geology* **11**, 166–81.

Murton, J.B. & French, H.M. (1994) Cryostructures in permafrost, Tuktoyaktuk coastlands, western Arctic Canada. *Canadian Journal of Earth Sciences* **31**, 737–47.

Mysak, L.A., Manak, D., & Marden, R. (1990) Sea ice anomalies observed in Greenland and Labrador seas during 1901–84 and their relation to an interdecadal arctic climate cycle. *Climate Dynamics* **5**, 111–33.

National Academy of Sciences (1991) *Opportunities in the Hydrologic Sciences*. National Research Council, Washington, DC.

National Academy of Sciences (2004) *Climate Data Records from Environmental Satellites: Interim Report*. Committee on Climate Data Records from NOAA Operational Satellites, National Research Council, Washington, DC, USA, 150 pp.

National Research Council (1954) *Classification for Snow*. Technical Memorandum 11, NRC, Ottawa.

National Weather Service (1994) Surface observations. *Observing Handbook* 7. U.S. Department of Commerce, National Oceanographic and Atmospheric Administration, Washington, DC.

Nesje, A. & Whillans, I.M. (1994) Erosion of Sognefjord, Norway. *Geomorphology* **9**, 33–45.

Nesje, A., Dahl, S.O., Valen, V., & Ovstedal, J. (1992) Quaternary erosion in the Sognefjord drainage basin, western Norway. *Geomorphology* **5**, 511–20.

Nghiem, S.V. & Tsai, W.-Y. (2001) Global snow cover monitoring with spaceborne Ku-band scatterometer. *IEEE Transactions on Geoscience and Remote Sensing* **39**, 2118–34.

Nguyen, A.T. & Herring, T.A. (2005) Analysis of ICESat data using Kalman filter and kriging to study height changes in East Antarctica. *Geophysical Research Letters* **32**, L23S03, doi:10.1029/2005GL024272.

Nilsson, C., Reidy, C.A., Dynesius, M., & Revenga, C. (2005) Fragmentation and flow regulation of the world's large river systems. *Science* **308**, 405–8.

Nixon, F.M. & Taylor, A.E. (1998) Regional active layer monitoring across the sporadic, discontinuous and continuous permafrost zones, Mackenzie Valley, Northwestern Canada. In: Lewkowicz, A.G. & Allard, M. (eds.) *Proceedings, Seventh International Conference on Permafrost, 23–27 June 1998*. Universite Laval, Centre d'Études Nordiques, Collection Nordicana No. 57, Yellowknife, Canada, pp. 815–20.

Norstrom, R.J. & Muir, D.C.C. (1994) Chlorinated hydrocarbon contaminants in arctic marine mammals. *The Science of the Total Environment* **154**, 107–28.

Norwegian Water and Energy Administration (NVE) (2002) Avrenningskart for Norge; armiddelverdier for avrenning 1961–90 (Runoff Map for Norway for the period 1961–90), Oslo, Norway.

O Cofaigh, C. (1996) Tunnel valley genesis. *Progress in Physical Geography* **20**, 1–19.

Odum, H.T. (1983) *Systems Ecology*. Wiley, Chichester.

Oerlemans, J. (ed.) (1989) On the response of valley glaciers to climatic change. In: *Glacier Fluctuations and Climate Change*. Kluwer, Dordrecht, pp. 353–71.

Oerlemans, J. (2001) *Glaciers and Climate Change*. Balkema, Rotterdam.

Oerlemans, J. (2005) Extracting a climate signal from 169 glacier records. *Science* **308**, 675–7.

Oeschger, H. et al. (1984) Climate processes and climate sensitivity. In: Hansen, J.E. & Takahashi, T. (eds.) Geophysical Monograph 29, American Geophysical Union, Washington, DC, pp. 299–306.

Ohmura, A. & Reeh, N. (1991) New precipitation and accumulation maps for Greenland. *Journal of Glaciology* **37**, 140–8.

Oke, T.R. (1987) *Boundary Layer Climates*, 2nd edn. Routledge, London.

Oppenheimer, M. (1998) Global warming and the stability of the West Antarctic Ice Sheet. *Nature* **393**, 325–32.

Osterkamp, T.E. (2005) The recent warming of permafrost in Alaska. *Global and Planetary Change* **49**, 187–202.

Osterkamp, T.E., Viereck, L., Shur, Y., Jorgenson, M.T., Racine, C., Doyle, A., & Boone, R.D. (2000) Observations of thermokarst and its impact on boreal forests in Alaska. *Arctic, Antarctic and Alpine Research* **32**, 303–15.

Østrem, G. (1959) Ice melting under a thin layer of moraine and the existence of ice in moraine ridges. *Geografiska Annaler* **41**, 228–30.

Østrem, G. (1966) The height of the glaciation limit in southern British Columbia and Alberta. *Geografiska Annaler* **48A**, 126–38.

Østrem, G. (1974) Present alpine ice cover. In: Ives, J.D. & Barry, R.G. (eds.) *Arctic and Alpine Environments*. Methuen, London, pp. 226–50.

Østrem, G. & Brugman, M. (1991) *Glacier and Mass Balance Measurements – A Manual for Field and Office Work*. Canadian National Hydrology Research Institute (NHRI) Science Report No. 4, Saskatoon.

Owen, L.A. & Derbyshire, E. (1989) The Karakoram glacial depositional system. *Zeitschrift fur Geomorphologie, Supplementband.* **76**, 33–73.

Owens, P.N. & Slaymaker, O. (2004) *Mountain Geomorphology*. Edward Arnold, London.

Painter, T.H., Dozier, J., Roberts, D.A., Davis, R.E., & Green, R.O. (2003) Retrieval of subpixel snow-covered area and grain size from imaging spectrometer data. *Remote Sensing of Environment* **85**, 64–77.

Papa, F., Legrésy, B., Mognard, N., Josberger, E., & Rémy, F. (2002) Snow depth estimations with the Topex-Poseidon altimeter and radiometer. *IEEE Transactions on Geoscience and Remote Sensing* **40**, 2162–70.

Parkinson, C., Cavalieri, D., Gloersen, P., Zwally, H., & Comiso, J. (1999) Arctic sea ice extents, areas and trends, 1978–96. *Journal of Geophysical Research (Oceans)* **C9**, 20837–56.

Partington, K.C. (1998) Discrimination of glacier facies using multitemporal SAR data. *Journal of Glaciology* **44**, 42–53.

Paterson, W.S.B. (1994) *The Physics of Glaciers*, 3rd edn. Pergamon Press, Oxford.

Penner, E. (1973) Frost heaving pressures in particulate materials. *Symposium on Frost Action on Roads 1*. OECD, Paris, pp. 379–85.

Peterson, J.A. & Peterson, L.F. (1994) Ice retreat from the neoglacial maxima in the Jayakesuma area, Republic of Indonesia. *Zeitschrift fur Gletscherkunde und Glazialgeologie* **30**, 1–9.

Petit, J.R. et al. (1999) Climate and atmospheric history of the past 420,000 years from the Vostok ice core, Antarctica. *Nature* **399**, 429–36.

Petit, J.R., Mounier, L., Jouzel, J., Korotkevitch, Y.S., Kotlyakov, V.I., & Lorius, C. (1990) Paleoclimatological and chronological implications of the Vostok core dust record. *Nature* **343**, 56–8.

Petit-Maire, N. (1999) Variabilité naturelle des environnements terrestres: les deux derniers extrèmes climatiques (18,000 ± 2,000 et 8,000 ± 1,000 ans BP). *Comptes Rendus de l'Académie des Sciences, Paris 2 (a)* **328**, 273–9.

Pewe, T.L. (1975) *Quaternary Geology of Alaska*. US Geological Survey Professional Paper 835.

Pewe, T.L. (1983) The periglacial environment in North America during Wisconsin time. In: Porter, S.C. (ed.) *Late Quaternary Environments of the United States, v.1: The Late Pleistocene*. Longman, London, pp. 157–89.

Pierson, T.C., Janda, R.J., Thouret, J.-C., & Borrero, C.A. (1990) Perturbation and melting of snow and ice by the 13 November 1985 eruption of Nevado del Ruiz, Columbia, and consequent mobilization, flow and deposition of lahars. *Journal of Volcanology and Geothermal Research* **41**, 17–66.

Pissart, A. (1987) Weichselian periglacial structures and their environmental significance: Belgium, the Netherlands and northern France. In: Boardman, J. (ed.) *Periglacial Processes and Landforms in Britain and Ireland*. Cambridge University Press, Cambridge, pp. 77–88.

Pomeroy, J.W. & Brun, E. (2001) Physical properties of snow. In: Jones, H.G. et al. (eds.) *Snow Ecology*. Cambridge University Press, Cambridge, pp. 45–126.

Pomeroy, J.W. & Schmidt, R.A. (1993) The use of fractal geometry in modeling intercepted snow accumulation and sublimation. *Proceedings Eastern Snow Conference* **50**, 1–10.

Porter, S.C. (1975) Equilibrium line altitudes of late Quaternary glaciers in the Southern Alps, New Zealand. *Quaternary Research* **5**, 27–47.

Porter, S.C. (1989) Some geological implications of average Quaternary glacial conditions. *Quaternary Research* **32**, 245–61.

Poser, H. (1994) Soil and climate relations in central and western Europe during the Wurm Glaciation, translated from Erdkunde, 1948, 2: 53–68. In: Evans, D.J.A. (ed.) *Cold Climate Landforms*. Wiley, Chichester, pp. 3–22.

Post, A. & Mayo, L.R. (1971) *Glacier Dammed Lakes and Outburst Floods in Alaska*. U.S. Geological Survey Hydraulic Investigations Atlas HA-455. Washington, DC, Government Printing Office.

Press, W.H., Flannery, B.P., Teukolsky, S.A., & Vetterling, W.T. (1989) *Numerical Recipes*. Cambridge University Press, Cambridge.

Prest, V.K. (1983) *Canada's Heritage of Glacial Features*. Geological Survey of Canada. Miscellaneous Report 28.

Price, A.G., Hendrie, L.K., & Dunne, T. (1978) Controls on the production of snowmelt runoff. In: Colbeck, S.C. & Ray, M. (eds.) *Modeling of Snow Cover Runoff*. U.S. Army Cold Regions Research and Engineering Laboratory, Hanover, pp. 257–68.

Prowse, T.D. (2000) *River-Ice Ecology*. National Water Research Institute, Environment Canada, Saskatoon.

Prowse, T.D. & Marsh, P. (1989) Thermal budget of river ice covers during break-up. *Canadian Journal of Civil Engineering* **16**, 62–71.

Prowse, T.D., Demuth, M.N., & Chew, H.A.M. (1990) The deterioration of freshwater ice due to radiation decay. *Journal of Hydraulic Research* **28**, 685–97.

Rahnstorf, S. (1997) Risk of sea change in the Atlantic. *Nature* **388**, 825–6.

Ramage, J.M. & Isacks, B.L. (2003) Interannual variations of snowmelt and refreeze timing on southeast-Alaskan icefields, U.S.A. *Journal of Glaciology* **49**, 102–16.

Rampton, V.N. (1982) *Quaternary Geology of the Yukon Coastal Plain*. Geological Survey of Canada Bulletin.

Rampton, V.N., Gauthier, R.C., Thibault, J., & Seaman, A.A. (1984) *Quaternary Geology of New Brunswick*. Geological Survey of Canada Memoir 416.

Ramsay, B. (1998) The interactive multisensor snow and ice mapping system. *Hydrological Processes* **12**, 1537–46.

Rango, A. & Martinec, J. (1979) Application of a snowmelt-runoff model using Landsat data, *Nordic Hydrology* **10**, 225–38.

Rango, A. & Martinec, J. (1982) Snow accumulation derived from modified depletion curves of snow coverage. *Symposium on Hydrological Aspects of Alpine and High Mountain Areas*, Exeter, IAHS Publication No. 138, pp. 83–90.

Raper, S.C.B., Wigley, T.M.L. & Warrick, R.A. (1996) Global sea level rise: past and future. In: Milliman, J.D. (ed.) *Rising Sea Level and Subsiding Coastal Areas*. Kluwer, Dordrecht.

Rasmussen, R.M., Vivekanandan, J., Cole, J., Myers, B., & Masters, C. (1999) The estimation of snowfall rate using visibility. *Journal of Applied Meteorology* **38**, 1542–63.

Rawlins, M.A., McDonald, K.C., Frolking, S., Lammers, R.B., Fahnstock, M., Kimball, J.S., & Vörösmarty, C.J. (2005) Remote sensing of snow thaw at the pan-Arctic scale using the SeaWinds scatterometer. *Journal of Hydrology* **312**, 294–311.

Ray, A.J. (1996) *I Have Lived Here Since the World Began*. Lester Publishing, Toronto.

Reid, L.M. & Dunne, T. (1996) *Rapid Evaluation of Sediment Budgets*. Catena Verlag, Reiskirchen.

Richardson, C. & Holmlund, P. (1996) Glacial cirque formation in northern Scandinavia. *Annals of Glaciology* **22**, 102–6.

Riggs, G. & Hall, D.K. (2004) Snow mapping with the MODIS Aqua instrument. *Proceedings of the 61st Eastern Snow Conference*, 9–11 June 2004, Portland, ME.

Rignot, E. & Kanagaratnam, P. (2006) Changes in the velocity structure of the Greenland Ice Sheet. *Science* **311**, 986–90.

Rind, D. & Peteet, D. (1985) Terrestrial conditions at the Last Glacial Maximum and CLIMAP sea surface temperature estimates: are they consistent? *Quaternary Research* **24**, 1–22.

Ritchie, J.C. (1987) *Postglacial Vegetation of Canada*. Cambridge University Press, Cambridge.

Robinson, D.A. (1997) Hemispheric snow cover and surface albedo for model validation. *Annals of Glaciology* **25**, 241–5.

Robinson, D.A. (1999) Northern Hemisphere snow cover during the satellite era. *Proceedings of the 5th Conference on Polar Meteorology and Oceanography*, Dallas, TX. American Meteorological Society, Boston, MA, pp. 255–60.

Robinson, D.A. & Frei, A. (2000) Seasonal variability of Northern Hemisphere snow extent using visible satellite data. *Professional Geographer* **51**, 307–14.

Robinson, D.A., Dewey, K.F., & Heim, R.R. (1993) Global snow-cover monitoring: an update. *Bulletin of the American Meteorological Society* **74**, 1689–96.

Roots, E.F. (1990) Environmental issues related to climate change in northern high latitudes. *Proceedings of a Symposium on the Arctic and Global Change*. Climate Institute, Washington, D.C.

Rose, J. (1987) Drumlins as part of a glacier bedform continuum. In: Menzies, J. & Rose, J. (eds.) *Drumlin Symposium*. Balkema, Rotterdam, pp. 103–16.

Rothrock, D.A., Yu, Y., & Maykut, G.A. (1999) Thinning of the arctic sea-ice cover. *Geophysical Research Letters* **26**, 3469–72.

Rothrock, D.A., Zhang, J., & Yu, Y. (2003) The arctic ice thickness anomaly of the 1990s: a consistent view from observations and models. *Journal of Geophysical Research* **108**, 3038 (28–37), doi:10.1029/2001JC001208.

Ruddiman, W.F. (2004) The role of greenhouse gases in orbital-scale climatic changes. *EOS* **85**, 1 and 6–7.

Ruddiman, W.F. (2005) How did humans first alter global climate? *Scientific American* March, 46–53.

Ruddiman, W.F. & Raymo, M.E. (1988) Northern Hemisphere climatic regimes during the last 3Ma: possible tectonic connections. *Philosophical Transactions of the Royal Society of London* **318B**, 411–30.

Rudoy, A.N., Galachov, V.P., & Damlin, A.L. (1989) Reconstruction of glacial discharge in the head of the Chuja River and alimentation of ice dammed lakes in the late Pleistocene. *Izvestiya Vsesoyuznogo Geografiicheskogo Obstichestva* **121**, 236–44 (quoted in Benn and Evans, 1998).

Russell, A.J. (1993) Supraglacial lake drainage near Söndre Stromfjörd, Greenland. *Journal of Glaciology* **39**, 431–3.

Ryder, J.M. (1971a) The stratigraphy and morphology of paraglacial alluvial fans in south-central British Columbia. *Canadian Journal of Earth Sciences* **8**, 279–98.

Ryder, J.M. (1971b) Some aspects of the morphometry of paraglacial alluvial fans in south-central British Columbia. *Canadian Journal of Earth Sciences* **8**, 1252–64.

Salomonson, V.V. & Appel, I.L. (2004) Estimating the fractional snow covering using the normalized difference snow index. *Remote Sensing of Environment* **89**, 351–60.

Sauber, J., Molnia, B., Carabajal, C., Luthcke, S., & Muskett, R. (2005) Ice elevations and surface change on the Malaspina Glacier, Alaska. *Geophysical Research Letters* **32(23)** L23S01, Paper No. 10.1029/2005GL023943.

Scambos, T., Sergienko, O., Sargent, A., MacAyeal, D., & Fastook, J. (2005) ICESat profiles of tabular iceberg margins and iceberg breakup at low latitudes. *Geophysical Research Letters* **32**, L23S09, Paper No. 10.1029/2005GL023802.

Schaerer, P. (1970) Variation of ground snow loads in British Columbia. *Proceedings of the 38th Western Snow Conference*, pp. 44–48.

Schellhubner, H.J. et al. (eds.) (2004) *Earth System Analysis for Sustainability*. MIT Press, Cambridge & London.

Schowengerdt, R.A. (1997) *Remote Sensing: Models and Methods for Image Processing*, 2nd edn. Academic Press, San Diego.

Schumm, S.A. (1973) Geomorphic thresholds and complex response of drainage systems. In: Morisawa, M. (ed.) *Fluvial Geomorphology*. Publications in Geomorphology 3, SUNY Press, Binghamton, pp. 299–310.

Schutz, B.E., Zwally, H.J., Shuman, C.A., Hancock, D., & DiMarzio, J.P. (2005) Overview of the ICESat Mission. *Geophysical Research Letters* **32**, L21S01, doi:10.1029/2005GL024009.

Seppala, M. (2004) Palsa. In: Goudie, A.S. (ed.) *Encyclopedia of Geomorphology*. Routledge, London and New York, vol. 2, pp. 756–8.

Serreze, M.C. & Barry, R.G. (2005) *The Arctic Climate System*. Cambridge University Press, Cambridge.

Serreze, M.C., Clark, M.P., & Frei, A. (2000) Characteristics of large snowfall events in the montane western United States as examined using snowpack telemetry (SNOTEL) data. *Water Resources Research* **37**, 675–88.

Shackleton, N.J. & Opdyke, N.D. (1973) Oxygen isotope and paleomagnetic stratigraphy of equatorial Pacific core V-28-238: oxygen isotope temperatures and ice volumes on a 100,000 and 1 million year scale. *Quaternary Research* **3**, 39–55.

Sharp, M. (1984) Annual moraine ridges at Skálafellsjökull, south-east Iceland. *Journal of Glaciology* **30**, 82–93.

Shaw, E.M. (1993) *Hydrology in Practice*, 3rd edn. Chapman and Hall, London.

Shaw, J. (1989) Drumlins, subglacial meltwater floods, and ocean responses. *Geology* **17**, 853–6.

Shepherd, A., Winham, D.J., Mansley, J.A.D., & Corr, H.F.J. (2001) Inland thinning of Pine Island Glacier, West Antarctica. *Science* **291**, 862–4.

Shimkus, K.M. & Trimonis, E.S. (1974) Modern sedimentation in the Black Sea. In: Degens, E.T. & Ross, D.A. (eds.) *The Black Sea: Geology, Chemistry and Biology*. American Association of Petroleum Geologists, Memoir vol. 20, pp. 249–78.

Shipp, S., Anderson, J., & Domack, E. (1999) Late Pleistocene-Holocene retreat of the West Antarctic Ice Sheet system in the Ross Sea: Part 1 – Geophysical results. *Geological Society of America Bulletin* **111**, 1486–516.

Siegert, M.J. (2001) *Ice Sheets and Late Quaternary Environmental Change*. Wiley, Chichester.

Shook, K. & Gray, D.M. (1997) Snowmelt resulting from advection. *Hydrological Processes* **11**, 1725–36.

Shreve, R.L. (1972) Movement of water in glaciers. *Journal of Glaciology* **11**, 205–14.

Siegert, M.J., Dowdeswell, J.A., & Melles, M. (1999) Late Weichselian glaciation of the Eurasian High Arctic. *Quaternary Research* **52**, 273–85.

Sikes, E.L. & Keigwin, L.D. (1994) Equatorial Atlantic sea surface temperature for the last 30 ka: a comparison of uranium 37, oxygen 18 and foraminiferal assemblage temperature estimates. *Paleoceanography* **9**, 31–45.

Skinner, B.J. & Porter, J.C. (1995) *The Blue Planet*. Wiley, New York.

Skofronick-Jackson, G., Kim, M.-J., Weinman, J.A., & Chang, D.-E. (2004) Physical model to determine snowfall over land by microwave radiometry. *IEEE Transactions on Geoscience and Remote Sensing* **42**, 1047–58.

Slaymaker, O. (1974) Alpine hydrology. In: Ives, J.D. & Barry, R.G. (eds.) *Arctic and Alpine Environments*. Methuen, London, pp. 134–55.

Slaymaker, O. (1987) Sediment and solute yields in British Columbia and Yukon: their geomorphic significance re-examined. In: Gardiner, V. (ed.) *International Geomorphology, Part I*. Wiley, Chichester, pp. 925–45.

Slaymaker, O. (1988) The distinctive attributes of debris torrents. *Journal of Hydrological Science* **33**, 567–73.

Slaymaker, O. (1990) Climate change and erosion processes in mountain regions of western Canada. *Mountain Research and Development* **10**, 171–82.

Slaymaker, O. (ed.) (2000) *Geomorphology, Human Activity and Global Environmental Change*. Wiley, Chichester.

Slaymaker, O. (2001) Why so much concern about climate change and so little attention to land use change? *The Canadian Geographer* **45**, 71–8.

Slaymaker, O. (2004) Mass balances of sediments, solutes and nutrients: the need for greater integration. *Journal of Coastal Research, Special Issue* **43**, 109–23.

Slaymaker, O. (2005) The cryosphere and global environmental change: some geomorphic perspectives. *Transactions of the Japanese Geomorphological Union* **26**, 359–70.

Slaymaker, O. & Spencer, T. (1998) *Physical Geography and Global Environmental Change*. Addison Wesley, Longman, Harlow.

Smith, B.E., Bentley, C.R., & Raymond, C.F. (2005) Recent elevation changes on the ice streams and ridges of the Ross Embayment from ICESat crossovers. *Geophysical Research Letters* **32**, L21S09, doi:10.1029/2005GL024365.

Smith, L.C., MacDonald, G.M., Velichko, A.A., Beilman, D.W., Borisova, O.K., Frey, K.E., Kremenetski, K.V., & Sheng, Y. (2004) Siberian Peatlands: a net carbon sink and global methane source since the early Holocene. *Science* **303**, 353–6.

Smith, S.L. & Burgess, M.M. (1999) Mapping the sensitivity of Canadian permafrost to climate warming. *International Association of Hydrological Sciences, Publication* **256**, 71–80.

Soldatova, I.I. (1992) Causes of variability of ice appearance dates in the lower reaches of the Volga. *Soviet Meteorology and Hydrology* **2**, 62–6.

Soldatova, I.I. (1993) Secular variations in river break-up dates and their relation with climate variation. *Meteorologia I Gidrologia* **9**, 89–96 (in Russian).

Sorenson, C.J. & Knox, J.C. (1974) Paleosols and paleoclimate related to late-Holocene forest-tundra border migrations: Mackenzie and Keewatin, NWT. In: Raymond, S. & Schledermann, P. (eds.) *International Conference on Prehistory and Paleoecology of Western North American Arctic and Subarctic*. Calgary, Archeological Association, University of Calgary, pp. 187–203.

Spikes, V.B., Hamilton, G.S., Arcone, S.A., Kaspari, S., & Mayewski, P.A. (2004) Variability in accumulation rates from GPR profiling on the west Antarctic plateau. *Annals of Glaciology* **39**, 238–44.

Spikes, V.B., Hamilton, G.S., Arcone, S.A., Kaspari, S., & Mayewski, P.A. (2005) *US International Trans-Antarctic Scientific Expedition (US ITASE): GPR Profiles and Accumulation Mapping*. USA, National Snow and Ice Data Center, Boulder, CO, Digital media.

Stalker, A.M. (1961) *Buried Valleys in Central and Southern Alberta*. Geological Survey of Canada Paper 60–32.

Starkel, L.A. (1987) Long-term and short-term rhythmicity in terrestrial landforms and deposits. In: Rampiro, M.R., Sanders, J.E., Newman, W.S., & Konigsson, L.K. (eds.) *Climate: History, Periodicity and Predictability*. Van Nostrand Reinhold, New York, pp. 323–32.

Stea, R.R. (1994) Relief and palimpsest glacial landforms in Nova Scotia, Canada. In: Warren, W.P. & Croot, D.G. (eds.) *Formation and Deformation of Glacial Deposits*. Balkema, Rotterdam, pp. 141–58.

Steffen, K. & Box, J. (2001) Surface climatology of the Greenland Ice Sheet: Greenland Climate Network 1995–9. *Journal of Geophysical Research* **106**, 33951–64.

Steffen, W., Andreae, M.A., Bolin, B., Cox, P.M., Crutzen, P.S., Cubasch, V., Held, H., Nakicenovic, N., Talaus-McManus, L., & Turner, B.L. (2004a) Earth system dynamics in the Anthropocene. In: Schellnhuber, H.J. et al. (eds.) *Earth System Analysis for Sustainability*.

Massachusetts Institute of Technology Press, MIT, pp. 313–40.

Steffen, W., Sanderson, A., & Tyson, P.D. (2004b) *Global Change and the Earth System: A Planet Under Pressure.* The IGBP Book Series. Springer, Berlin.

Stock, J. & Dietrich, W.E. (2003) Valley incision by debris flows: evidence of a topographic signature. *Water Resources Research* **39**, doi:10.1029/2001WR001057.

Stone, K.H. (1963) Alaskan ice-dammed lakes. *Annals of the Association of American Geographers* **53**, 332–49.

Street, R.B. & Melnikov, P.I. (1990) Seasonal snow cover, ice and permafrost. In: Tegart, W.J.M., Sheldon, G.W., & Griffiths, D.C. (eds.) *Climate Change: the IPCC Impacts Assessment*, Chapter 7. Australian Government Publishing Service, Canberra, pp. 7-1–7-33.

Studinger, M., Bell, R.E., & Tikku, A.A. (2004) Estimating the depth and shape of subglacial Lake Vostok's water cavity from aerogravity data. *Geophysical Research Letters* **31**, L12401, 10.1029/2004GL019801.

Sturm, M. & Benson, C.S. (1985) A history of jökulhlaups from Strandline Lake, Alaska, USA. *Journal of Glaciology* **31**, 272–80.

Sturm, M., Benson, C.S., & MacKeith, P. (1986) Effects of the 1966–8 eruptions of Mount Redoubt on the flow of Drift Glacier, Alaska, USA. *Journal of Glaciology* **32**, 355–62.

Sturm, M., Holmgren, J., & Liston, G.E. (1995) A seasonal snow cover classification system for local to global application. *Journal of Climate* **8**, 1261–83.

Sturm, M., McFadden, J.P., Liston, G.E., Chapin, F.S., Holgren, J., & Walker, M. (2001) Snow-shrub interactions in Arctic tundra: a feed-back loop with climate implications. *Journal of Climatology* **14**, 336–44.

Sugden, D.E. & John, B.S. (1976) *Glaciers and Landscape.* Edward Arnold, London.

Sugden, D.E., Bentley, M.J., Fogwill, C.J., Hulton, N.R.J., McCulloch, R.D., and Purves, R.S. (2005) Late glacial glacier events in southern-most South America: a blend of northern and southern signals. *Geografiska Annaler* **87A**, 273–88.

Tabor, R.W. (1971) Origin of ridge-top depressions by large-scale creep in the Olympic Mountains, Washington. *Geological Society of America, Bulletin* **82**, 1811–22.

Tait, A.B., Hall, D.K., Foster, J.L., & Armstrong, R.L. (2000) Utilizing multiple datasets for snow-cover mapping. *Remote Sensing of Environment* **72**, 111–26.

Tempelman-Kluit, D. (1980) Evolution of the physiography and drainage in southern Yukon. *Canadian Journal of Earth Sciences* **17**, 1189–203.

The Greenland Summit Ice Cores CD-ROM (1997) Available from the National Snow and Ice Data Center, University of Colorado at Boulder, and the World Data Center-A for Paleoclimatology, National Geophysical Data Center, Boulder, CO.

Thomas, R.H. (1993) Ice sheets. In: Gurney, R., Foster, J.L., & Parkinson, C.L. (eds.) *Atlas of Satellite Observations Related to Global Change.* Cambridge University Press, Cambridge, pp. 385–400.

Thomas, R.H. (2004) Greenland: recent mass balance observations. In: Bamber, J.L. & Payne, A.J. (eds.) *Mass Balance of the Cryosphere: Observations and Modelling of Contemporary and Future Changes.* Cambridge University Press, Cambridge, pp. 393–436.

Thomas, R., Csatho, B., Davis, C., Kim, C., Krabill, W., Manizade, S., McConnell. J., & Sonntag, J. (2001) Mass balance of higher-elevation parts of the Greenland ice sheet. *Journal of Geophysical Research* **106**, 33707–16.

Thompson, L.G., Moseley-Thompson, E., Bolzan, J.F., & Koci, B.R. (1989) A 1500-year record of tropical precipitation in ice cores from the Quelccaya ice cap, Peru. *Science* **229**, 971–73.

Thompson, R.S. (1990) Late Quaternary vegetation and climate in the Great Basin. In: Betancourt, J.L., Van Devender, T.R., & Martins, P.S. (eds.) *Packrat Middens: The Last 40,000 Years of Biotic Change.* University of Arizona Press, Tucson, AZ, pp. 200–39.

Thompson, S.K. (1992) *Sampling.* Wiley, New York.

Thorarinsson, S. (1939) The ice dammed lakes of Iceland with particular reference to their values as indicators of glacier oscillations. *Geografiska Annaler* **21A**, 216–42.

Thorarinsson, S. (1957) The jökulhlaup from the Katla area in 1955 compared with other jökulhlaups in Iceland. *Jökull* **7**, 21–25.

Thorarinsson, S. (1958) *The öræfajökull Eruption of 1362*. Acta Naturalia Islandica, Reykjavik, vol. 2.

Thouret, J.C. (1990) Effects of the November 13, 1985 eruption on the snow pack and ice cap of Nevado del Ruiz volcano, Columbia. *Journal of Volcanology and Geothermal Research* **41**, 177–201.

Tikhomirov, B.A. (1961) The changes in biogeographical boundaries in the north of USSR as related with climatic fluctutations and activity of man. *Botanisk Tidsskrift* **56**, 285–92.

Titus, J.G. & Narayanan, V. (1995) *The Probability of Sea Level Rise*. U.S. Environmental Protection Agency, Office of Planning, Policy and Evaluation. Climate Change Division. EPA 230-R-95-008-EPA, Washington, DC.

Tooley, M. (1978) *Sea-Level Changes: North-West England During the Flandrian Stage*. Clarendon Press, Oxford.

Trabant, D.C. & Meyer, D.F. (1992) Flood generation and destruction of Drift Glacier by 1989–90 eruption of Redoubt Volcano, Alaska. *Annals of Glaciology* **16**, 33–8.

Tranter, M. & Jones, H.G. (2001) The chemistry of snow: processes and nutrient cycling. In: Jones, H.G. et al. (eds.) *Snow Ecology*. Cambridge University Press, Cambridge, pp. 127–67.

Troll, C. (1948) The sub-nival or periglacial cycle of denudation, translated from Erdkunde, 1948, 2: 1–21. In: Evans, D.J.A. (ed.) *1994 Cold Climate Landforms*. Wiley, Chichester, pp. 23–44.

Tsang, L., Chen, C., Chang, A.T.C., Guo, J., & Ding, K. (2000) Dense media radiative transfer theory based on quasicrystalline approximation with applications to passive microwave remote sensing of snow. *Radio Science* **35**, 731–49.

Turner, B.L., Kasperson, R.K., Meyer, W.B., Dow, K.M., Golding, D., Kasperson, J.X., Mitchell, R.C., & Ratick, S.J. (eds.) (1990) *The Earth as Transformed by Human Action*. Cambridge University Press, Cambridge.

Ulaby, F.T. & Stiles, W.H. (1980) The active and passive microwave response to snow

parameters 2. Water equivalent of dry snow. *Journal of Geophysical Research* **85**, 1045–9.

Ussing, N.V. (1903) Om Jyllands Hedesletter og Teoriene om deres Dannelse. *Oversigt over det Kongelige Danske Videnskabernes Selskabs Forhandlinger* 2, 99–165.

Ustin, S. (ed.) (2004) *Manual of Remote Sensing. Volume 4, Remote Sensing for Natural Resource Management and Environmental Monitoring*, 3rd edn. Wiley, Chichester.

Velitchko, A.A. (ed.) (1984) *Late Quaternary Environments of the Soviet Union*. Longman, London.

Vinje, T. (2001) Anomalies and trends of sea ice extent and atmosphere circulation in the Nordic Seas during the period 1864–1998. *Journal of Climate* **14**, 255–67.

Vinnikov, K.Ya, Groisman, P.Ya, & Lugina, K.M. (1990) Empirical data on contemporary global climate changes (temperature and precipitation). *Journal of Climate* **3**, 662–77.

Vinnikov, K.Ya, Robock, A., Stouffer, R.J., Walsh, J.E., Parkinson, C.L., Cavalieri, D.J., Mitchell, J.F.B., Garrett, D., & Zakharov, V.F. (1999) Global warming and Northern Hemisphere sea ice extent. *Science* **286**, 1934–7.

Vitousek, P.M., Mooney, H.A., Lubchenco, J., & Melillo, J.M. (1997) Human domination of earth's ecosystems. *Science* **277**, 494–9.

von Bubnov, S. (1963) *Fundamentals of Geology*. Oliver and Boyd, Edinburgh and London.

Vörösmarty, C.J., Fekete, B., Maybeck, M., & Lammers, R.B. (2000) The global system of rivers: its role in organizing continental land mass and defining land-to-ocean linkages. *Global Biogeochemical Cycles* **14**, 599–621.

Waitt, R.B. (1980) About forty last-glacial Lake Missoula jokulhlaups through southern Washington. *Journal of Geology* **88**, 653–79.

Waitt, R.B. (1985) Case for periodic, colossal jokulhlaups from glacial Lake Missoula. *Geological Society of America Bulletin* **95**, 1271–86.

Walder, J.S. & Driedger, C.L. (1995) Frequent outburst floods from South Tahoma Glacier, Mount Rainier, USA: relation to debris flows, meteorological origin and implications for subglacial hydrology. *Journal of Glaciology* **41**, 1–10.

Walker, A.E. & Silis, A. (2002) Snowcover variations over the Mackenzie River basin, Canada derived from SSM/I passive microwave satellite data. *Annals of Glaciology* **34**, 8–14.

Walker, D.A., Billings, W.D., & Molenaar, J.G.de (2001) Snow-vegetation interactions in tundra environments. In: Jones, H.G. et al. (eds.) *Snow Ecology*. Cambridge University Press, Cambridge, pp. 266–324.

Walker, D.A., Bockheim, J.G., Chapin III, H.S., Eugster, W., King, J.Y., McFadden, J.P., Michaelson, G.J., Nelson, F.E., Oechel, W.C., Ping, C.L., Reeburgh, W.S., Regli, S., Shiklomanov, N.J., & Vourlitis, G.L. (1998) A major arctic soil pH boundary: implications for energy and trace gas fluxes. *Nature* **394**, 469–72.

Walsh, J.E. & Chapman, W.L. (2001) Twentieth century sea ice variations from observational data. *Annals of Glaciology* **33**, 444–8.

Walther, G.-R., Post, E., Convey, P., Menzel, A., Parmesan, C., Beebee, T.J.C., Fromentin, J.-M., Hoegh-Goldberg, O., & Bairlein, F. (2002) Ecological responses to recent climate change. *Nature* **416**, 389–95.

Wan, Z., Zhang, Y., Zhang, Q., & Li, Z.-L. (2004) Quality assessment and validation of the MODIS global land surface temperature. *International Journal of Remote Sensing* **25**, 261–74.

Warrick, R.A. & Oerlemans, J. (1990) Sea level rise. In: Houghton, J.T. et al. (eds.) *Climate Change: The IPCC Assessment*. Cambridge University Press, Cambridge, pp. 257–81.

Warrick, R.A., Le Prevost, C., Meier, M.F., Oerlemans, J., & Woodworth, P.L. (1996) Changes in sea level. In: Houghton, J.T., Meira Filho, L.G., Callander, B.A., Harris, N., Kattenberg, A., & Maskell, K. (eds.) *Chapter 7, Climate Change 1995: The Science of Climate Change*. Cambridge University Press, Cambridge, pp. 363–405.

Washburn, A.L. (1979) *Geocryology*. Arnold, London.

Wasson R.J. (1994) Annual and decadal variation of sediment yield in Australia, and some global comparisons. *International Association of Hydrological Sciences Publication* **224**, 269–79.

WCED (1987) *Our Common Future*. Oxford University Press, Oxford.

Webb, P.N. & Harwood, D.M. (1991) Late Cenozoic history of the Ross Embayment, Antarctica. *Quaternary Science Reviews* **10**, 215–33.

Weber, F., Nixon, D., & Hurley, J. (2003) Semi-automated classification of river ice types on the Peace River using RADARSAT-1 synthetic aperture radar (SAR) imagery. *Canadian Journal of Civil Engineering* **30**, 11–27.

Weertman, J. (1974) Stability of the junction of an ice sheet and an ice shelf. *Journal of Glaciology* **13**, 3–11.

Welch, D.M. (1970) Substitution of space for time in a study of slope development. *Journal of Geology* **78**, 234–9.

Weller, G. & Lange, M. (1999) *Impacts of Global Climate Change in the Arctic Regions*. Centre for Global Change and Arctic System Research. University of Alaska, Fairbanks.

Weller, G. & Wendler, G. (1990) Energy budgets over various types of terrain in polar regions. *Annals of Glaciology* **14**, 311–14.

Westman, W.E. (1978) *Ecology, Impact Assessment and Environmental Planning*. Wiley, New York.

Wharton, R.A., McKay, C.P., Simmons, G.M., & Parker, B.C. (1985) Cryoconite holes on glaciers. *Biological Science* **35**, 499–503.

WHO (1967) The Constitution of the World Health Organization. *WHO Chronicle* **1**, 29.

Wigley, T.M.L. & Raper, S.C.B. (1992) Implications for climate and sea level of revised IPCC emissions scenarios. *Nature* **357**, 293–300.

Wigley, T.M.L. & Raper, S.C.B. (1993) Future changes in global mean temperature and sea level. In: Warrick, R.A. et al. (eds.) *Climate and Sea Level: Observations, Projections and Implications*. Cambridge University Press, Cambridge, pp. 111–33.

Wilson, J.T. (1938) Glacial geology of part of northwestern Quebec. *Transactions of the Royal Society of Canada*, **Section 4(32)**, 49–59.

Winebrenner, D.P., Arthern, R.J., & Shuman, C.A. (2001) Mapping Greenland accumulation rates using observations of thermal emission at 4.5 cm wavelength. *Journal of Geophysical Research* **106**, 33919–34.

Wingham, D.J., Ridout, A.J., Scharroo, R., Arthern, R.J., & Shum, C.K. (1998) Arctic elevation change from 1992–6. *Science* **282**, 456–8.

Winther J.-G. & Solberg, R. (2002) *International Symposium on Remote Sensing in Glaciology, Maryland, USA, 4–8 June 2001.* Annals of Glaciology, vol. 34, International Glaciological Society.

WMO (1970) *WMO Sea-Ice Nomenclature, Terminology, Codes and Illustrated Glossary.* WMO/OMM/BMO 259, TP 145. World Meteorological Organization, Geneva.

WMO (2005) *Sea Ice Glossary.* Joint WMO/IOC Technical Commission for Oceanography and Marine Meteorology (JCOMM) Expert Team on Sea Ice (ETSC).

Woo, M.-K. (1996) Hydrology of northern North America under global warming. In: Jones, J.A.A., Liu, C., Woo, M.-K., & Kung, H.-T. (eds.) *Regional Hydrological Responses to Climate Change.* Kluwer, Dordrecht, pp. 73–86.

Woo, M.-K. & Marsh, P. (2005) Snow, frozen soils and permafrost hydrology in Canada, 1999–2002. *Hydrological Processes* **19**, 215–29.

Woo, M.-K. & Thorne, R. (2003) Streamflow in the Mackenzie basin, Canada. *Arctic* **56**, 328–40.

Wookey, P.A., Parsons, A.N., Welker, J.M., Potter, J.A., Callaghan, T.V., Lee, J.A., & Press, M.C. (1993) Comparative responses of phenology and reproductive development in simulated environmental change in sub-arctic and high arctic plants. *Oikos* **67**, 490–502.

Worsley, P. (2004) Periglacial geomorphology. In: Goudie, A.S. (ed.) *Encyclopedia of Geomorphology (vol. 2).* Routledge, London, pp. 772–6.

Wright, J.V. (2001) *A History of the Native People of Canada v. 1: 10,000–1,000 BC.* Canadian Museum of Civilization Corporation, Ottawa.

Wright, R. & Anderson, J.B. (1982) The importance of sediment gravity flow to sediment transport and sorting in a glacial marine environment: eastern Weddell Sea, Antarctica. *Geological Society of America Bulletin* **93**, 951–63.

Wu, X., Budd, W.F., Lytle, V.I., & Massom, R.A. (1999) The effect of snow on Antarctic sea ice simulations in a coupled atmosphere–sea ice model. *Climate Dynamics* **15**, 127–43.

Wyrwoll, K.-H. (1977) Causes of rock-slope failure in a cold area: Labrador–Ungava. *Geological Society of America Reviews in Engineering Geology* **3**, 59–67.

Yamada, T. (1998) *Glacier Lake and its Outburst Flood in the Nepal Himalaya.* Monograph No. I. Tokyo, Data Center for Glacier Research, Japanese Society of Snow and Ice.

Yang, D., Kane, D., Ahang, Z., Legates, D., & Goodison, B. (2005) Bias corrections of long term (1973–2004) daily precipitation data over northern regions. *Geophysical Research Letters* **32**, L19501, doi:10.1029/2005GL024057.

Yang, D., Kane, D.L., Hinzman, L.D., Zhang, X., Zhang, T., & Ye, B. (2002) Siberian Lena River hydrologic regime and recent change. *Journal of Geophysical Research* **107**, 4694 (14-1–14-10).

Yang, D., Ye, B., & Kane, D.L. (2004a) Streamflow changes over Siberian Yenisei River basin. *Journal of Hydrology* **296**, 59–80.

Yang, D., Ye, B., & Shiklomanov, A. (2004b) Discharge characteristics and changes over the Ob River watershed in Siberia. *Journal of Hydrometeorology* **5**, 595–610.

Ye, B., Yang, D.Q., & Kane, D.L. (2003) Changes in Lena River stream flow hydrology: human impacts versus natural variations. *Water Resources Research* **39**, 1200–14.

Yershov, E.D. (2004) *General Geocryology.* Cambridge University Press, Cambridge.

Zeman, L.J. & Slaymaker, O. (1978) Mass balance model for calculation of ionic input loads in atmospheric fallout and discharge from a mountainous basin. *Bulletin of the Hydrological Sciences* **23**, 103–18.

Zhang, T., Frauenfold, O., Serreze, M., Etisinger, A., Oelke, C., McCreight, J., Barry, R., Gilichinsky, D., Yang, D., Ye, H., Ling, F., & Chudmova, S. (2005) Spatial and temporal variability in active layer thickness over the Russian Arctic drainage basin. *Journal of Geophysical Research* **110**, No. D16, D6101, doi:10.1029/2004JD00564.

Zhang, T., Heginbottom, J.A., Barry, R.G., & Brown, J. (2000) Further statistics on the distribution of permafrost and ground ice in the Northern Hemisphere. *Polar Geography* **24**, 126–31.

Zwally, H.J., Schutz, B., Abdalati, W., Abshire, J., Bentley, C., Brenner, A., Bufton, J., Dezio, J., Hancock, D., Harding, D., Herring, T., Minster, B., Quinn, K., Palm, S., Spinhirne, J., & Thomas, R. (2002) ICESat's laser measurements of polar ice, atmosphere, ocean, and land. *Journal of Geodynamics* **34**, 405–45.

Zwally, H.J., Schutz, R., Bentley, C., Bufton, J., Herring, T., Minster, J., Spinhirne, J., & Thomas, R. (2003) *GLAS/ICESat L1B Global Elevation Data V018*, 15 October to 18 November 2003. National Snow and Ice Data Center, Boulder, CO. Digital media – updated current year.

Zwally, H.J., Giovinetto, M.B., Li, J., Cornejo, H.G., Beckley, M.A., Brenner, A.C., Saba, J.L., & Yi, D. (2005) Mass changes of the Greenland and Antarctic ice sheets and shelves and contributions to sea-level rise: 1992–2002. *Journal of Glaciology* **51**, 509–27.

INDEX

Note: Page numbers in bold refer to figures and tables